Microprocessor and Microcomputer Basics

Microprocessor and Microcomputer Basics

JEFFERSON C. BOYCE

Prentice-Hall, Inc. Englewood Cliffs, New Jersey 07632

Library of Congress Cataloging in Publication Data

Boyce, Jefferson C
Microprocessor and microcomputer basics.

(Prentice-Hall series in electronic technology)
Bibliography: p.
Includes index.
1. Microprocessors. 2. Microcomputers. I. Title.
QA76.5.B67 001.6'4'04 78-26455
ISBN 0-13-581249-6

Editorial/production supervision by Steven Bobker
and Phyllis Springmeyer
Interior design by Phyllis Springmeyer
Cover design by Edsal Enterprises
Manufacturing Buyer: Gordon Osbourne

10 9 8 7 6 5 4 3 2 1

Printed in the United States of America

Prentice-Hall International, Inc., *London*
Prentice-Hall of Australia Pty. Limited, *Sydney*
Prentice-Hall of Canada, Ltd., *Toronto*
Prentice-Hall of India Private Limited, *New Delhi*
Prentice-Hall of Japan, Inc., *Tokyo*
Prentice-Hall of Southeast Asia Pte. Ltd., *Singapore*
Whitehall Books Limited, *Wellington, New Zealand*

To Betty —

For 35 inspirational years

Contents

Preface

Just as the transistor revolutionized the field of electronics, so has the *microprocessor* revolutionized the digital computer. The *microprocessor* promises to put computer capability in every home at a cost no greater than a modern television set. Further, it is reducing the cost of industrial control, automating supermarket checkout, controlling automobile engines for maximum performance and minimum pollution, ad infinitum.

This book proposes to reveal the innermost secrets of the *microprocessor* and how it is used in modern technology. Since *microprocessor* use requires background in the actual hardware (parts and pieces) comprising the *microprocessor* and the software (instructions that tell the hardware what to do), this book presents an integrated coverage of both. Sufficient detail is presented to bring a noncomputer-oriented person to a level of capability where practical uses for *microprocessors* and *microcomputers* can be recognized and existing designs evaluated in light of technical requirements.

Microprocessor and Microcomputer Basics is not a "cookbook" that gives step-by-step instructions for the construction and operation of a *microprocessor* or *microcomputer*. Such publications already abound. Instead, an attempt is made in this book to lay the basic groundwork so that the reader may *understand* the *microcomputer*. Although information from many manufacturers is included, a rigorous attempt has been made to refrain from discussing specific hardware until a firm footing in fundamentals has been established. Therefore, actual *microcomputers* are not included until relatively late in the book.

After reading and studying this book carefully, performing the problems and exercises in each chapter, and investigating modern applications of *microprocessors* and *microcomputers*, the reader should be able to:

1. Use the terminology of the trade.
2. Explain the fundamentals of digital computers.
3. Explain the fundamentals of *microprocessors* and *microcomputers*.
4. Apply *microprocessor* technology to *microcomputer* applications.
5. Apply *microcomputer* technology to practical uses.

Perusal of the table of contents will reveal how the objectives may be attained. Chapters 1 through 4 prepare the reader for the new field of *microprocessors* and *microcomputers* by reviewing the principles of digital logic, computer arithmetic, and the conventional general-purpose digital computer. Chapter 5 points out the differences between the general-purpose digital computer and the *microprocessor*. The *microprocessor* is developed to the major block diagram level to introduce its salient points.

A detailed discussion of typical *microprocessor* hardware is provided in Chapter 6. Each of the major blocks developed in Chapter 5 is investigated in detail. Chapter 7 concentrates on *microprocessor* software (i.e., those instructions required to cause the *microprocessor* hardware to perform its basic operations). Hardware information from Chapter 6 is integrated to provide a complete understanding of *microprocessor* capabilities.

Chapter 8 discusses typical input/output methods and devices, while Chapters 9 and 10 are devoted to *microcomputer* memory devices, both "internal" and "external." A typical *microcomputer* is examined in Chapters 11 and 12. Hardware-oriented information is discussed in Chapter 11, while the software aspects are integrated in Chapter 12.

The final chapter describes actual "in-use" applications of *microprocessors* and *microcomputers*. Both hardware and software aspects are covered so that the reader may appreciate the close tie that exists between them. A large-scale *microprocessor* is also described in Chapter 12 as a preview of possible future developments. Thus, within these pages lies the means of progressing from a bare knowledge of digital logic fundamentals to a workable knowledge of *microprocessors* and *microcomputers*.

Microprocessor and Microcomputer Basics has been the result of a natural progression from its predecessors, *Digital Computer Fundamentals* and *Digital Logic and Switching Circuits*. Much of the material gathered during previous research pointed the way toward this new and expanding field. Commodore Business Machines, Inc.; INTEL Corporation; MOS Technology; SIGNETICS Corporation; Texas Instrument, Incorporated (Semiconductor Group), and many other manufacturers supplied copious technical information and assistance. They are largely responsible for the information that is included, but the author accepts responsibility for the organization and interpretation of the data. The cooperation of those at Prentice-Hall who have taken part in this effort is also acknowledged. Last, but surely not least, the encouragement of my loving wife must be stressed. As in previous endeavors, without the understanding and devotion to everyday tasks that she assumed, this book would not have evolved.

Jefferson C. Boyce

CHAPTER 1

Introduction to Microprocessors and Microcomputers

We have entered the "micro" era, a time in which the marriage of two technologies will greatly influence our way of life. The love affair between digital computer technology and semiconductor technology has spawned the *microprocessor*.* Just as the birth of a child influences and changes the lives of the parents, so will the birth of the microprocessor affect the technologies that conceived and nurtured it. Furthermore, the parents' dreams for a child's future are as strongly reflected in technology's dreams for the microprocessor. Before we look to the future, however, let us delve into the past and see what brought these two powerful technologies together.

1-1 SHORT HISTORY

The ancestry of the digital computer can be traced to ancient times. One of the earliest devices built to aid man in his computing tasks was the *abacus*, a digital device that finds mention in history prior to the

*A microprocessor is, by generally accepted definition, a large-scale integrated circuit assembly that contains about 70% of the computing capability of a very small digital computer. Technological advances will soon see a *complete* computer on a single integrated-circuit assembly, and differentiation between microprocessor and computer will be in physical size only.

birth of Christ. A skilled abacus operator can still successfully compete with mechanical desktop calculators, although the electronic calculator easily outstrips either of them. Following the abacus, there was little additional progress in the field of mechanical aids to mathematics until the appearance of Blaise Pascal's desk calculator in 1642. Pascal's device used simple gears to add and subtract. Other mathematicians improved Pascal's early machine and achieved multiplication, but the lack of mechanical precision hampered progress.

The next major milestone associated with development of digital machines occurred in the early 1800s. Charles Babbage envisioned a mechanical device which incorporated many of the principles of the modern digital computer. His "Difference Engine" was developed to calculate and print mathematical tables. Again, imperfect materials, a shortage of precision tools, withdrawal of government support, and a lack of understanding among his associates resulted in abandonment of his project after several incomplete models had been constructed.

Steady progress in the late 1800s and early 1900s saw techniques developed for the application of mechanical and electrical principles to the record storing, account handling, and bookkeeping processes. The 1940s witnessed development of an *electromechanical* computer. Pulse techniques which evolved in connection with World War II radar development were wedded with increased use of applied mathematics. Automation requirements to meet wartime production needs resulted in machines that could perform routine tasks without human intervention. Electronics took over, and soon the production machines were *programmed* to make decisions concerning quality, quantity, and so forth. The general development of vacuum-tube electronic computers in World War II was quickly followed by increased activity in design and application of "intelligent machines."

While the application of vacuum tubes resulted in a tremendous increase in speed of operation, the instructions to the machine still tended to be stored external to the actual computer and inserted sequentially as needed. In addition, the high electrical power requirements and poor reliability of the vacuum-tube computer began to restrict its capability. Two important events that were to shape the future of the digital computer occurred in the mid-1940s. In 1946, John von Neumann made a now-classic proposal for computer design. He proposed placing the program of instructions for the computer in storage *internally* along with the data being processed. This was to shape the entire future architecture of digital computers. In 1947, the work of William Shockley, John Bardeen, and Walter H. Brattain culminated in the demonstration of a solid-state amplifying device—the transistor. In a very small fraction of the space and requiring an even smaller fraction of the operating power, the transistor removed the barrier of physical size and power requirements. Soon computers which formerly required *rooms* to house their components shrank to desk size and computers which were beyond comprehension during vacuum tube days began to appear. The digital computer and semiconductor had met.

1-2 SEMICONDUCTOR TECHNOLOGY

First, let us trace the evolution of semiconductor technology. Actually, there have been only two major breakthroughs in the semiconductor field (as far as computers are concerned) since the discovery of the transistor. The push to make more powerful and versatile computers in less space and at lower cost eventually resulted in the successful manufacture of the integrated circuit (IC) in the early 1960s. Many transistors and diodes were manufactured on the same tiny chip of silicon, thus reducing size and cost when compared with discrete component circuits. It had taken many years to learn how to effectively isolate transistors and diodes from each other when they existed on the same chip, but the time was well spent. The now more powerful computers could be put to work designing even more complex and compact IC assemblies.

As component density became greater and greater, it was soon apparent that the techniques for the original ICs could not be applied to the new problem. An intermediate step between ICs and the now-commonplace large-scale-integration (LSI) assemblies did result, however, in the manufacture of medium-scale-integration (MSI) assemblies. Many considered MSI nothing more than the big brother of the original IC and continued to set their sights on bigger game.

Continued development and new techniques finally resulted in capabilities to pack *hundreds* of components into extremely small spaces, and in about 1967 the first practical LSI assembly emerged. It was this high-density packaging that has allowed the microprocessor to become a reality and to usher in the era of low-cost computing. Figure 1-1 shows the evolution of LSI. It should be noted that each of the assemblies in Figure 1-1 occupies approximately the same amount of space on the tiny silicon chip. Further advances in semiconductor technology are increasing component density by factors of up to 1000. These new techniques are called Very Large Scale Integration (VLSI) or Extra Large Scale Integration (ELSI).

1-3 DIGITAL COMPUTER TECHNOLOGY

Next let us look at the growth of digital computer technology. Early electromechanical computers had a storage capacity of 132 words, and it took 3 seconds to add two numbers. The next-generation machines were constructed with vacuum tubes. The example shown in Figure 1-2

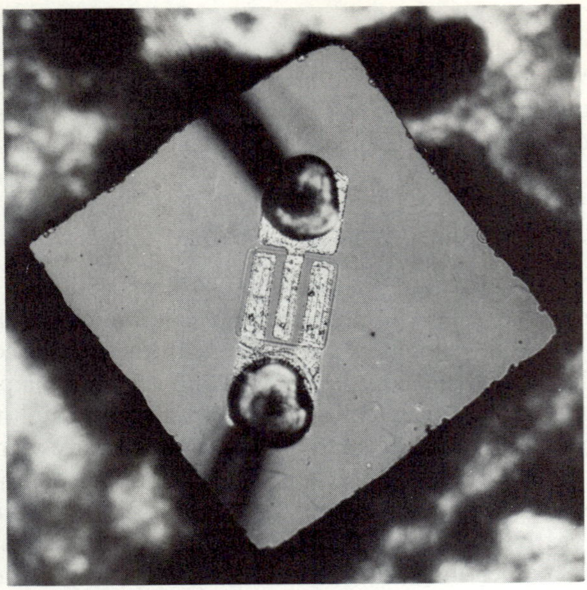

(a)

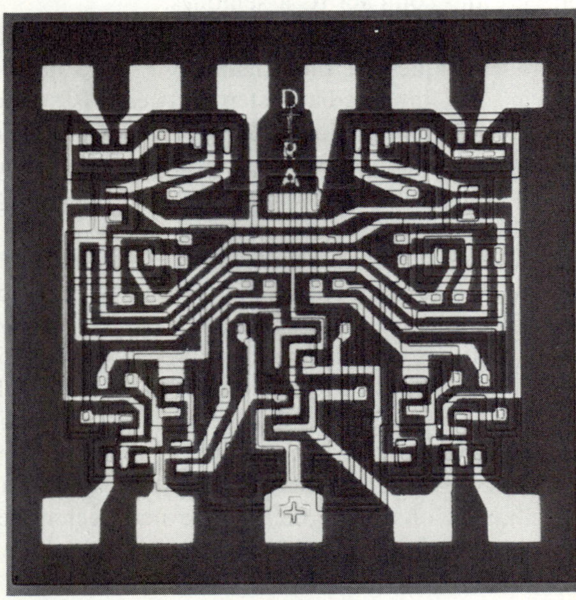

(b)

Figure 1-1 Evolution of Large-Scale Integration (LSI): (a) Early Transistor (Discrete Semiconductor Device); (b) Small-Scale Integration (Integrated Circuit Capable of Storing 1 Bit of Information); (c) Medium-Scale Integration (Integrated Circuit Capable of Storing 256 Bits of Information); (d) Large-Scale Integration (Integrated Circuit Capable of Storing 65,536 Bits of Information) (All Courtesy Fairchild Camera and Instrument Corp)

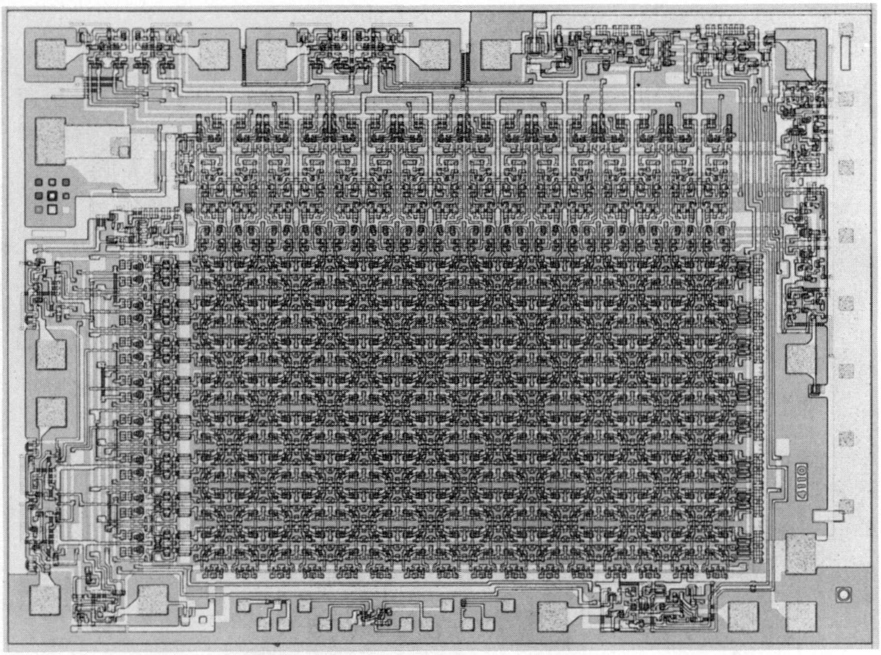

(c)

(d)

Figure 1-1 (Continued)

Figure 1-2 ENIAC, One of the First All-Electronic Computers, Circa Late 1940s (Courtesy UNIVAC Division, Sperry Rand Corp)

could add two 10-digit numbers in $\frac{1}{5000}$ of a second, quite a jump from the earlier 3-second add time (storage capacity was still limited, however). Occupying 15,000 square feet of floor space, weighing 30 tons, and containing approximately 18,000 vacuum tubes, this behemoth ushered in the era of electronic computers.

Transistor-based computers greatly reduced space and power requirements, and the early transistorized computer could easily perform additions in mere microseconds.* Storage capacity was in the general range of 1000 to 4000 words internally. Peripheral (external) storage devices, such as the magnetic tape units, greatly extended the storage capacity of second-generation and all later digital computers. The development of integrated-circuit techniques again reduced the size and power requirements, and the digital computer in Figure 1-3 is representative of today's *minicomputer*. Mini describes size only, because this machine possesses the following characteristics: (a) operating time in nanoseconds,† (b) storage

*A microsecond is a millionth of a second.

†A nanosecond is a billionth of a second.

Figure 1-3 Modern Minicomputer (Courtesy Hewlett-Packard)

capacity up to 31,000 words internally, and (c) a full system weight of only 110 pounds.

Finally, the modern large-scale computer of Figure 1-4, with a capability of nanosecond operating times, a storage capacity limited only by the imagination, and a multi-million-dollar price tag, shows the results of less than half a century of technological advances in the digital computer field. As new techniques and devices appear, speed of operation is decreasing, price is decreasing, and storage capacity is increasing.

1-4 MICROPROCESSOR EVOLUTION

As the digital computer grew, one of the problems that began to develop was that of trying to apply the computer to more and more everyday tasks, such as supermarket checkout, small machine control, and

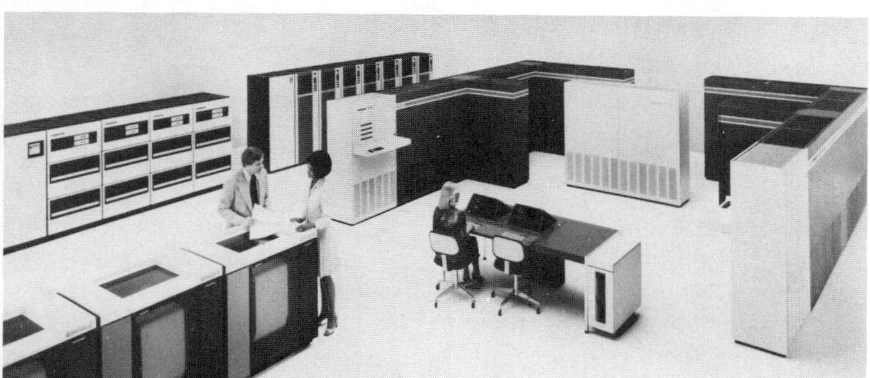

Figure 1-4 The Burroughs B6800, A Large Modern Computer (Courtesy Burroughs Corp)

traffic light controllers. The immensity of the large-scale computer made such picayune applications out of the realm of possibility. Even the minicomputer was "overdesigned" for such simple tasks. With the knowledge that semiconductor technology could put thousands of circuits on very small chips, the designers approached the everyday application of computers from a different viewpoint.

Why not take the many "hard-wired logic" control and data-processing applications and replace the logic circuits with a general purpose, *programmable*, digital computer? Then *one* computer could perform many tasks, requiring only a change in programming to convert from one task to another. By designing the computer specifically for applications requiring small amounts of storage and minimal mathematical capabilities, perhaps the semiconductor people could come up with a low-cost "computer-on-a-chip." After all, if they could put out a calculator chip, why not a computer chip?

As is often the case, however, the designers' requirements and the industry's capabilities were years apart. Semiconductor techniques could meet some of the requirements, but the true computer-on-a-chip was still a design goal, not an immediate reality. Many different concepts evolved, but generally the result was an LSI chip that could perform only the primitive operations of a digital computer, and then only when properly programmed. Memory was very limited, sometimes no more than a few

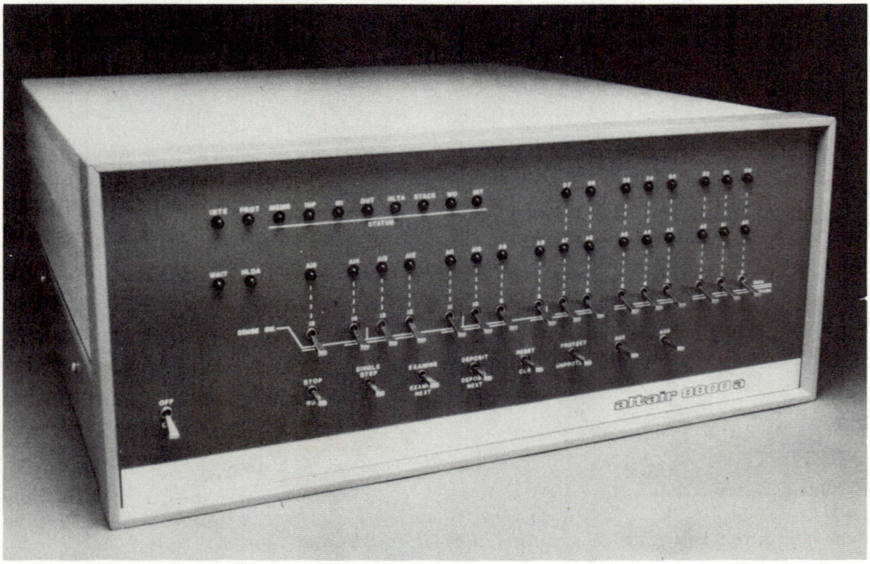

Figure 1-5 Early Microcomputer, The Altair 8800 (Courtesy Pertec Computer Corp)

characters, so more ICs were needed for storage. No provisions were made to match the processor to the external world, and more ICs were required for the interface operation. Timing and data routing were external to the processor chip. All in all, so many supporting ICs were needed that some designers began to doubt the wisdom of their ideas as their microcomputer* began to look like a minicomputer. An early microcomputer is shown in Figure 1-5.

Once again, though, the semiconductor people came through. Multifunction LSI support chips replaced dozens of ICs, and the microprocessor-based microcomputer shrank from a number of printed circuit boards to only one (see Figure 1-6).

If one is willing to settle for small-scale control applications, the true computer-on-a-chip exists today; Figure 1-7 shows such a chip. It includes the actual processor that performs operations and routes data. It stores its own program, interfaces with the outside world, and includes its own

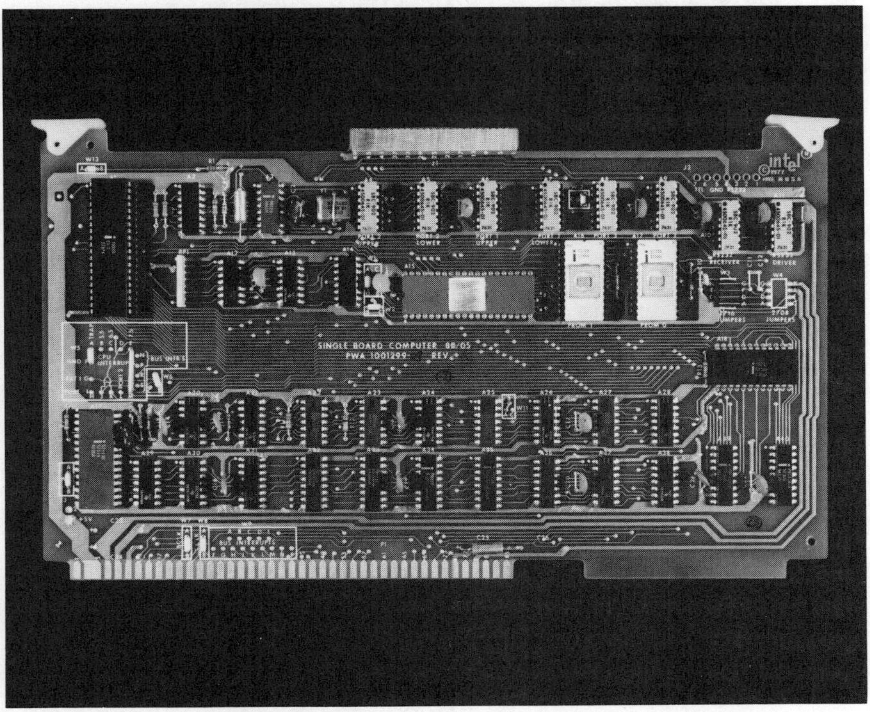

Figure 1-6 Single-Board Microcomputer (Courtesy INTEL Corp)

*A microcomputer is a microprocessor-based computer which has *complete* computer capability, including memory, timing, and the ability to communicate with the outside world.

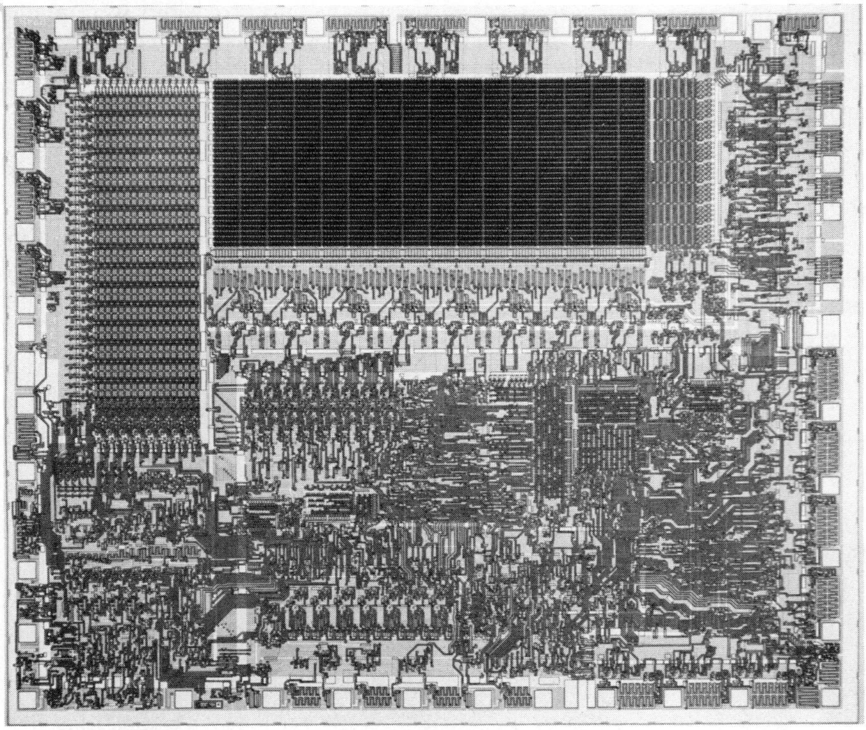

Figure 1-7 Computer on a Chip (Courtesy INTEL Corp)

timing. And the end is not yet in sight. Semiconductor technology is growing, and more and more capability is being packaged in smaller and smaller spaces. As the processor chips become more versatile, more applications are generated, which ultimately require more versatile processors. Thus, the microprocessor, offspring of two burgeoning technologies, is accepting the challenge of its new world. As the fledgling grows to maturity, it will also join with some new technology and "perpetuate the race." As with the human race, each new generation meets its own challenge and passes on to the next some of its wisdom, together with some of its problems.

1-5 APPLICATIONS

The almost instant success of the microprocessor is generally due to its versatility. Its applications are endless, as can be seen in some of the everyday uses shown below.

If you were to take an imaginary tour of the world, it would be impossible to travel far without encountering the microprocessor and microcomputer. Even before you leave home, this powerful new servant to mankind is in evidence. Remember that delicious dinner last night? If you have an electronic oven, chances are that dinner was prepared under microprocessor control (Figure 1-8). It may have even been placed in the oven hours before, in frozen form, and the microprocessor left to its program processes to have it table-ready at the proper time.

Are you taking a taxi to the airport or the dock? A microprocessor-based fare meter could easily be computing your fare as you travel, while another microprocessor is hard at work getting maximum mileage and performance from the engine, consistent with minimum environmental pollution. Will you get to your destination on time? If the microprocessor has anything to say about it, you probably will. In many cities the complete traffic flow is microprocessor/microcomputer-controlled, assuring minimum delay day or night.

Figure 1-8 Microprocessor-Controlled Electronic Oven (Courtesy Litton Microwave Cooking)

At the airport you will find your baggage being routed by microcomputer and your ticket being issued by a locally based microprocessor or microcomputer communicating with a larger, centrally located computer elsewhere. Your aircraft will be equipped with antiskid wheel brakes controlled by a microcomputer. A microprocessor will keep tabs on the various engine, electrical, and mechanical systems that keep the aircraft operating and will ensure safe, efficient operation throughout the flight. Computer-controlled navigation devices will give the aircraft's exact location at all times, assuring on-time arrival at your destination.

Should you be traveling by ship, you will be a bit more on your own. The microprocessor/microcomputer has made inroads in sea transportation, but it is not yet as common as in land and air transportation. Loran C, a navigation system used by ships at sea, takes advantage of microprocessor techniques to display exact location at all times. If you need to send a message home, microprocessor-based communication systems will route the message most efficiently to assure its transmission to its destination.

As you travel throughout the world, you cannot help but marvel at the ingenious applications of the microprocessor. Each country you visit is investigating this fantastic new device and applying it to improve their way of life. In fact, by the time you return home from your trip, you may even note microprocessor invasion of your hometown.

Next time you stop for gas at a service station, look carefully. Maybe that pump (Figure 1-9) is microprocessor-controlled, calculating your charges at the same time it is dispensing gas to another customer and forwarding daily sales information to a central computer. Are you a little short of cash after your trip? Stop by the corner self-service bank, identify yourself with your computer card and identification number, and the computer-controlled bank will have you solvent again in no time.

If you are a patron of public transportation in the larger cities, the microprocessor is collecting fares, making change, issuing tickets, and operating the vehicles. Large sports stadiums control their scoreboards and display boards by microcomputer. And if you are more the stay-at-home type, television games use microprocessor techniques to give you the maximum challenge.

The improvements in manufacturing techniques have reduced costs to the point where amateur computing has become a major hobby (Figure 1-10). National amateur computing organizations have been formed and conventions are held often to update the amateur computer buff to the newest equipment and techniques.

The day of the "computer in every home" is nearly upon us. Even if the actual computer is not resident in the home, every home will have access to computer capability through a data link directly into the home.

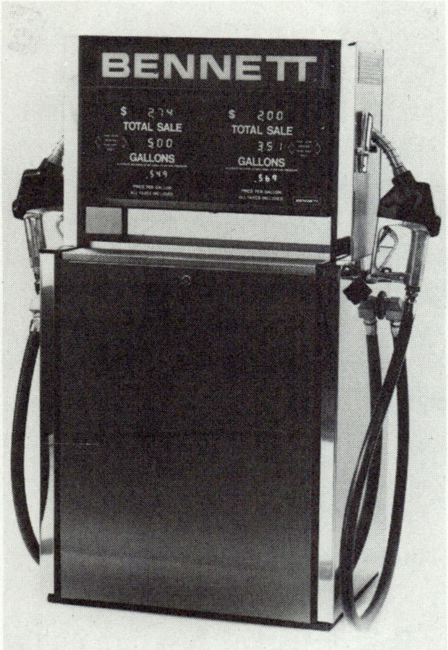

Figure 1-9 Microprocessor-Controlled Gasoline Pump and Microprocessor Board. (Courtesy Bennett Pump Company)

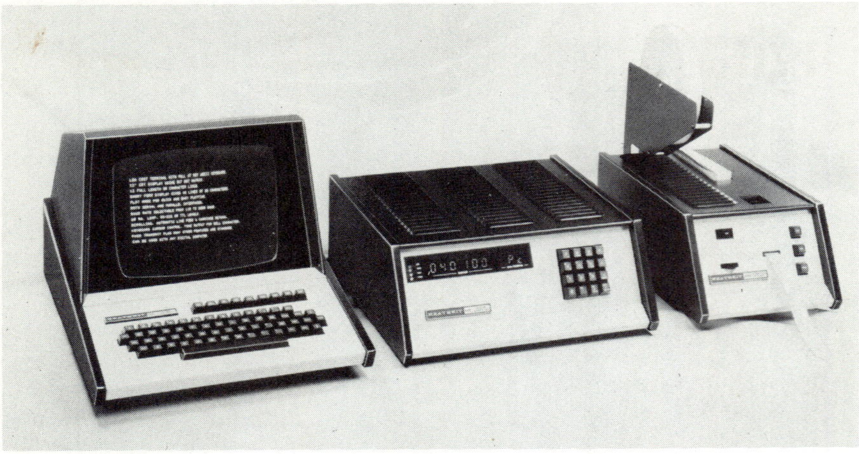

Figure 1-10 Amateur Computer (Courtesy Heath Company)

Think of the advantages! Need a recipe for Baked Alaska? Just turn on the computer (or computer terminal), key in the code for recipes, and ask for the Baked Alaska recipe! How about income tax? At the end of the year, assuming that you have been inputting the necessary data concerning income, expenses, and so on, all that is necessary is to ask for the income tax program, identify yourself, and come back in a few minutes. The job will be done—except for payment, of course!

It is really foolish to attempt to second-guess all the possible applications of the microprocessor/microcomputer. Only the imagination of the user seems to be the limit. The remainder of this book will attempt to provide you with enough information so that you can imaginatively apply the microprocessor/microcomputer.

CHAPTER 2

Digital Logic

Even though use of a microprocessor does not require delving into the innermost workings of the tiny silicon chip, a basic knowledge of the principles of digital logic makes microprocessor operation easier to understand. The microprocessor has evolved from very basic digital logic functions and by understanding these functions it is possible to systematically investigate the microprocessor chip's individual operations. It is assumed that the reader already has background in digital logic and this chapter merely serves as a review. If additional logic background is required, consult any of the appropriate books listed in the bibliography.

Most digital equipment consists of two general types of logic circuits: combinational and sequential. In a *combinational logic circuit*, the output depends on the *input combinations* present when the output is examined. But a *sequential logic circuit* contains memory elements, so the output depends not only on the state of the input at any given instant but also on the *entire input history* of the circuit. A microprocessor contains both types of circuits.

2-1 COMBINATIONAL LOGIC CIRCUITS

Combinational circuits make decisions which range from the very simple to the extremely complex. Understanding how these decisions are made requires some knowledge of the binary number system and the mathematics of logic (Boolean algebra). It also entails application of such "tools" as *truth tables, symbology*, and *waveforms*.

The circuits that perform decision-making operations, generally known as *gates*, are found in profusion in microprocessor circuits. AND gates, OR gates, NOT circuits, NOR gates, and NAND gates generally fit into the category of combinational circuits.

AND Gates. In microprocessors, combinational logic circuits perform specified operations on the inputs and provide outputs that are related to the inputs in accordance with predefined rules. For example, one of the logic circuits used is called an AND gate. The *AND gate* provides a *useful* output only when *all* its inputs are in *useful* form simultaneously.

AND gates are used extensively in microprocessors to perform logical, arithmetic, and control operations. Applications such as the routing of information from one part of the microprocessor to another are typical. Consider that information to be transferred is applied to one input of an AND gate and its output is routed to a storage device. The information should not be transferred, however, until an arithmetic operation is performed and the storage device is ready to accept the information. These conditions may be concisely described in the following manner:

> Information is to be made available to the storage device if and only if the arithmetic operation is complete AND the storage device is ready to accept the information AND information is present at the input.

A complete description of microprocessor operation entails thousands of such statements and obviously is quite unwieldy. Furthermore, mathematical manipulation of such statements (a requirement for understanding microprocessor operation) is practically impossible. The algebraic technique of assigning letters as abbreviations for English-language statements is applied to digital logic manipulation.

Let X = useful output
A = arithmetic operation complete
Y = storage device ready
D = data (information)

Thus, $X = A$ AND Y AND D, and the complete description of what is required to transfer the results of an arithmetic operation to a storage device is shown in relatively simple algebraic form. The branch of algebra applicable to manipulation of logic quantities is called *Boolean algebra* (after George Boole, 1815–1864.) The basic ideas of Boolean algebra are shown in this chapter, and some of the laws and identities of Boolean algebra are given in Appendix A.

The Boolean algebra expression given above describes the operation of a specific AND function, but without additional description the full meaning may not be apparent. For example, what are the output conditions when the input data are present, the arithmetic operation is complete, but the storage device is not ready? Or when input data are present and the storage device is ready but the arithmetic operation is not complete? Or any of a number of other possibilities?

Truth Tables. Investigation of all possible combinations of a number of conditions may be accomplished in tabular form using *truth tables*, a technique borrowed from formal logic. The *truth value* of English-language phrases and combinations of phrases is investigated with truth tables. Consider the phrase "the arithmetic operation is complete." If the arithmetic operation is complete, the phrase is true (T); if the operation is *not complete*, it is false (F). (For the purpose of this discussion, T is defined as useful form and F is defined as not useful form.) No other conditions can occur. The arithmetic operation *must* be either true or false. *It cannot be both together nor can it be nonexistent*. All other input conditions may be viewed in this same manner.

Each condition, however, must be examined with all possible combinations of every other condition. Since each condition has *two* possible forms, the total number of possible combinations may be expressed as

$$N=2^m$$

where $N=$ total number of possible combinations
$m=$ total number of variables

EXAMPLE 2-1 How many possible combinations exist with three variables?

SOLUTION

$N=2^m=2^3=8$

Three variables are used in Example 2-1. The eight possible combinations of the three variables may be tabularly arranged in a number of different ways, but the most common form is shown in Table 2-1a. It is now merely necessary to apply the AND gate rule to each of the combinations to determine the outputs. Only *one* combination of inputs supplies a useful (T) output, that combination of inputs where *all* inputs are useful (T) at the same time. Note the ease with which this information is derived.

Table 2-1 AND Gate Truth Tables

Inputs A	*Inputs* MR	*Inputs* D	*Outputs* X	Row	*Inputs* A	*Inputs* MR	*Inputs* D	*Outputs* X
F	F	F	F	0	0	0	0	0
F	F	T	F	1	0	0	1	0
F	T	F	F	2	0	1	0	0
F	T	T	F	3	0	1	1	0
T	F	F	F	4	1	0	0	0
T	F	T	F	5	1	0	1	0
T	T	F	F	6	1	1	0	0
T	T	T	T	7	1	1	1	1

(a) (b)

Perhaps by this time the reader has recognized that the means of identifying variable conditions is binary in nature; that is, the conditions exist in only one of two states. It becomes convenient to use binary notation (1–0) in truth tables rather than the formal logic T–F notation because of the many correlations that exist with numerical operation in the microprocessor. One (1) is commonly used to represent true (T) and zero (0) replaces false (F). The truth table of Table 2-1a now appears as shown in Table 2-1b, which is the form used in most applications.

How can it be assured that *all* possible combinations of variables being investigated are used? It has been shown that eight combinations exist with three variables, and the truth tables of Table 2-1 have eight combinations. Investigation of the truth table verifies that each row is different, so all possible combinations must be included. The sequence of rows is unimportant, although the arrangement shown in Table 2-1 is preferred. Random sequencing could be used, but such a method is quite inefficient, especially when a large number of variables are used. The easiest method of arrangement is to assign an identification number to each row and arrange the binary value of each row to correspond to its identification number. The rows are arranged in ascending numerical order, top to bottom.

Number System Concepts. Correlation of row-identification numbers and the binary value of each row requires the ability to recognize binary equivalents of decimal numbers, and vice versa. Fundamental to this skill is the understanding of the concepts of number systems and conversions between decimal and binary, or vice versa. Development of these concepts follows.

Any number, in any number system, can be expressed as a *summation of products*. For example, the decimal number 1234.56 may also be expressed in the form

$$1 \times 10^3 + 2 \times 10^2 + 3 \times 10^1 + 4 \times 10^0 + 5 \times 10^{-1} + 6 \times 10^{-2}$$

Each position is related to its adjacent position by a power of 10, such as 10^3, 10^2, 10^1 and so on. The value of a digit placed in a specific position is equal to the positional value of the position multiplied by the digit. Thus, a 2 placed in the 10^1 position is equal to 20, while a 2 placed in the 10^3 position is equal to 2000.

Binary numbers are also formed using positional notation. The binary number 1010.10 may also be expressed in the form

$$1 \times 2^3 + 0 \times 2^2 + 1 \times 2^1 + 0 \times 2^0 + 1 \times 2^{-1} + 0 \times 2^{-2}$$

In binary numbers each position is related to its adjacent position by a power of 2, such as 2^3, 2^2, 2^1, and so on. The value of a digit placed in a specific position is equal to the positional value of the position multiplied by the digit. Thus, a 1 placed in the 2^1 position is equal to 2, while a 1 placed in the 2^3 position is equal to 8.

Correlation between the decimal number system and the binary number system (for decimal 0 through 15) can be seen in Table 2-2.

OR Gate. Another type of gate used in microprocessors is the OR gate. The *OR gate* provides a *useful* output when *any one or more* of its inputs is *useful.* An OR gate could be used to route data into the

Table 2-2 Decimal-Binary Number Comparison

Decimal	*Binary*
0	0000
1	0001
2	0010
3	0011
4	0100
5	0101
6	0110
7	0111
8	1000
9	1001
10	1010
11	1011
12	1100
13	1101
14	1110
15	1111

microprocessor's control circuitry. For example, the microprocessor might accept information if it comes from a storage device, a keyboard, or a group of switches. These conditions may be described in the following manner:

> Input to the microprocessor occurs if information is available from the storage device OR the keyboard OR the switch panel.

If X = useful output

S = data available from the storage device

K = data available from the keyboard

B = data available from the switches

then $X = S$ OR K OR B. Once again, the substitution of letter variables for English-language statements greatly simplifies presentation of information when discussing microprocessors.

Inverters. Often the electronic implementation of operations in microprocessors results in outputs being false when they should be true, and vice versa. The third type of circuit, the *inverter*, allows false inputs or outputs to be made true, and true inputs or outputs to be made false. It will be seen shortly that such capability results in much greater versatility in logic circuits.

Now that the three basic gates of digital logic have been discussed, the methods of describing them should be shown. Each of the gates, when used in diagrams showing their interconnections, has a unique symbol for identification purposes. Furthermore, when a Boolean algebra expression describing the function of gates is used, a mathematical symbol to correlate with the identification symbols may be employed. Figure 2-1 shows the symbolic representation of the AND gate. OR gate, and NOT circuit (inverter).

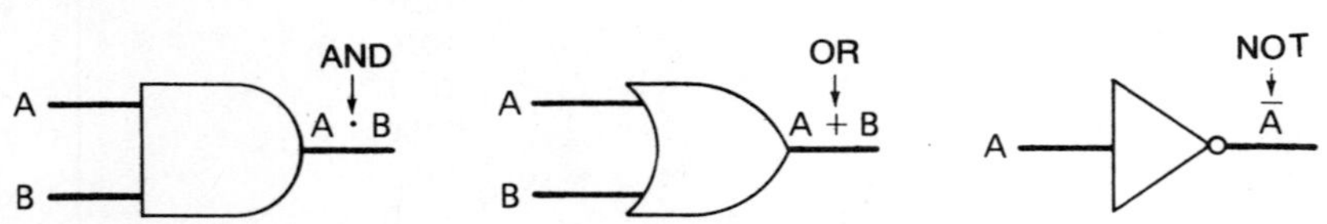

Figure 2-1 AND, OR, and NOT Gates

NOR and NAND Gates. Two specific cases exist where basic circuits are combined into other basic circuits. These cases are the NOR gate and the NAND gate. Manufacturing techniques may make it more economical to implement a microprocessor or its peripheral equipment using either NOR or NAND gates rather than OR, AND, and NOT circuits. Since the three basic circuits, together with all sequential circuits, may be implemented with NOR and NAND gates, a considerable cost saving can result during manufacture of microprocessor devices using the "derived" gates.

The *NOR gate* performs the same function as an OR gate followed by an inverter, or NOT operation. It has been shown that the output of an OR operation is true when any of the inputs are true. If an inverter follows the OR gate, the output will be false when any of the inputs to the OR gate are true. The only condition that results in a true output is when both inputs are false. This information is summarized in Figure 2-2a.

If the desired output of the NOR gate is to be true, the logical function is fulfilled *only* when inputs A and B are false, $\overline{A}\cdot\overline{B}$. Thus, a means now exists (the NOR gate) to obtain the logical AND operation for false inputs.

The *NAND gate* performs the same function as an AND gate followed by an inverter, or NOT operation. It has been shown that the output of an AND operation is true when and only when all inputs are true. If an inverter follows the AND operation, the output will be false when and only when all inputs to the AND gate are true. All other input conditions result in a false output. This information is summarized in Figure 2-2b.

If the desired output of the NAND gate is to be true, the logical function is fulfilled when either A or B is false, or $\overline{A}+\overline{B}$. Thus, a means now exists (the NAND gate) to obtain the logical OR operation for false inputs.

One final "convention" should be mentioned before proceeding. Microprocessor circuits are often investigated using devices that display the electrical form of logic quantities. Oscilloscopes (Figure 2-3a), logic analyzers (Figure 2-3b), and logic probes (Figure 2-3c) are typical. These devices are interested in the *relative magnitude* of the logic signal. The most common convention is to consider the true or 1 condition as HIGH and the false or 0 condition as LOW. When viewed on an oscilloscope or logic analyzer, a HIGH signal appears toward the top of the cathode-ray tube, while a LOW signal appears toward the bottom. Thus, a HIGH signal is higher than a LOW signal. Logic probes employ either a light-on/light-off technique or separate lights for HIGH/LOW indications. In this book both 1–0 and HIGH–LOW conventions will be used to describe logic conditions, so the reader should be prepared to recognize the correlation between both conventions.

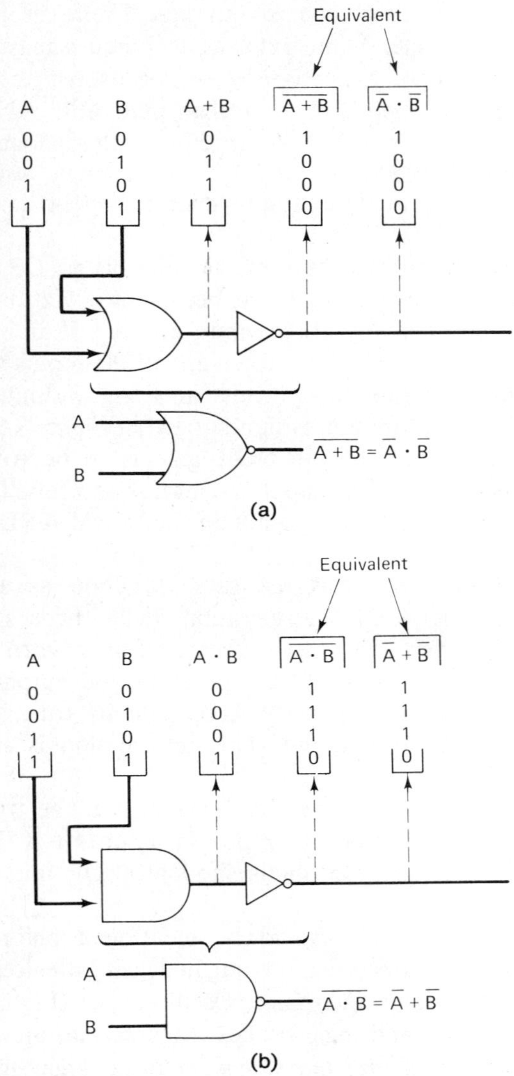

Figure 2-2 NOR and NAND Gates: (a) NOR; (b) NAND

Boolean Algebra. Application of combinational logic circuits to actual microprocessor functions not only demonstrates the use of logic circuits, but also shows some of the operations of Boolean algebra. One general group of decision-making circuits *compares* inputs and supplies an

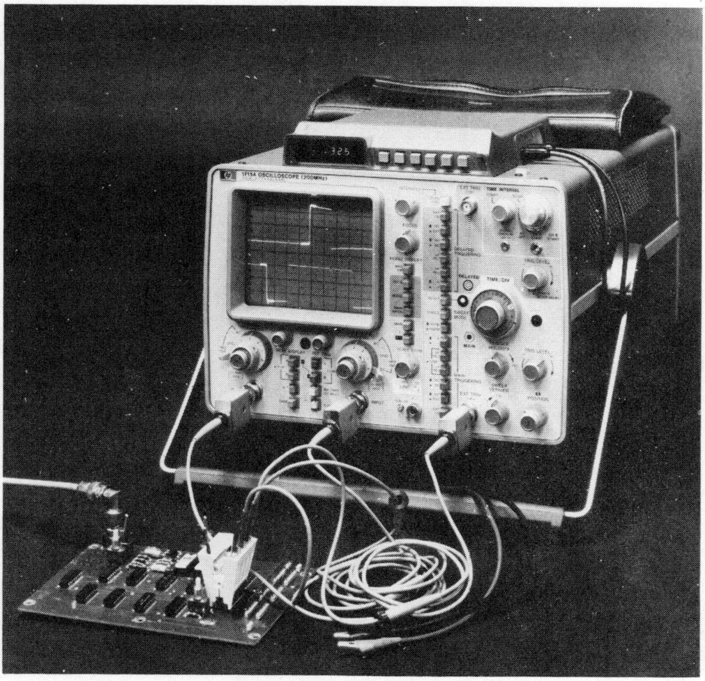

(a)

Figure 2-3 Digital Test Equipment (a) Oscilloscope; (b) Logic Analyzer; (c) Logic Probes (Courtesy Hewlett-Packard)

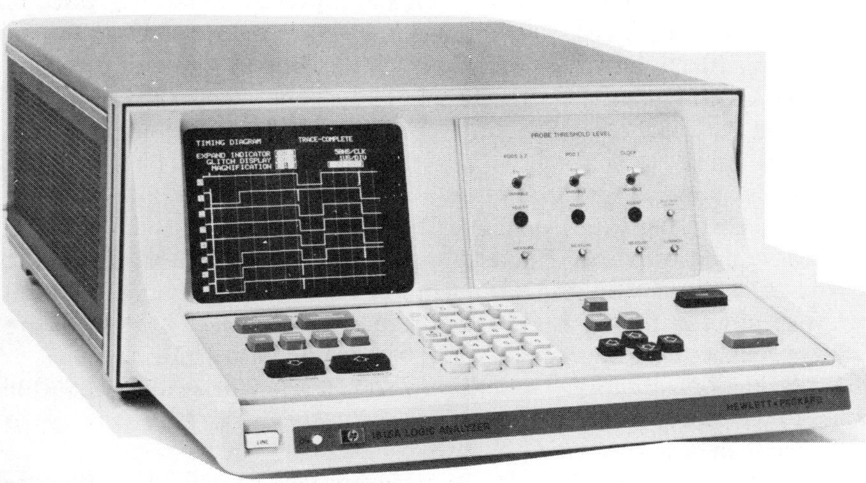

(b)

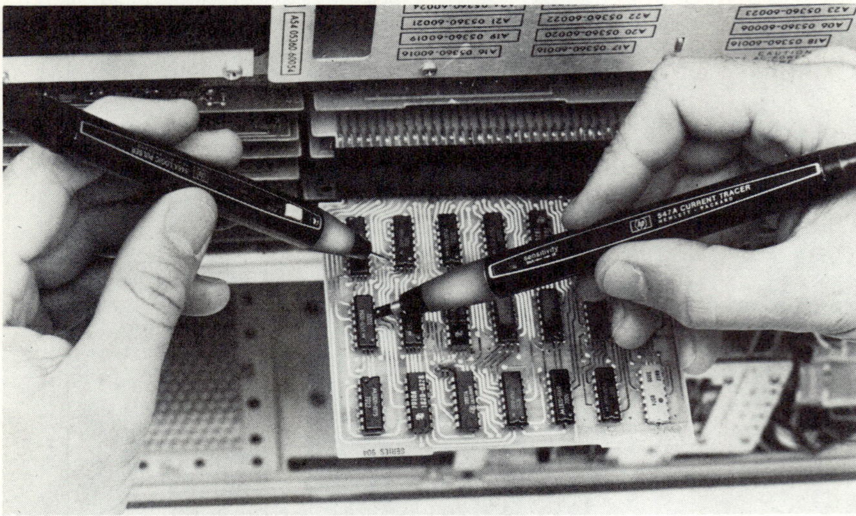

(c)

Figure 2-3 (Continued)

output based on the relationship of these inputs. For example, it is a common requirement in microprocessors to decide whether two binary numbers are equal or not equal. In simple form, the *Exclusive-OR* circuit (Figure 2-4) can perform this function.

The Exclusive-OR operation is a function that appears so often in digital circuitry that it has been assigned a special operating symbol. It is not, however, an independent operation and thus may also be expressed in terms of the basic gates. The expression *A Exclusively-OR B* is written $A \oplus B$. This function is true ($A \oplus B = 1$) when A or B is true, but *not* when both are true. Figure 2-4 shows the standard symbol for the Exclusive-OR operation, its truth table, algebraic expression, and a typical implementation using gates.

The expression for the Exclusive-OR circuit ($A\overline{B} + \overline{A}B = A \oplus B$) demonstrates how AND, OR, and NOT operations may be combined and written in algebraic form. The term $A \cdot \overline{B}$ is ORed with the term $\overline{A} \cdot B$. Each of the terms consists of an AND operation with one input in the true form and one input in the false or NOTed form. These *Boolean expressions* are used throughout microprocessor data sheets and instruction manuals to describe the operations to be performed on inputs. The reader should be able to correlate Boolean expressions with the logic circuits used to implement the expressions.

Another application of Boolean algebra can be seen if an inverter (NOT circuit) is added to the Exclusive-OR circuit. Recalling that an

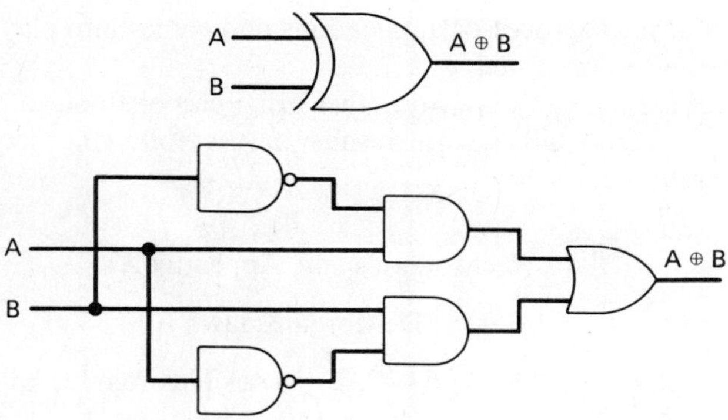

A	B	A ⊕ B
0	0	0
0	1	1
1	0	1
1	1	0

Figure 2-4 Exclusive-OR Function

inverter NOTs its input, the output of the NOTed Exclusive-OR can be written $\overline{A\cdot\overline{B}+\overline{A}\cdot B}$. One of the laws of Boolean algebra (*DeMorgan's law*) states that inversion or *negation* of a Boolean expression or statement may be accomplished by negating each variable/constant, changing each AND to OR, and each OR to AND. That is, $\overline{A\cdot B}=\overline{A}+\overline{B}$; $\overline{A+B}=\overline{A}\cdot\overline{B}$. One must be certain that the difference between the expressions $\overline{A+B}$ and $\overline{A}+\overline{B}$ is recognized. Note that in one case A and B are combined in an OR relationship and *then* the output is inverted. In the other case A and B are each inverted first and then the inverted variables are combined in the OR relationship. *The two expressions are not equal*, as can be shown by constructing truth tables for each expression and comparing the results (see Table 2-3). The basic DeMorgan's law is a very powerful tool. It is

Table 2-3 De Morgan's Laws Truth Table

Not identical: $\overline{A\cdot B}$ / $\overline{A}\cdot\overline{B}$; $\overline{A}+\overline{B}$ / $\overline{A+B}$

A	$\overline{A}$	B	$\overline{B}$	A · B	$\overline{A\cdot B}$	$\overline{A}+\overline{B}$	A + B	$\overline{A+B}$	$\overline{A}\cdot\overline{B}$
0	1	0	1	0	1	1	0	1	1
0	1	1	0	0	1	1	1	0	0
1	0	0	1	0	1	1	1	0	0
1	0	1	0	1	0	0	1	0	0

Identical: $\overline{A\cdot B}$ and $\overline{A}+\overline{B}$; Identical: $\overline{A+B}$ and $\overline{A}\cdot\overline{B}$

used throughout microprocessor applications and comes into play any time a NOT operation is performed.

Applying DeMorgan's law and other principles of Boolean algebra to the NOTed Exclusive-OR circuit results in the following sequence of algebraic operations:

$\overline{A\cdot\overline{B}+\overline{A}\cdot B} = \overline{A\cdot\overline{B}}\cdot\overline{\overline{A}\cdot B}$ (DeMorgan's law, Appendix A)

$= (\overline{A}+\overline{\overline{B}})\cdot(\overline{\overline{A}}+\overline{B})$ (DeMorgan's law)

$= (\overline{A}+B)\cdot(A+\overline{B})$ ($\overline{\overline{B}}=B$, $\overline{\overline{A}}=A$ per Theorem 13, Appendix A).

$= \overline{A}A+\overline{A}\,\overline{B}+AB+B\overline{B}$ (logic variables may be multiplied algebraically as in conventional algebra)

$= 0+\overline{A}\,\overline{B}+AB+0$ ($\overline{A}A=0, \overline{B}B=0$ per Theorem 6, Appendix A)

$= \overline{A}\,\overline{B}+AB$ ($0+A=A$ per Theorem 8, Appendix A)

$= AB+\overline{A}\,\overline{B}$ (Commutative Law, Appendix A)

When the truth table for the Exclusive-NOR operation is investigated, it is seen that a means now exists to determine the *equality* of two binary digits. When A and B are both 1 or both 0, the logic value of the negated Exclusive-OR operation is 1. If the Exclusive-NOR operation is implemented as in Figure 2-5, it is possible to obtain an indication of not

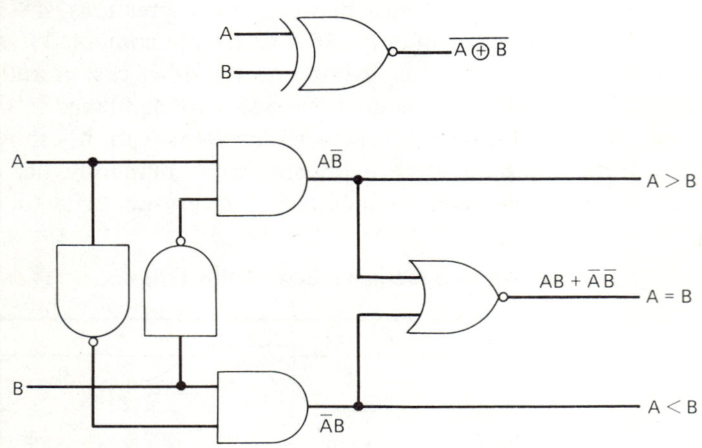

A	B	A > B	A = B	A < B
0	0	0	1	0
0	1	0	0	1
1	0	1	0	0
1	1	0	1	0

Figure 2-5 Exclusive-NOR Function

only $A = B$ and $A \neq B$, but also the relative magnitude ($A < B$ or $A > B$) of two binary digits. The truth table accompanying Figure 2–5 verifies the statements above.

Combinational logic circuits are also applied to the recognition of a prescribed set of conditions. It will be seen in subsequent chapters that the microprocessor is told what it is to do next by a group of binary digits (*bits*), called an *instruction word*. Each instruction consists of a unique combination of 1s and 0s, and combinations of gates are provided to recognize each unique instruction and supply control information to the rest of the microprocessor. The *instruction decoder*, which performs this function, is discussed in Chapter 4.

Since the input to a microprocessor is quite often in a form other than that which the microprocessor can accommodate, combinational logic circuits called *converters* are employed to change information from one form to another. Such circuits are also found within the microprocessor, as well as at the output of the microprocessor.

2-2 SEQUENTIAL CIRCUITS

Flip-Flops. Storage/timing operations are performed by *sequential*, or time-dependent circuits. The term *flip-flop* (*FF*) is commonly applied to logic circuits that perform sequential operations. A flip-flop is a *bistable* electronic circuit, a circuit that has *two and only two stable states*. It may be likened to a light switch, which is either on or off, but never in an "in-between" state. The ON state is often called the SET state, and the OFF state the RESET or CLEAR state. A flip-flop's graphic symbol is shown in Figure 2-6a. This FF (called the RS FF) has two inputs, R and S, and two outputs Q and $\overline{Q}$ (NOT Q). The operation of a FF may be described in the following manner:

> If the FF is in the SET state, the Q output is HIGH or true and the $\overline{Q}$ output is LOW or false. Conversely, in the RESET state, the $\overline{Q}$ output is HIGH and the Q output is LOW.

Application of an *activating** input to the SET input places the RS FF in the SET state, while the application of an activating input to the RESET input results in the RS FF assuming the RESET state. As drawn, the RS FF of Figure 2-6 requires a HIGH-going input as an activating

*An activating input is one that causes the logic device to perform its intended function.

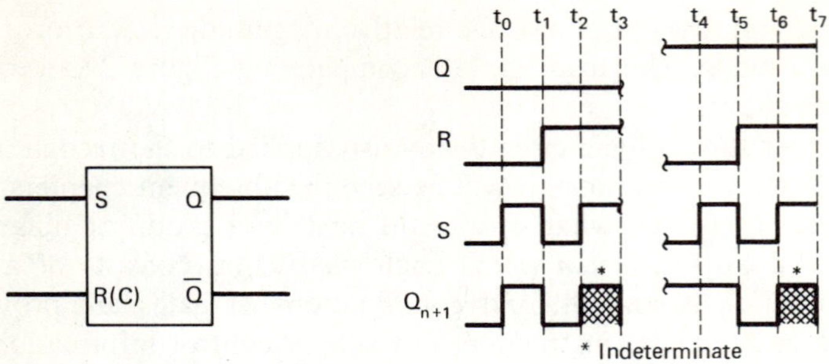

(a) Symbol

(b) Waveforms

Note: Q_n is the state of the FF *before* the inputs shown and Q_{n+1} is the state *after* the inputs shown are applied.

Input		If Q output			
R	S	Was	It becomes	Was	It becomes
Inactive	Inactive	Inactive →	Inactive	Active →	Active
Inactive	Active	Inactive →	Active	Active →	Active
Active	Inactive	Inactive →	Inactive	Active →	Inactive
Active	Active	*		*	

(c) General operating characteristics

R	S	Q_n	Q_{n+1}
0	0	0	0
0	0	1	1
0	1	0	1
0	1	1	1
1	0	0	0
1	0	1	0
1	1	0	*
1	1	1	*

(d) Detailed state table

R	S	Q_{n+1}
0	0	Q_n
0	1	1
1	0	0
1	1	*

(e) Simplified state table

Figure 2-6 RS Flip-Flop

signal. HIGH-going is defined as a transition from a LOW logic level to a HIGH logic level, as shown in Figure 2-6b.

The type of diagram shown in Figure 2-6b is representative of conventional presentations encountered in instruction sheets furnished with microprocessors. Such diagrams (and practically all microprocessor operations) are time-oriented. Time is usually considered to start at the left side of the page and to progress to the right. Time intervals are represented by subscripted letters, such as t_0, t_1, t_2, and so on.

Many other types of FF configurations exist in microprocessor circuits, depending on specific requirements. Such names as T (toggle),

Inputs							Outputs					
							R-S	Clocked R-S	T	D	J-K	Clocked J-K
S	R	T	D	J	K	CP	Q_{n+1}	Q_{n+1}	Q_{n+1}	Q_{n+1}	Q_{n+1}	Q_{n+1}
0	0						Q_n					
0	1						0					
1	0						1					
1	1						*					
0	0					0		Q_n				
0	0					1		Q_n				
0	1					0		Q_n				
0	1					1		0				
1	0					0		Q_n				
1	0					1		1				
1	1					0		Q_n				
1	1					1		*				
		0							Q_n			
		1							$\overline{Q}_n$			
			0							0		
			1							1		
				0	0						Q_n	
				0	1						0	
				1	0						1	
				1	1						$\overline{Q}_n$	
				0	0	0						Q_n
				0	0	1						Q_n
				0	1	0						Q_n
				0	1	1						0
				1	0	0						Q_n
				1	0	1						1
				1	1	0						Q_n
				1	1	1						$\overline{Q}_n$

*Indeterminate

Figure 2-7 Sequential Logic Characteristics

D (delay), and JK are encountered. Each of these performs a specific timing operation and responds according to specific rules. The common FF configurations are summarized in Figure 2-7, together with their rules of operation.

Counters. Flip-flops may be combined to form many different types of counting, timing, and information-handling circuits. A *counter* is a device, composed of properly connected sequential (and sometimes combinational) elements, which performs the very basic function of counting.

Counters are binary in nature, but may be modified to count in other than the natural binary sequence—for example, base 8, base 10, and so on.

Counters in microprocessors are very seldom constructed from single FFs any longer. A simple four-stage counter is shown in Figure 2-8a simply for the purpose of reviewing the concepts of binary counting. Within modern microprocessors and microcomputers, complete counters are either packaged as a part of the microprocessor, or appear as a separate integrated-circuit assembly. The counter of Figure 2-8 is an UP counter which starts counting from some predetermined point (often all FFs are cleared) and counts in ascending order. The count is accumulated in the binary system and proceeds in the order shown in Table 2-2. When

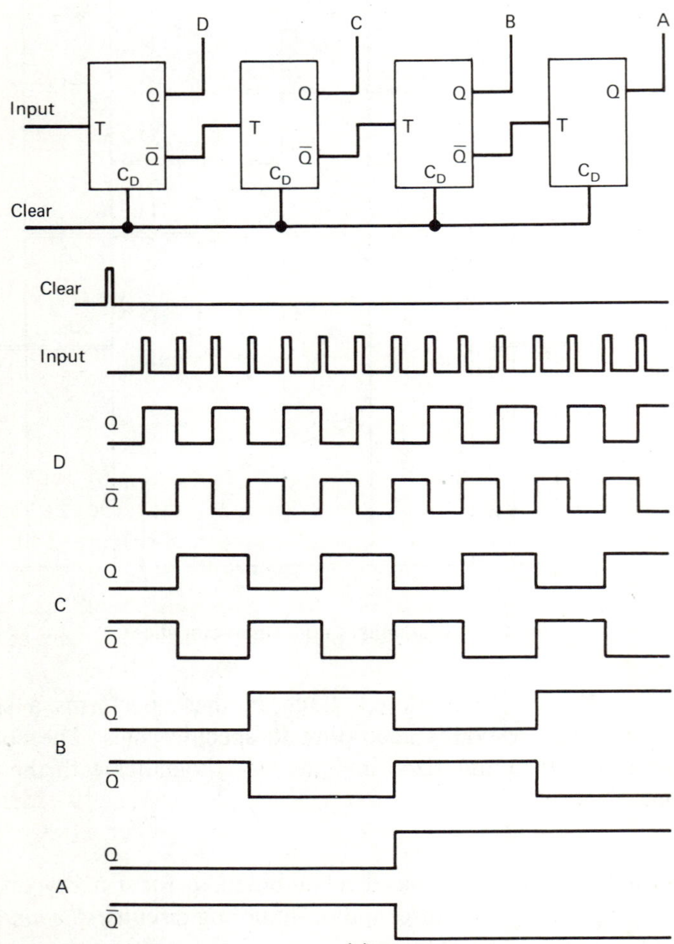

(a)

Figure 2-8 Counters: (a) Binary UP Counter; (b) Decimal Counter

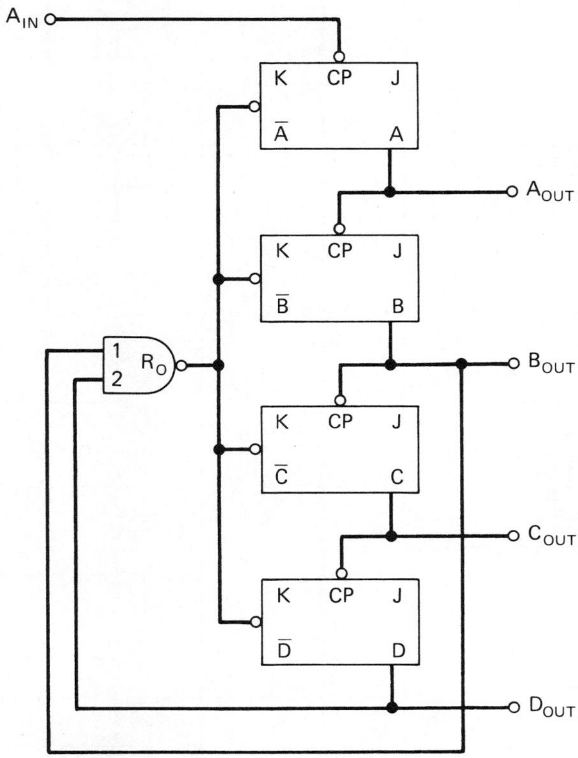

(b)

Figure 2-8 (Continued)

the maximum count is reached, the counter recycles and repeats. The waveforms of Figure 2-8a should be correlated with the FF rules of operation for both this and the decimal counter of Figure 2-8b. If an application is encountered that requires a recycle at some count earlier than the full capability of the counter, combinational circuits are used to detect that count and force a recycle. Figure 2-8b shows how the basic UP counter may be modified to recycle at count 10, thus forming a *decimal* counter.

Counters may be operated in modes other than UP counting. DOWN counting is common, where the counter starts from a predetermined point (often all FFs SET) and counts in a descending order. Both UP and DOWN capability may be combined with proper gating circuits, as may preloading the counter to any desired starting count.

Registers. Requirements exist within the microprocessor for temporary storage of information. This function may be accomplished by logic circuits called *registers*. A register is a group of logic circuits, usually

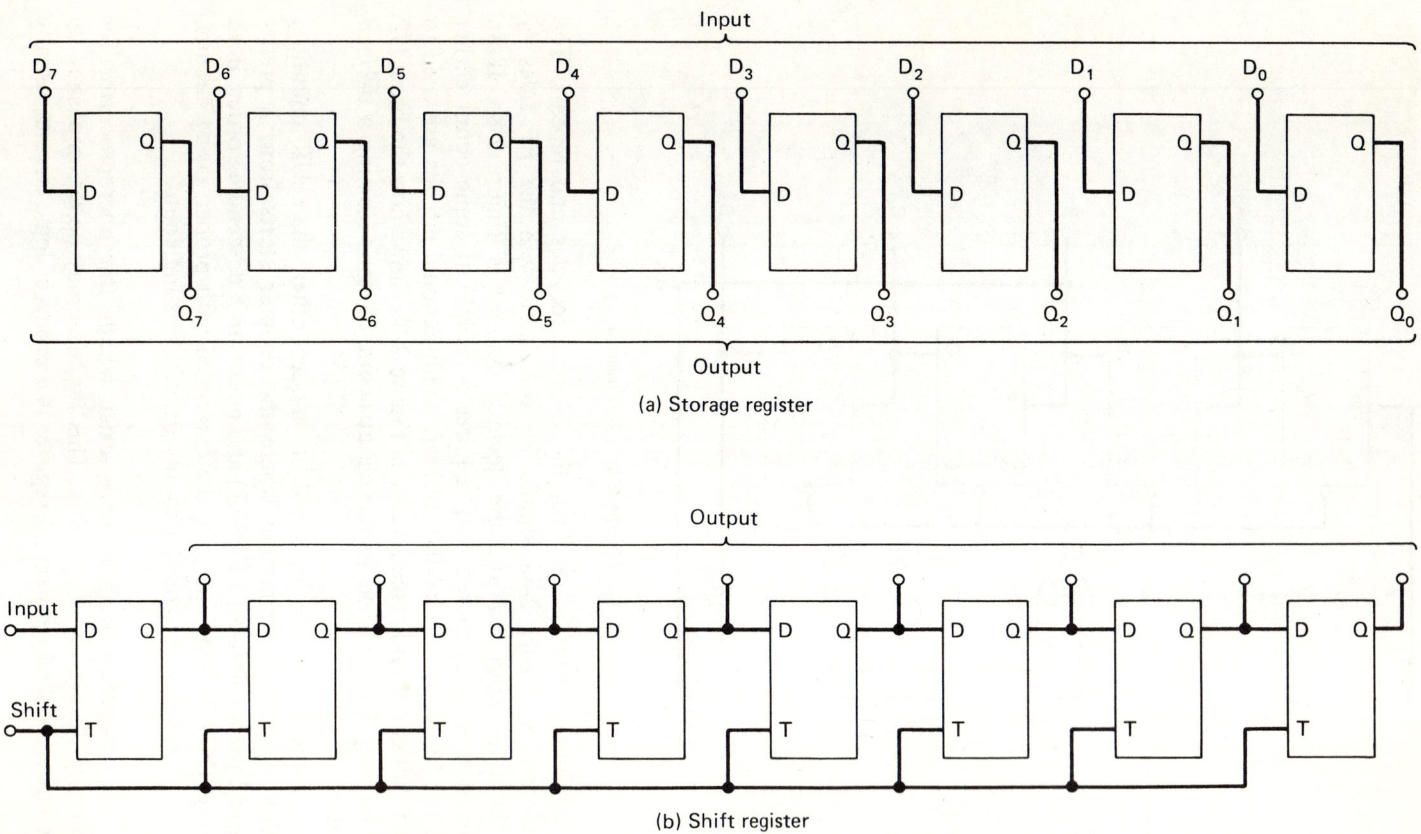

Figure 2-9 Registers: (a) Storage Register; (b) Shift Register

FFs, arranged in a manner that allows storage and processing of information in binary form. The number of FFs in a register is determined by the requirements of the specific application.

A *storage register retains information*, while a *shift register processes information*. Information may be entered into and taken from registers sequentially bit by bit (*serial* form) or all bits at once (*parallel* form).

As stated before, information in a microprocessor is contained in groups of bits called *words*. Often it is necessary to temporarily store a word awaiting time for the microprocessor to process it. Assuming that the word contains 8 bits, then 8 FFs connected as in Figure 2-9a will perform the storage function. Additional combinational circuits are used to transfer the information from the register to the microprocessor.

Other applications require information conversion from serial form to parallel form. The register connection of Figure 2-9b will accept information bit by bit, store it until needed, and then provide it to the microprocessor, all 8 bits at a time. As with the storage register, additional combinational circuits are used during information transfer.

Registers, like counters, are seldom found constructed from separate FFs any longer. Separate integrated-circuit assemblies, or direct inclusion on the microprocessor chip, are the most common forms of modern registers.

2-3 SUMMARY

This chapter has served merely to review the basic principles of digital logic. Combinational circuits and their applications, together with sequential circuits and their applications, have been mentioned, and some of the basic logic-circuit interconnections shown. The questions that follow represent the general knowledge that the reader should have before proceeding. If difficulty is encountered with any of the questions and review of this chapter does not alleviate the difficulty, it is recommended that the reader consult any of the books listed in the bibliography for assistance. Comprehension of material to follow may be difficult if the reader is unable to grasp the basic concepts of digital logic.

QUESTIONS

1. Define combinational circuits and explain their uses in microprocessors.
2. Define sequential circuits and explain their uses in microprocessors.
3. What is an AND gate? How would it be used in a microprocessor?
4. Explain the use of truth tables in describing logic functions.
5. What is an OR gate? How would it be used in a microprocessor?
6. What is an inverter? How would it

be used in a microprocessor?
7. What is a NOR gate? How would it be used in a microprocessor?
8. What is a NAND gate? How would it be used in a microprocessor?
9. Draw a logic diagram of an Exclusive-OR circuit and show its truth table and Boolean algebra description.
10. Explain DeMorgan's law and how it may be used in microprocessors.
11. Explain the operation of an RS flip-flop.
12. Explain the operation of a four-stage binary counter.
13. Show how a four-stage binary counter could be modified to count base 10; base 12.
14. Explain the difference between a storage register and a shift register.

ADVANCED QUESTIONS

1. Show how NOR gates may be combined to perform the OR function; the AND function; the NOT function; and the NAND function.
2. Show how NAND gates may be combined to perform the AND function; the OR function; the NOT function; and the NOR function.
3. Using logic diagrams and/or Boolean algebra, explain the Distributive Law; the Laws of Absorption; and the Laws of Expansion (see Appendix A).
4. Explain the operation of a D flip-flop.
5. Explain the operation of a JK flip-flop.
6. Can a single register be used for both storage and shifting purposes? Explain how.

CHAPTER 3

Computer Arithmetic

Although the microprocessor is basically a processor of information, much of its capability is closely allied to arithmetic/logical capability. We have already seen in Chapter 2 how the logical operations AND, OR, and NOT may be accomplished with single-digit binary numbers. In this chapter we shall investigate logic operations concerned with multidigit binary numbers, as well as the accomplishment of arithmetic operations with multidigit binary numbers.

Microprocessors perform arithmetic and logic functions in some form of binary operations. The only real hindrance to the use of binary operations is the apparent difficulty that many experience when binary-based arithmetic is performed. If one realizes that binary-based arithmetic operates with a set of rules that parallel arithmetic in the decimal (or other) number systems, part of the mystery disappears. The rules for binary-based arithmetic are developed in this chapter.

3-1 BINARY ADDITION

Binary addition operates with the same rules as decimal addition, except that it is simpler. The binary addition rules shown below completely define the operations to be performed.

A		B		S	C
Addend		Augend		Sum	Carry
0	plus	0	=	0	0
0	plus	1	=	1	0
1	plus	0	=	1	0
1	plus	1	=	0	1

Examples of binary addition are shown in Examples 3-1, 3-2, and 3-3.

EXAMPLE 3-1

Decimal	*Binary*
5	101
+2	+ 10
7	111

EXAMPLE 3-2

10	1010
+ 9	+1001
19	10011

EXAMPLE 3-3

27	11011
+15	+ 1111
42	101010

The Half-Adder Circuit. Logic circuits may be used to implement the binary addition operation. By equating numeric 0 with binary 0 and numeric 1 with binary 1, the techniques of Boolean algebra can be used to determine the interconnection of logic circuits required to perform binary addition. According to the rules of binary addition, the addition of an addend digit A and an augend digit B results in a sum digit and perhaps also a carry digit. Examining the sum (S) requirement first, it can be seen that a sum of 1 exists when A is 0 and B is 1 or when A is 1 and B is 0—that is, $A\overline{B}+\overline{A}B$. A carry ($C$) of 1 appears only when both A is 1 and B is 1 (AB). Thus, a logic circuit is required that will accept two inputs, A and B, and provide two outputs, $S=A\overline{B}+\overline{A}B$ and $C=AB$.

C is easily obtained from a simple two-input AND gate. The S output occurs from an *Exclusive-OR* circuit (see Figure 2-4). An assembly that supplies both *S* and *C* outputs for the two inputs *A* and *B* is called a *half-adder*. The half-adder may be used to add *two* binary digits but is limited in application because of the possibility of a carry occurring in a less-significant position. A full-adder is required, and used, in practically all binary arithmetic operations due to the carry limitation of the half-adder.

The Full-Adder Circuit. A *full-adder* is capable of accepting not only the addend and the augend, but also a carry input. It furnishes a sum and carry output. Using the basic rules of binary addition, Table 3-1 is

Table 3-1 Full-Adder Truth Table

A	B	C_i	S	C_o
0	0	0	0	0
0	0	1	1	0
0	1	0	1	0
0	1	1	0	1
1	0	0	1	0
1	0	1	0	1
1	1	0	0	1
1	1	1	1	1

developed to show all possible combinations of the addend, augend, and carry input with resultant sum and carry outputs. One possible implementation of the logic equations resulting from Table 3-1 is shown in Figure 3-1. In microprocessor applications one full-adder is used for each binary digit. If each binary number consists of 8 bits, eight full-adders would be found in the microprocessor. Storage and shift registers, along with gating circuits, are associated with the full-adders so that information may be stored and manipulated prior to performance of the actual addition operation. Detailed hardware operation is discussed in subsequent chapters.

3-2 SIGNED NUMBERS

Concepts. Perhaps it has been noted that addition examples shown so far have considered only unsigned numbers. In practical use, both positive and negative numbers are encountered, and the microcomputer must be capable of handling them in various combinations. When working with signed numbers, computer words often contain a *sign bit* adjacent to the most-significant digit. The most common convention assigns a binary 0 as a positive sign and a binary 1 as a negative sign. For example $+10_{10}$ exists as the 4-bit binary word (*plus* sign bit) 01010,

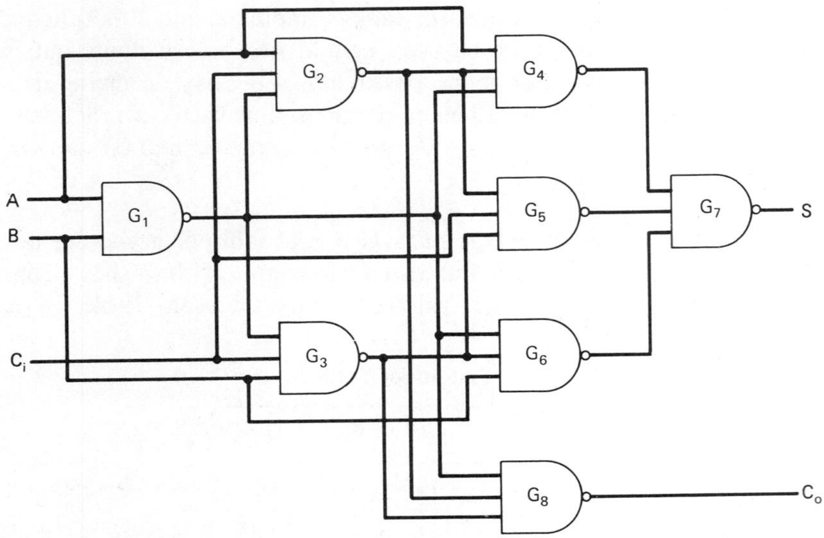

Sum = $\bar{A}\cdot\bar{B}\cdot C + \bar{A}\cdot B\cdot\bar{C} + A\cdot\bar{B}\cdot\bar{C} + A\cdot B\cdot C$

Carry out = $\bar{A}\cdot B\cdot C + A\cdot\bar{B}\cdot C + A\cdot B\cdot\bar{C} + A\cdot B\cdot C = A\cdot C + A\cdot B + B\cdot C$

Figure 3-1 Full-Adder Logic Diagram

thereby requiring 5 bits. However, *negative* numbers may be stored in a variety of forms, as follows:

Decimal form	*Sign-true magnitude form*		*One's-complement form*		*Two's-complement form*	
-10_{10}	1	1010	1	0101	1	0110
	sign	magnitude	sign	magnitude	sign	magnitude

The sign-true magnitude form defines a negative number by the 1 bit adjacent to the most-significant digit. Number *magnitude* remains in the same form as the equivalent positive number.

Complemented numbers are frequently used in the subtraction and mixed-sign addition processes. Every number system has two types of complements: *radix-minus-one* and *true*. The *radix-minus-one complement* is formed by subtracting each digit of a number from the radix of the number system minus one. *Radix* is a term used interchangeably with *base*. Consider the familiar decimal number system (radix 10). The radix-minus-one complement of 7 is 2 ($9-7=2$), while the radix-minus-one complement of 66 is 33 ($99-66=33$). The radix-minus-one complement in the decimal system is more commonly called the *nine's complement*. In the binary number system (radix 2), the radix-minus-one complement is formed by changing each 1 in a number to 0 and each 0 to 1. Thus, the radix-minus-one complement of 1 is 0, and the radix-minus-one complement of 10101 is 01010. The radix-minus-one complement in the binary system is called the *one's complement*.

True complements are obtained by subtracting each digit of a number from the radix of the number system.* The true (ten's) complement of 7 is 3 (10−7=3), while the ten's complement of 66 is 34 (100−66=34). Binary *true (two's) complements* are also easily formed. One merely determines the one's complement and adds 1 to the result. For example, the two's complement of 10101 is obtained by one's-complementing (01010) and adding 1 (01011).

Addition of Signed Numbers. Addition of signed binary numbers follows the same rules used for addition of signed decimal numbers.

> *Rule 1:* If both the addend and augend have the same signs, the sum is the total of both numbers with the sign retained.

Addition of two signed positive numbers presents no problem as long as the sum is no greater than the absolute value storable in the computer's registers and memory. A 5-bit register, when used with signed numbers, can accommodate values between 0 (00000) and +15 (01111). A sum greater than +15, therefore, must provide some indication of "overflow." Example 3-4 shows an addition that falls within the representable range of such a register, while Example 3-5 shows an example of an overflow condition.

EXAMPLE 3-4

Decimal	Binary (sign, magnitude)	
+ 8	0 1000	numbers in *sign-true magnitude* form
+ 6	0 0110	
+14	0 1110	← sum is in *sign-true magnitude* form

(sign ┐ ┌ magnitude)

* *True complements* may also be obtained by subtracting each digit of a number from the radix-minus-one of the number system and then adding 1 to the least-significant digit of the resulting number. For example, (9−7)+1=3; and (99−66)+1=34.

EXAMPLE 3-5

```
                 sign┐ ┌—magnitude
  +11              0 1011 }
+  7               0 0111 }——— numbers in sign-true magnitude form
 ————              ——————
  +18              1 0010 ←——— sum is in sign-true magnitude
                               form (considering all 5 bits),
                               but the sign is different; there-
                               fore, the result is invalid; also,
                               if a carry exists into the sign
                               position with no carry out, it
                               is further indication of in-
                               validity (an overflow indica-
                               tion would probably occur)
```

Furthermore, if two negative numbers are summed, it is also apparent that difficulties exist (see Example 3-6).

EXAMPLE 3-6

```
     − 5                 1 0101
  (+)− 9                 1 1001
   ————                  ——————
    −14                1 0 1110 ←——magnitude correct
              overflow—┘   ———sign bit incorrect
```

Because of the sign-bit discrepancy (and other problems soon to be discovered), negative numbers are seldom manipulated in sign-true magnitude form within the ARITHMETIC/LOGIC section of the microprocessor. If stored in memory in sign-true magnitude form, conversion to a form compatible with the ARITHMETIC/LOGIC section's operation must be performed *before* and *after* arithmetic manipulation. The use of *complementary addition* solves many of the problems encountered when negative numbers must be used.

Microprocessors commonly operate in the two's-complement mode. As described earlier, two's complements are formed by obtaining the one's complement and adding 1. The problem of Example 3-6 is developed using both ten's and two's complements in Example 3-7.

EXAMPLE 3-7

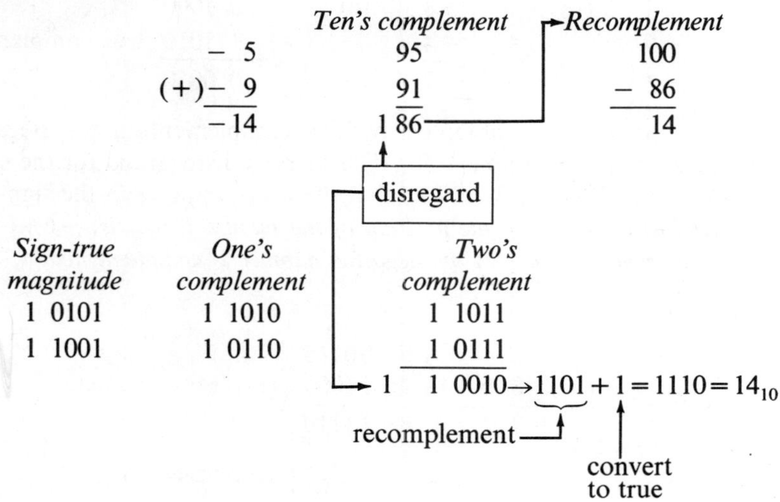

The two's-complement result is correct. The sign bit represents a negative number, the result of adding two negative numbers. Furthermore, although an overflow occurs, it is a valid operation, since a carry propagating *out* of the sign-bit position was accompanied by a carry *into* the sign-bit position. Magnitude representation is in the two's-complement form and, when converted to true form as shown, reveals the proper value. It should be remembered, however, that conversion probably does not occur until it is necessary to make this information available to the human user. The microprocessor may be constructed to always recognize a number prefixed with a "one" sign bit as a negative number and respond accordingly.

> *Rule 2:* If the addend and augend have unlike signs, *subtract* the smaller from the larger. The sum is the *difference* in absolute value of the two quantities with the sign of the larger attached.

This rule requires *subtraction*, and such a capability has not yet been discussed. However, an alternative compatible with designs using *only* adders is available and in common use. Verification of the *complementary addition* equivalence of subtraction is shown by demonstration of Rule 2. Consider first the addition shown in Example 3-8a.

EXAMPLE 3-8

$$
\text{(a)}\quad \begin{array}{rr} & +8 \\ (+) & -6 \\ \hline & +2 \end{array}
\qquad
\text{(b)}\quad \begin{array}{rr} & +8 \\ (+) & +4 \\ \hline + & \boxed{1}\ 2 \end{array}
\qquad
\text{(c)}\quad \begin{array}{rrl} & 0\ 1000 & \\ (+) & 1\ 1010 & \text{true complement} \\ \hline \boxed{1} & 0\ 0010 & \end{array}
$$

The same result is obtained by true-complementing the negative number and adding (Example 3-8b). The carry is disregarded for the same reason shown in Example 3-7. Note that the result appears in the sign-true magnitude form. *Reversing the position of the numbers, that is,* $-6+(+8)$, *shows no difference as long as the negative number is complemented.*

EXAMPLE 3-9

$$
\begin{array}{r} +6 \\ (+)-8 \\ \hline -2 \end{array}
\qquad
\begin{array}{r} +6 \\ (+)+2 \\ \hline 8 \end{array}
\qquad
\begin{array}{rl} 00110 & \\ 11000 & \text{(two's complement)} \\ \hline 11110 & \\ \downarrow & \\ 10001 & \text{recomplement} \\ +1 & \\ \hline 10010 = -2_{10} & \end{array}
$$

Note: 8 complements to -2

The hardware required to perform complementary addition (subtraction) is contained within the microprocessor. From a conceptual viewpoint, it is

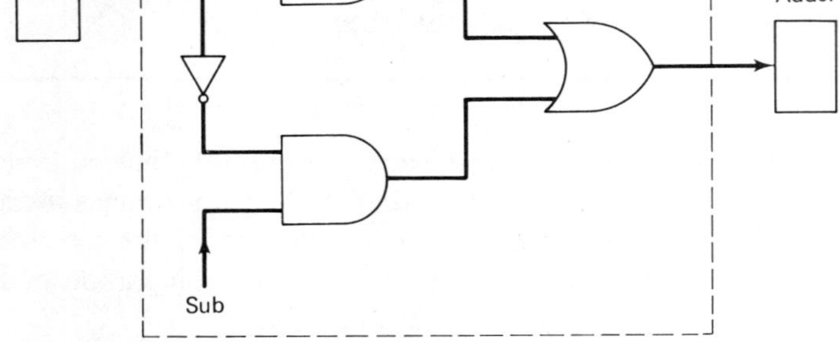

Figure 3-2 Concept of Gating

merely necessary to select the digit to be supplied to the adder for addition or its complement if subtraction is to be performed. Figure 3-2 demonstrates a simple gating network that routes the necessary logic signals for 1 bit of the complete word from a register to an adder input.

Table 3-2 shows a number of examples that summarize addition of signed binary numbers in a typical microprocessor.

Table 3-2 Addition and Subtraction of Signed Binary Numbers

	sign magnitude	
+8	0 1 0 0 0	+numbers in sign-true magnitude form
(+)+6	0 0 1 1 0	
+14	0 1 1 1 0	
−5	1 1 0 1 1	−numbers in two's complement form
(+)−9	1 0 1 1 1	
−14	☒ 1 0 0 1 0 (discard; sign)	complement→{1 1 0 1+1=1 1 1 0=14_{10} {convert to true magnitude form} true magnitude form
+8	0 1 0 0 0	+numbers in sign-true magnitude form
(+)−6	1 1 0 1 0	−number in two's complement form
+2	☒ 0 0 0 1 0 (discard; sign)	sign-true magnitude form
−6	1 1 0 1 0	−number in two's complement form
(+)+8	0 1 0 0 0	+number in sign-true magnitude form
+2	☒ 0 0 0 1 0 (discard; sign)	sign-true magnitude form
−9	1 0 1 1 1	−number in two's complement form
(+)+5	0 0 1 0 1	
−4	1 1 1 0 0 (sign)	complement→{0 0 1 1 +1=0 1 0 0=-4_{10} {convert to true magnitude form} true magnitude form

(continued)

Table 3-2 Continued

+5 (+) −9 −4	same as above (see previous page)		
+8 (−)+6 +2	+8 =(+)−6 +2	−8 (−)+6 −14	−8 =(+)−6 −14
+8 (−)−6 +14	+8 =(+)+6 +14	−8 (−)−6 −2	−8 =(+)+6 −2
+6 (−)+8 −2	+6 =(+)−8 −2	−6 (−)+8 −14	−6 =(+)−8 −14
+6 (−)−8 +14	+6 =(+)+8 +14	−6 (−)−8 +2	−6 =(+)+8 +2
Subtraction: change sign of subtrahend, complement and add			

Rules. Assuming that the arithmetic section of the microprocessor performs arithmetic by the two's-complement method, the following rules describe its operation.

1. \+ numbers are in sign-true magnitude form.
2. − numbers are in two's-complement form.
3. The sign is correct as it appears in the sign position of the result.
4. The result is invalid if a *carry propagates into the sign position* but not out, or if there is *no carry into the sign position with a carry out*.
5. Subtraction is performed by changing the sign of the subtrahend and adding the subtrahend to the minuend using the rules for addition of signed numbers.

Table 3-2 shows that each subtraction is actually an addition. Thus, subtraction in the microprocessor uses existing adders, and the inputs are modified by the microprocessor when subtraction is required.

3-3 BINARY MULTIPLICATION AND DIVISION

Concepts of Multiplication. Binary multiplication is a very simple process that is performed in accordance with the rules*

$$0 \times 0 = 0$$
$$0 \times 1 = 0$$
$$1 \times 0 = 0$$
$$1 \times 1 = 1$$

When implemented in the microprocessor, however, binary multiplication requires a number of data manipulations. Since both binary and decimal multiplication use similar concepts, the more familiar decimal procedures are shown first.

A typical decimal multiplication problem is 123×32. One viewpoint on multiplication is that the multiplier (32) tells how many times the multiplicand (123) is to be added to itself. Thus, 123 is written 32 times and the result obtained by *successive addition*. The concept of *partial products* is also useful. Since 32 may also be written as $30+2$, the multiplication problem becomes $123 \times (30+2)$. One partial product is obtained ($123 \times 2 = 246$), and the second partial product ($123 \times 30 = 3690$) is summed with it ($246 + 3690 = 3936$) to form the final answer.

In examining the "paper-and-pencil" method of performing multiplication, an interesting point comes to light, as shown in Example 3-10.

EXAMPLE 3-10

```
  123
  ×32
 ----
  246   first partial product
 369    second partial product
 ----
 3936
```

The second partial product was shifted *one* position to the left to compensate for the fact that the actual multiplication was by 30, *not* 3. The zero in the least-significant position of the second partial product is conveniently left off and the second partial product is shifted left so that

*Note that these rules are the same as the logical AND rules.

its least-significant digit starts in the ten's column. This is actually equivalent to multiplying by 3×10.

Sample Problem. Binary multiplication is accomplished in the same manner as decimal multiplication, but since the "multiplication table" is less complex, the actual multiplication operation is much simpler. Both successive addition and partial product methods are used. The popularity of the "shift and add" partial-product method has made it a natural choice for discussion in this text.

The rules of binary multiplication may be simplified as follows:

1. Copy the multiplicand to obtain the partial product if the multiplier digit is a 1.
2. The partial product is 0 if the multiplier digit is a 0.

A sample binary multiplication problem using partial products is shown in Example 3-11.

EXAMPLE 3-11

```
      1010      multiplicand
    ×  101      multiplier
    ------
      1010 ┐
     0000  ├    partial products
    1010   ┘
    ------
    110010      final product
```

The actual operations performed during the multiplication of Example 3-11 follow exactly the simplified binary multiplication rules. The least-significant bit of the multiplier is examined, and since it is a 1, the multiplicand is copied directly. Examination of the next multiplier bit shows that it is 0, so the second partial product becomes all zeros. Since this bit is in the 2^1 position, it is necessary to displace either the first partial product one position to the right or the second partial product one place to the left, as performed in decimal multiplication.* Most microprocessors possess the capability to add only two numbers at a time, so the first two partial products must be summed and stored before proceeding. The next multiplier bit (1) results in a direct copy of the multiplicand, with the usual

*Remember, this operation is equivalent to multiplying by a power of 2 (binary) or a power of 10 (decimal).

displacement of the partial product. Addition is once again performed and the result stored. This final partial product is summed with the stored partial products, supplying the completed product. The process of examining the multiplier digit, forming the partial product, shifting, and adding is a common method of performing multiplication in digital computers.

Determination of the sign of the product presents no problem. When two quantities to be multiplied have the same sign, the product sign is positive. If their signs are different, the sign of the product is negative.

It should be noted that the shift-and-add method of multiplication uses existing adders and registers in the microprocessor and requires no additional hardware. Owing to the fact that many shifts and adds must be performed, however, multiplication requires a longer time period than does addition or subtraction. Unless separate hardware is provided, multiplication is a process that is performed in the computer by means of software (computer programs) rather than hardware. In other words, a separate routine (series of steps) is used each time a multiplication is required.

Concepts of Division. Division is a more complex operation than multiplication. A sequence of shift and conditional add operations is usually performed in multiplication, where the multiplier bits determine whether the multiplicand should or should not be added to already calculated partial products. Conversely, division is usually performed by a sequence of shift and conditional *subtract* operations. The decision concerning subtracting the divisor from the dividend is based on whether or not this subtraction yields a valid result. Table 3-3 shows the concepts associated with *restoring division* methods. Reference to Example 3-12 provides some insight into the techniques employed by the microprocessor to implement restoring division. The subtraction operation is often performed by the adders in the microprocessor using two's-complement notation. Remember, however, that division, like multiplication, is often software-controlled and quite time consuming. When shorter operating times are required, the methods discussed during explanation of the multiplication process may be used.

Rules for Performing Restoring Division. The procedures used in Table 3-3 and Example 3-12 may be summarized as follows:

1. Make a trial subtraction of the divisor from the partial dividend.
2. If the divisor can be subtracted from the partial dividend without a negative difference, perform the subtraction and enter a 1 in the quotient and continue with division process if possible.
3. If the subtraction cannot be made, enter a 0 in the quotient and do not

alter the partial dividend. Shift the divisor 1 bit right and start again at step 1.

4. If the signs of the divisor and the dividend are the same, the quotient is positive.
5. If the signs of the divisor and the dividend are different, the quotient is negative.

Table 3-3 "Restoring Division" Concepts

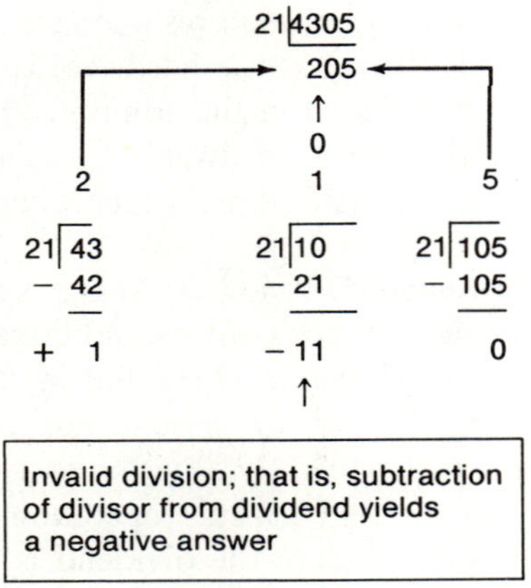

(a) Decimal Division

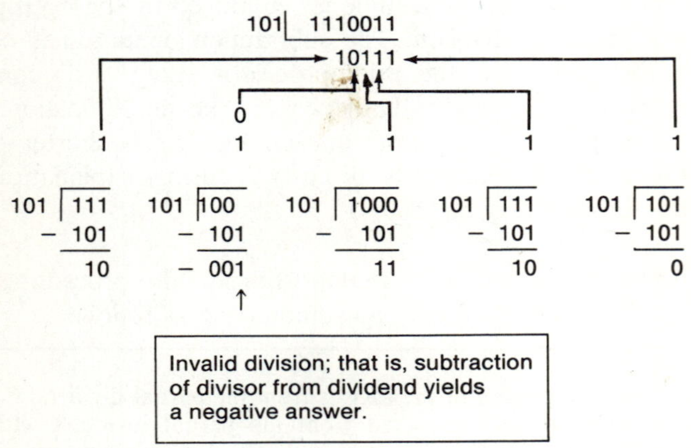

(b) Binary Division

EXAMPLE 3-12

```
          10111
    101 |1110011
         101
         ---
          100
          000     trial divisor quotient negative
          ----
          1000
           101    shift divisor right
           ----
           111
           101
           ---
            101
            101
            ---
            000
```

3-4 ARITHMETIC/LOGIC UNIT (ALU) OPERATION

Arithmetic Function. Included in most microprocessors is a functional section called the ARITHMETIC/LOGIC unit (ALU). Depending on design parameters, the ALU can perform not only addition and subtraction, but also several logic functions. A common ALU implementation is shown in Figure 3-3. It accepts two 4-bit binary words and, depending on control signals, performs arithmetic or logic operations on the two words. The operation of the ALU shown in the logic diagram of Figure 3-3 is described in detail in Table 3-4. Note that with the Mode Control (M) HIGH, logic functions are selected, while with M LOW, arithmetic operations result. The C_n input further controls the actual arithmetic operations that are performed.

A typical arithmetic operation using the ALU of Figure 3-3 will show the practicality of the logic circuit. Consider the simple operation of adding 101 and 111. Each of the numbers is placed in a register and supplied to the ALU. Consider 101 to be stored in register A and 111 in Register B. The inputs and resulting output (F) is shown below.

A	0101
B	0111
F	1100

The readers who possess the advanced digital logic background may trace the "1s and 0s" from inputs (A, B) to output (F) in order to verify the

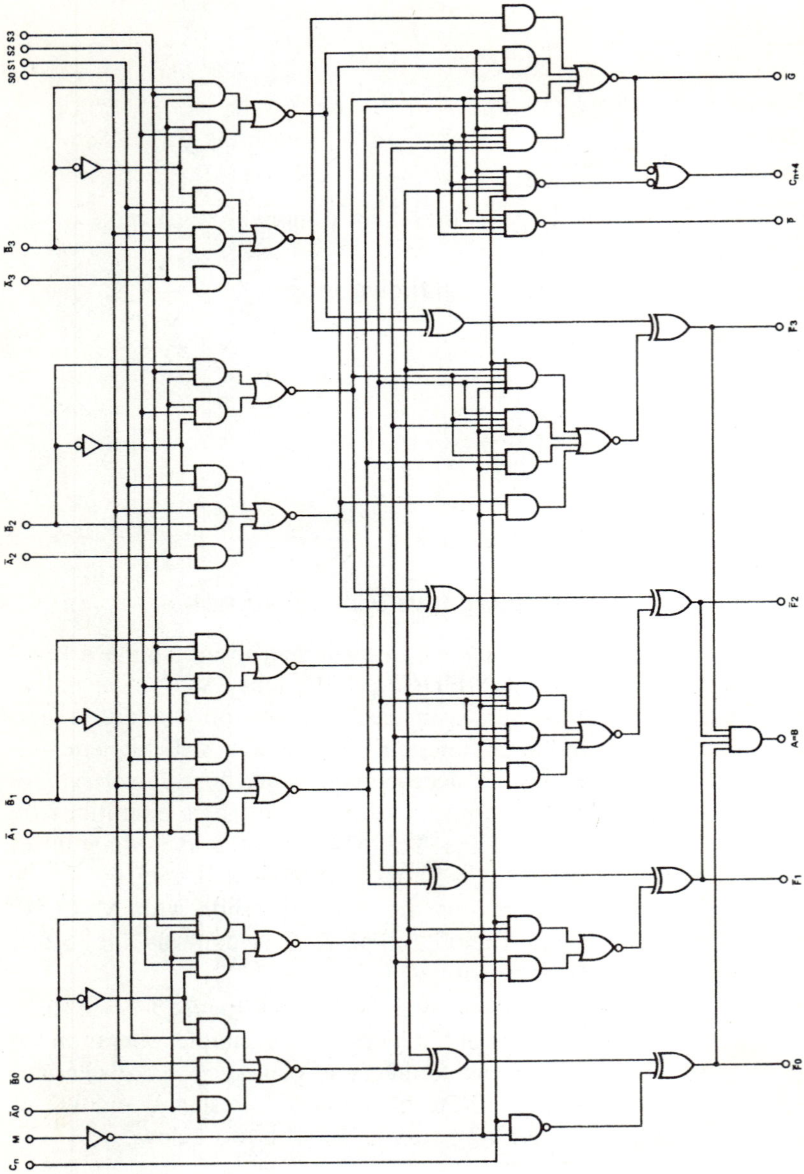

Figure 3-3 ALU Chip Diagram (Courtesy SIGNETICS)

Table 3-4 ALU Truth Tables

Function Select				*Inputs/Outputs Active Low* Logic Function	Arithmetic* Function	*Inputs/Outputs Active High* Logic Function	Arithmetic* Function
S3	S2	S1	S0	$(MC=H)$	$(MC=L, \bar{C}_n=L)$	$(MC=H)$	$(MC=L, \bar{C}_n=H)$
L	L	L	L	$\bar{A}$	A minus 1	$\bar{A}$	A
L	L	L	H	$\overline{AB}$	AB minus 1	$\overline{A+B}$	$A+B$
L	L	H	L	$\bar{A}+B$	$A\bar{B}$ minus 1	$\bar{A}B$	$A+\bar{B}$
L	L	H	H	Logic "1"	minus 1	Logic "0"	minus 1
L	H	L	L	$\overline{A+B}$	A plus $(A+\bar{B})$	$\overline{AB}$	A plus $A\bar{B}$
L	H	L	H	$\bar{B}$	AB plus $(A+\bar{B})$	$\bar{B}$	$(A+B)$ plus $A\bar{B}$
L	H	H	L	$\overline{A\oplus B}$	A minus B minus 1	$A\oplus B$	A minus B minus 1
L	H	H	H	$A+\bar{B}$	$A+\bar{B}$	$A\bar{B}$	$A\bar{B}$ minus 1
H	L	L	L	$\bar{A}B$	A plus $(A+B)$	$\bar{A}+B$	A plus AB
H	L	L	H	$A\oplus B$	A plus B	$\overline{A\oplus B}$	A plus B
H	L	H	L	B	$A\bar{B}$ plus $(A+B)$	B	$(A+\bar{B})$ plus AB
H	L	H	H	$A+B$	$A+B$	AB	AB minus 1
H	H	L	L	Logic "0"	A plus A	Logic "1"	A plus A
H	H	L	H	$A\bar{B}$	AB plus A	$A+\bar{B}$	$(A+B)$ plus A
H	H	H	L	AB	$A\bar{B}$ plus A	$A+B$	$(A+\bar{B})$ plus A
H	H	H	H	A	A	A	A minus 1

*Expressed as two's complement

detailed operation of the A plus B operation of the ALU. From a practical viewpoint, however, microprocessor data sheets do not commonly present this level of detail. As the microprocessor is examined in more detail, it will be found that the *function* is the important point to be understood, *not* the detailed circuit operation. Thus, to use a microprocessor most effectively, one merely needs to understand how the arithmetic operations (addition, subtraction, etc.) function.

An $A = B$ output is also provided so that comparator operations may be performed. Furthermore, by use of the C_{n+4} output, relative magnitude indications also may be obtained. The ALU is placed in the subtract mode for these comparator operations, and the comparison results are determined by use of the information in Table 3-4.

The capabilities provided by an ALU therefore replace the adders, comparators, logic operators, and arithmetic control portion of the earlier general-purpose digital computers. It merely remains to furnish registers to manipulate the data into and out of the ALU. Modern microprocessors contain numerous registers that may be used for this purpose.

Logical Functions. Logical operations are also performed in the ALU. Such operations may be verified by analysis of the logic diagram of Figure 3-3 and the information in Table 3-4. However, an example may aid the reader in this analysis.

As just demonstrated, arithmetic operations are performed on *all* bits in the register supplying the ALU so as to obtain a *composite* answer. When performing logical operations, however, the ALU obtains *separate* answers for each *pair* of bits in the same location in the registers. In the ALU of Figure 3-3, for example, at the same time the two most-significant bits (A_3, B_3) are combined in the desired logic operation to obtain a result, the *same* logic operation is being performed on the other pairs of bits (A_2 and B_2, A_1 and B_1, A_0 and B_0). The end result is (in Figure 3-3) four *separate* answers. Two examples follow:

AND

	3	2	1	0
A	0	1	0	1
B	1	1	1	1
F	0	1	0	1

←bit positions→

EXCLUSIVE-OR

	3	2	1	0
A	1	0	1	0
B	1	1	1	1
F	0	1	0	1

four separate results for each operation

Sufficient information is available in Figure 3-3 and Table 3-4 so that those readers who desire may trace the AND operation through the ALU from input to output. Just as with arithmetic operations, only a *functional* understanding of the logical operations is required for microprocessor use.

3-5 BINARY-CODED-DECIMAL (BCD) ARITHMETIC

Concepts of BCD Arithmetic. Many of today's electronic calculators, plus some of the microprocessors, use BCD arithmetic methods. The usual criterion for use of BCD in favor of direct binary codes is that of predominantly manual input, such as in calculators. Input and output are relatively simple in BCD, and minimum conversion hardware is required.

BCD addition and subtraction are discussed here only as an introduction to BCD operations. Other arithmetic operations can be derived from addition and subtraction as in binary operation and are left to texts specifically directed to BCD arithmetic.

In BCD a decimal digit is represented in 4 bits. Sixteen possible combinations of 4 bits exist, yet only 10 are required for decimal representation. The remaining six combinations are not used in BCD, although they do occur and must be compensated for in arithmetic operations.

Typical Problem in 8–4–2–1 BCD. Table 3-5 shows the 8–4–2–1 BCD code used to represent the decimal numbers 0 through 15. Note that the only *valid* representations are 0 through 9. No corrections are required. Since four binary digits are used to represent each decimal digit, a correction must be applied when any decimal number greater than 10 is required. Consider the BCD form of 12_{10}, which is 0001 0010. However, the 8–4–2–1 representation of 12_{10} is 1100, which is invalid in BCD without correction. Since the difference between the maximum decimal number that can be shown with a 4-bit BCD representation (9) and a 4-bit 8–4–2–1 representation (15) is 6, the binary equivalent of 6 (0110) must be added to any *invalid* 4-bit 8–4–2–1 representation.

Thus, 1100+0110=1 0010. The carry that results exceeds the four-digit BCD limit and must be applied to the next-most-significant BCD position. Leading zeros are added to complete the BCD ten's position, and the BCD representation of 12_{10} becomes 0001 0010.

Table 3-5 BCD 8-4-2-1 Code—Addition and Carry Concepts

Decimal sum (A+B)	*Uncorrected sum* S_u	8-4-2-1	*Corrected sum* S	*Correction*
0	0000		0000	
1	0001		0001	
2	0010		0010	
3	0011		0011	
4	0100		0100	
5	0101	Valid	0101	
6	0110		0110	
7	0111		0111	
8	1000		1000	
9	1001		1001	
10	1010		0001 0000	(+6)
11	1011		0001 0001	(+6)
12	1100	Add 6	0001 0010	(+6)
13	1101		0001 0011	(+6)
14	1110		0001 0100	(+6)
15	1111		0001 0101	(+6)

Rules:

1. No correction when $S_u \leqslant 1001$
2. $S = S_u + 0110$ when $S_u > 1001$

Table 3-6 BCD Addition Examples

	Decimal	8-4-2-1	
	58	0101	1000
	29	0011	1001
Uncorrected sum (S_u)	7(17)	0111	0001
Correction			+0110
carry	10	+0001	
Corrected sum (S)	87	1000	0111
		8	7

An example of the principles of BCD 8–4–2–1 addition shown in Table 3-5 is detailed in Table 3-6. Note the applications of Rule 1 in the ten's position and Rule 2 in the unit's position. (See Table 3–5.)

BCD Adder Operation. The circuit of Figure 3-4 performs BCD-corrected addition and subtraction on 4 bits (one BCD digit). For addition, the control input (SUBTRACT) is LOW. The first 4-bit adder adds the B_0 through B_3 inputs to the A_0 through A_3 inputs, the binary sum is generated on outputs S_0 through S_3, and the binary carry appears on output C_4. Whenever the binary sum exceeds nine, that is, when S_3 AND (S_2 OR S_1) OR C_4 is a true expression, a decimal carry is generated by the gating structure shown. The carry flip-flop is set and a binary 6 is forced onto the B inputs of the second 4-bit adder by the quad multiplexer. The outputs D_0 through D_3 represent the BCD corrected sum $D = A$ plus B.

For subtraction, the control input (SUBTRACT) is HIGH, inverting B_0 through B_3 inputs to the first adder. The quad multiplexer feeds the $\overline{Q}$ output of the carry flip-flop into the Carry In of the first adder, which performs $\overline{\text{Carry}}$ plus A plus $\overline{B}$, one of the binary subtraction algorithms. The Carry Out (C_4) signal is inverted before it is routed through the quad multiplexer into the JK input of the Carry-Borrow flip-flop. Whenever this flip-flop is being set, the binary result at S_0 through S_3 requires correction by subtracting 6 or adding 10. This is performed in the second 4-bit adder by routing the $\overline{C_4}$ signal into the C_0 (weight 2) and the B_2 (weight 8) inputs. The outputs D_0 through D_3 represent the BCD corrected result, $D=A-B$.

3-6 SUMMARY

In this chapter it has been shown that the common arithmetic and logical operations can be performed using simple logic circuits. Although the methods employed in the microprocessor may seem both

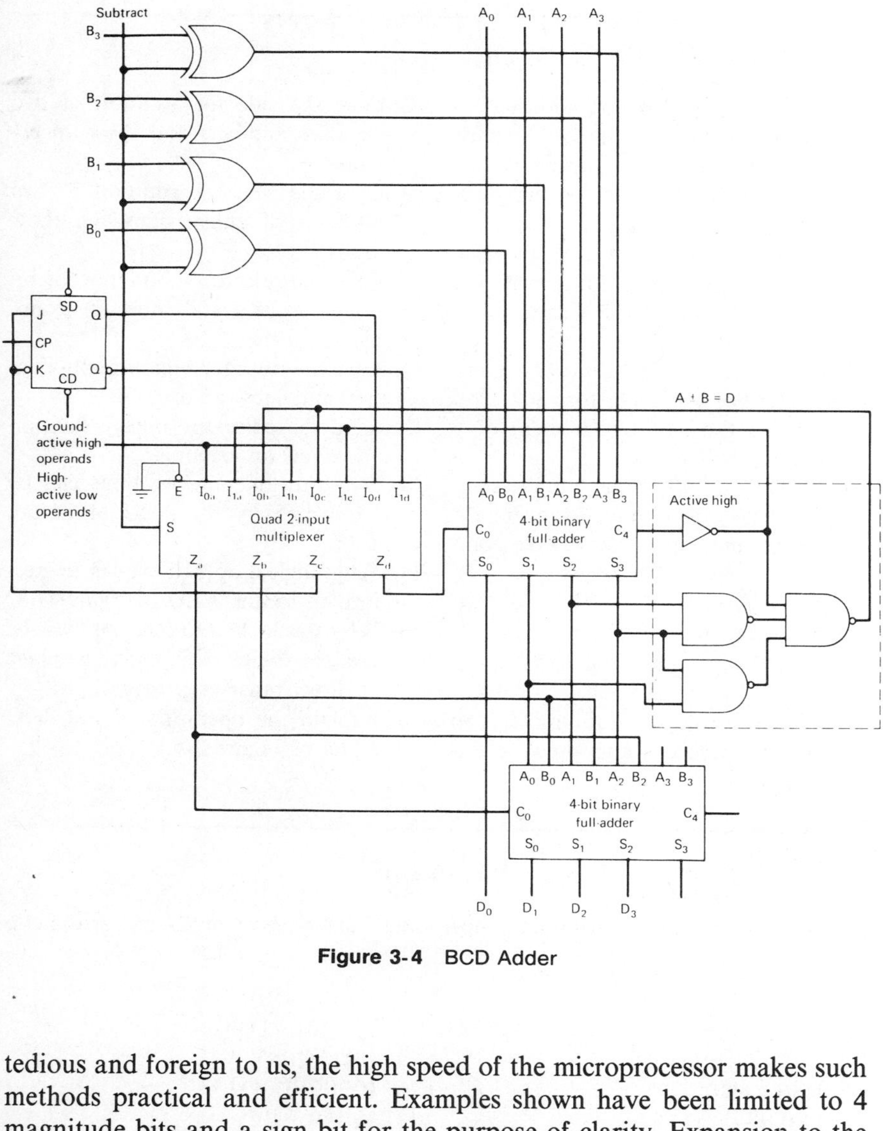

Figure 3-4 BCD Adder

tedious and foreign to us, the high speed of the microprocessor makes such methods practical and efficient. Examples shown have been limited to 4 magnitude bits and a sign bit for the purpose of clarity. Expansion to the more common 8-bit words merely increases the number of logic circuits required. Complexity is increased by adding more of the same kind of circuits, not because of the addition of any new circuits.

More complex operations, such as square roots, trigonometric operations, and so on, use the basic ALU circuits and merely use program steps to repetitively apply basic operations to perform the complex operations. The programming approach to complex operations is shown in later chapters.

QUESTIONS

1. List the rules of binary addition.
2. Draw a logic diagram of a device that will perform addition of two binary digits. Assume no carry input is required.
3. Draw a logic diagram of a device that will perform addition of two binary digits. Assume a carry input is required.
4. How is the radix-minus-one complement of a number formed?
5. How is the true complement of a number formed?
6. Write the following numbers in sign-true magnitude form, one's-complement form, and two's complement form.

a.	+10	e.	−30
b.	−7	f.	+33
c.	+9	g.	−8
d.	−15	h.	−6

7. Give the rule for addition of two positive numbers and show an example.
8. Give the rule for addition of two negative numbers and show an example.
9. Give the rule for addition of two numbers of unlike signs and show an example.
10. Give the rule for subtraction of binary numbers and show an example.
11. Give the rules for binary multiplication and show an example.
12. Give the rules for binary division and show an example.
13. What arithmetic operations can be performed by the ALU of Figure 3-3?
14. What logical operations can be performed by the ALU of Figure 3-3?
15. Why would BCD representation be used in digital equipment in place of direct binary representation?
16. Explain the operation of the BCD adder of Figure 3-4.

PROBLEMS

Perform the following arithmetic operations, using complementary arithmetic where appropriate.

1. 1101101 + 1011010
2. 11011 + 1010
3. 100111 + 1011101
4. 1110 + 1110
5. 110110 − 10111
6. 1001001 − 101110
7. 110101 − 1011
8. 1110101 − 111111
9. 1011 × 101
10. 1000010 × 111
11. 101001 ÷ 101
12. 101101 ÷ 110

Perform the following arithmetic operations, using 8–4–2–1 BCD.

13 14 + 14
14. 27 + 10
15. 66 + 36
16. 109 + 90

CHAPTER 4

Fundamentals of Digital Computers

Understanding the microprocessor necessitates recognition of two major factors. First, the microprocessor can be approached from a *functional* viewpoint, with the emphasis on specific operations that the device performs. Second, it must be recognized that the microprocessor is time-oriented. The interrelationships of the functions performed and the time it takes to achieve that performance are of the utmost importance.

The operation of the microprocessor is closely associated with that of the digital computer. Therefore, the digital computer's operations will be outlined in this chapter so that the microprocessor's operations will be more meaningful when encountered in subsequent chapters.

4-1 GENERAL COMPUTER CONCEPTS

Functional Operation. Five basic functions are performed by *any* information-processing device, whether it be a digital computer or a microprocessor. These functions are:

1. The INPUT function, which interfaces between the outside world and the device.
2. The OUTPUT function, which interfaces between the device and the outside world.

3. The STORAGE (MEMORY) function, which stores information and instructions about what to do with the information.
4. The ARITHMETIC/LOGIC (ALU) function, which performs the mathematical and logical operations of the device.
5. The CONTROL function, which sequences and controls the device.

These are *functional* categories only, and in most cases are not directly equatable to actual hardware. Often a function shares hardware locations directly with a number of different physical sections of the device. For example, in the *functional block diagram* of a digital computer (Figure 4-1), the functions *inside* the division labeled *Central Processor Unit* (CPU) are commonly located within the physical part of the computer so labeled. Note that the INPUT and OUTPUT functions are only partially contained within the CPU. Other portions of these functions are located in other pieces of hardware that also provide INPUT and OUTPUT functions. However, because of the complexity of the computer, it is usually advantageous to learn its operation from the functional standpoint before becoming hardware-oriented.

The flow of information within the computer may be seen in Figure 4-1. Except in special cases, all information that the computer possesses must be entered via the INPUT function. (In microcomputers, some of the computer's information may be permanently contained within the machine.) Depending on the type of information, it may be used either to tell the computer what to do (*instructions*) or it may be used as data in arithmetic or logical operations (*data*). As an instruction, the information tells the computer where the next information is stored or what to do with the next packet of information obtained. As data, it will be routed to the ARITHMETIC/LOGIC function for processing and back to STORAGE

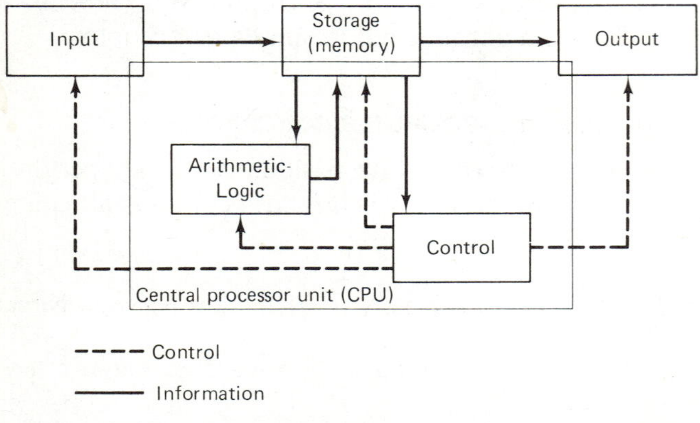

Figure 4-1 Digital Computer Functional Diagram

to await its next use. The OUTPUT function displays the results of the internal processing of information when called upon by the CONTROL function. The flow of information throughout the computer is from INPUT to STORAGE to OUTPUT, with detours along the way to the ARITHMETIC/LOGIC function for processing operations and to the CONTROL function to determine what is to be done next.

Sequential Operation. Control signals, occurring at the proper time and in the proper sequence, are developed by the CONTROL function. There is no need, for example, to have the ARITHMETIC/LOGIC function operating when it is not needed. Therefore, it is *enabled* only during the time it is required, as are all other functions of the computer. The *timing diagram* of Figure 4-2 shows a very simplified version of the major control signals required to enable and disable the basic computer functions of Figure 4-1. It should be understood that these timing signals are very broad in coverage and that they will be modified greatly as more is learned about the computer.

Information is accepted from the input device(s) during the period T_0 to T_1, when the input device(s) and STORAGE function are enabled. It is during this time that the list of instructions (the program) is given to the

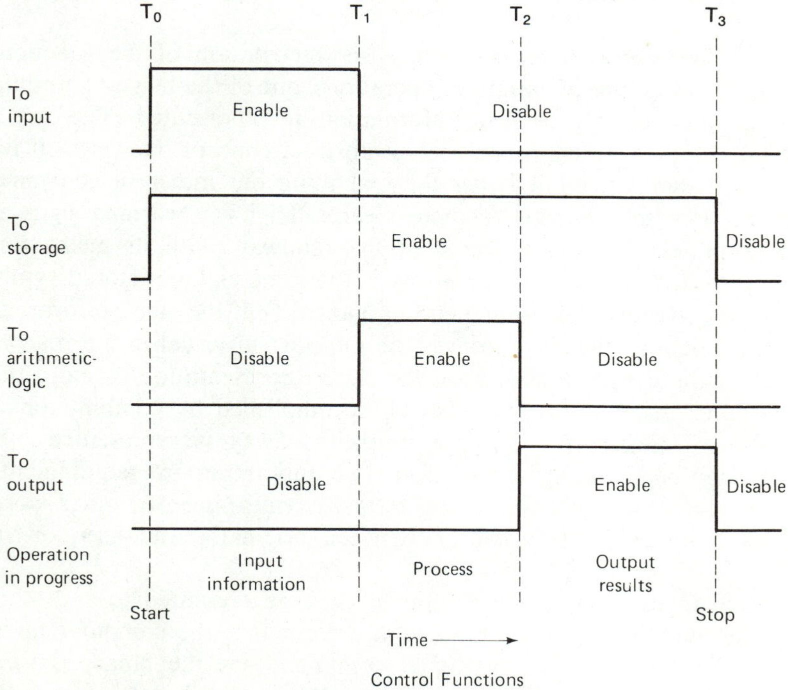

Figure 4-2 Gross Timing Diagram, Digital Computer

computer. The computer is helpless without direction to tell it what to do and how to do it. The list of instructions informs the computer of the actions to be performed and gives the sequence in which they are to be performed. Preparation of the program is the task of a *programmer*, who, in conjunction with a *systems analyst*, prepares a *detailed, sequential list* of steps which direct the operations necessary to perform the information processing. With microprocessors the program is often prepared directly by the system designer because of the close tie between hardware and software.

The information to be processed may also be presented to the computer during the period T_0 and T_1. Instructions and data are placed in STORAGE and called upon as needed during the T_1–T_2 time period. With both STORAGE and ARITHMETIC/LOGIC functions activated, instructions and data are processed as required by the *stored program*, and intermediate/final results are returned to STORAGE. Upon completion of information processing, the ARITHMETIC/LOGIC function is disabled and the OUTPUT function is enabled. The stored program then controls transfer of results to the output device(s) during T_2–T_3. All functional sections are disabled upon completion of the stored program. As the computer is examined in greater detail, more complete sequential timing operations are shown.

Information Representation. After recognition of the functional and sequential nature of computer operations, one of the biggest stumbling blocks is the concept of how information is represented. The use of symbols to represent information is an abstract concept. However, if it is understood that a symbol is *not* the real thing but merely a convenient representation, the concept becomes clearer. Highway warning signs are good examples. The actual curve in the roadway, with its guard rails, roadside trees, and scenery, in no way is the same as the pictorial symbol advising the motorist of the impending hazard. Yet the sign performs the required function; that is, it causes the motorist to visualize the roadway curving in the direction shown on the sign. Representation of more than one piece of information may also be accomplished by combinations of symbols. A rough-surfaced curve in the roadway is represented by a "curve" sign *and* a "rough road" sign. The only important requirement is that the user of the symbols recognized their meanings. In other words, symbols *communicate* information between originator and user, or vice versa.

Each instruction to the computer must be presented to and stored within the memory of the computer in a form that the computer understands. The symbols used in digital computers are the binary (base 2) numbers (1 and 0). It may seem impractical to use only two symbols for information representation, but electrical/electronic implementation of computer functions is difficult with more than two symbols. Therefore, all information presented to and stored within the computer is in binary form. In addition, the information must be in a format that is easily and unambiguously interpretable by the computer. These problems are attacked during initial design of the computer, and the solutions vary from computer to computer. In all cases, of course, binary symbols are used. The format of the information generally consists of selection of a *fixed number of binary symbols* (called a *computer word*) to be presented to the memory of the computer for each instruction in the program. *Information to be processed* (commonly called *data*) is also arranged in a similar format. Formats vary from extremely simple to quite complex.

The computer, then, is given information in groups of binary symbols called *computer words*. If the information is comprised of data to be processed, such as letters, numbers, or symbols, the package of information is called a *data word*. If the computer is being told what to do, the package of information is called an *instruction word*. Figure 4-3 is a simplified representation of a data word and an instruction word that could be used with a computer that handles information in groups of 16 binary symbols.

Note that the instruction word consists of two parts: the operation and the operand. The *operation* portion of the word (*op code) informs the*

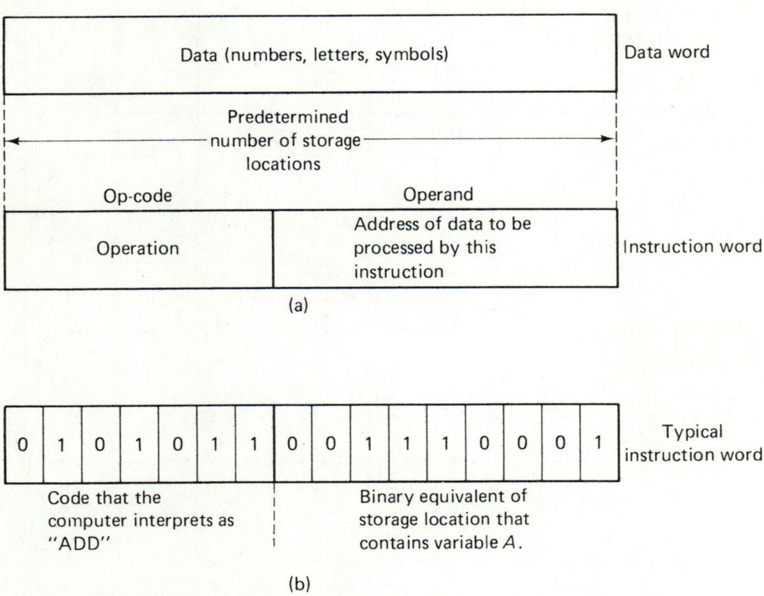

Figure 4-3 Computer Words: (a) Concept; (b) Typical Instruction Word

computer what to do, and the *operand* supplies a coded indication that *tells where any data to be operated upon are located in memory*. Computer words are supplied in the order in which they are to be used, and the computer automatically selects the next computer word in sequence unless told to do otherwise. The basic cycle of operation is one of retrieving the computer word from memory, evaluating it, obtaining the information at the indicated operand location, performing the operation with the obtained information, and progressing to the next computer word for another instruction.

Information Storage. The classical example of computer information storage employs the common post office box. Almost all post offices have private boxes for their patrons. The boxes are arranged in matrix (row and column) form, and are sequentially numbered as shown in Figure 4-4. A package of information (a letter, for example) is placed in a specific location (the box number, or address) for later retrieval. In like manner, information is stored in computer memory. A package of information (the computer word) containing a given amount of intelligence is assigned an identifiable location in memory and may be retrieved by "addressing" that

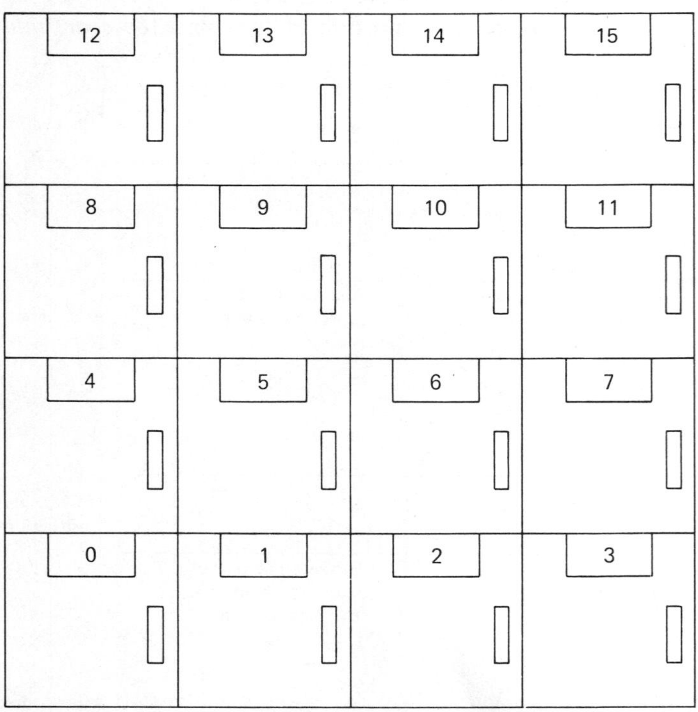

Figure 4-4 Post Office Box Analogy of Information Storage

location. It should be pointed out at this time that there is no way of knowing whether the information is instruction or data except by knowledge of its assigned location. Assignment of memory locations is primarily the job of the programmer, and the computer must be told where to go for instructions and where to go for data. The actual mechanism of information storage is discussed in Section 4-3, and again in considerable detail in Chapter 9.

Stored Program Capability. By combining the concepts of sequential operation, information representation, and storage, the final concept of *stored program capability* may be seen. If the computer is considered as a device to perform arithmetic calculations, the major advantage over a simple calculator or hand calculations becomes apparent.

A simple calculator, capable of performing addition, subtraction, multiplication, and division, does not have the advantages of the computer. Such a calculator responds to the wishes of the operator merely by mechanizing the basic arithmetic operations. Multiplication of the sum of two numbers $(A+B)$ by the difference of two other numbers $(C-D)$ is a good example.

EXAMPLE 4-1: Solve $(A+B)(C-D)$, where $A=1$, $B=2$, $C=5$, and $D=3$.

SOLUTION

Step **1.** Operate CLEAR key.
2. Enter A.
3. Operate + key.
4. Enter B.
5. Operate = key.
6. Store sum.
7. Operate CLEAR key.
8. Enter C.
9. Operate − key.
10. Enter D.
11. Operate = key.
12. Operate X key.
13. Enter stored sum.
14. Operate = key.
15. Record the product.

Note that all the steps performed by the calculator could have been performed by paper and pencil. The simple calculator, however, adds accuracy and speed to hand calculations, while obtaining the same result.

The *stored program computer*, however, solves the problem of Example 4-1 without the requirement for multiple operations by the operator. A *program* is developed by the programmer that causes the computer to perform each of the required operations in sequence *without* operator intervention. Each of the operations in the calculator solution has an equivalent step in the program. Of course, each step in the program is coded in binary form and stored in the memory in that form. A typical example is the step that enters A. Actually, this step is accomplished by adding the value of A to zero, since in the previous step the computer was "cleared" of all previous data in the arithmetic circuits. So the "enter A" step and the "operate + key" step are combined into a single step.

Thus, the command to the computer is one that identifies the ADD operation *and* the location of the quantity A. If the simplified computer word of Figure 4-3 is considered, the operation section is the code for ADD and the operand is the code for the address (location) of A. Figure 4-3b is a possible 16-bit (binary digit) computer word that may be fed to and stored in computer memory to cause A to be retrieved and added into the arithmetic circuits.

Each of the steps is coded in a manner similar to the example of Figure 4-3b and entered *in sequence* into the memory. Values for A, B, C, and D are entered next at locations called out in the respective sections of the computer word. It is only necessary then to select the address of the first word in the program; the computer *sequentially* performs the steps necessary to obtain the solution to the problem, and, if the program is so written, makes the solution available to the user.

The stored program capability, then, relieves the operator of anything more than mechanical functions and supplies problem solutions in much shorter times. Example 4-1 is a relatively simple problem, yet the advantages of stored program operations are quite apparent. When a complex problem, such as space travel trajectories which require the interactions of many variables, is encountered, there is no doubt that the stored program digital computer is a welcome addition to science, industry, and the community.

Overall Information Flow. Now that sequential operation, information representation and storage, and stored programs are understood, computer operations can be explored in more detail. By viewing the functional sections of the computer in a slightly different manner, the constant action of the control function is quite apparent. Figure 4-5 separates the storage function into three blocks for the purpose of establishing a straight-line flow of information from input to output. This is *not* to imply that the results of processing data are stored in different parts of STORAGE (MEMORY), since in many cases instructions and results are intermixed. It should also be noted that data and instructions are separated during the input and storage functions. Such an arrangement is to be expected, since the sequential nature of computer operations requires each

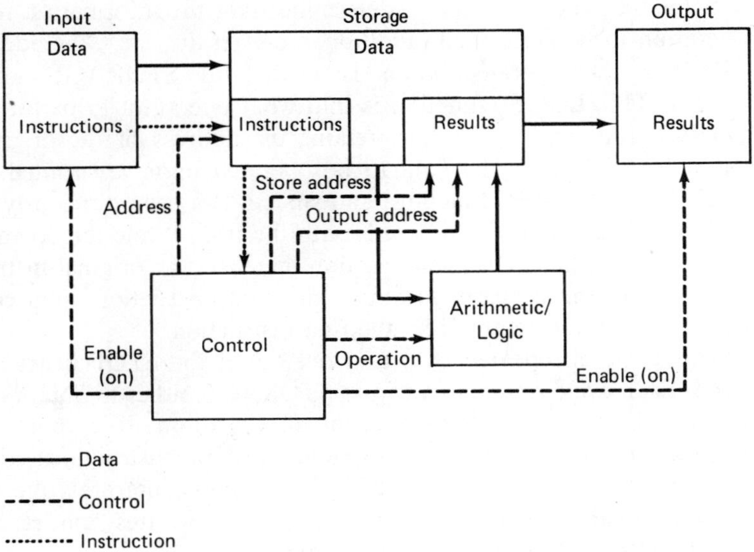

Figure 4-5 Computer Operation, Functional Diagram

step of the program to immediately follow the preceding one. (There are ways around this problem, but they will not be discussed until more details of computer operation are developed.) In a similar manner, instructions are shown being routed *only* to the CONTROL function. More detailed discussion in subsequent chapters shows occasional trips by instructions to and from the ARITHMETIC/LOGIC function.

Control lines extend from the CONTROL section to all other sections in Figure 4-5, just as they did in the conceptual diagram of Figure 4-1. Note, however, that more detail is provided in the expanded diagram, expecially in the STORAGE section. The addition of just these few extra control lines materially expands the coverage of the diagram.

Generally, instructions are loaded into the computer first, followed by data. The CONTROL function provides the "enabling" command to the INPUT function and maintains the input devices in an active state until all information is loaded. The list of instructions (*program*) includes address assignment for each instruction so that the computer operator knows the location of each instruction. Within the CONTROL function are circuits that allow selection of *any* memory address. When the operator selects the address of the first instruction in the program and activates the computer (operates the START control), the memory is enabled and the information present at that address is made available to the CONTROL function. The information is examined within the CONTROL function and decisions made as to what action is to follow.

If the CONTROL function determines that the information requires an operation to be performed (addition, store results, etc.), the address of the data to be operated upon is routed to STORAGE and the ARITHMETIC/LOGIC function is told what operation to perform. The CONTROL function, meanwhile, retains the address of the original instruction for later use. The data to be operated upon are routed to the ARITHMETIC/LOGIC function; and since it has been previously set up to perform the operation, it proceeds with the task. While the operation is being performed, the CONTROL section increases the original instruction address by one and prepares to obtain the next instruction upon completion of the operation required by the first instruction.

This cycle is repeated over and over until the STOP instruction is received. Each cycle results in a trip to STORAGE either to obtain data to be processed or to store the results of an operation. If output of total results (or intermediate results) is required, these instructions are also included in the program. The CONTROL function interprets the output instruction, enables the output device(s), and locates the results in STORAGE to be made available for output.

The timing diagram of Figure 4-2 can now be expanded to show the control functions more accurately (Figure 4-6). For the purpose of simplicity, it is assumed that all input operations are conducted at one time, as are the output operations. As will be noted in subsequent chapters, it is not necessary to input *all* instructions and data at one time. Once initial instructions are provided and stored in MEMORY, those instructions may also include frequent excursions to the input devices for additional instructions or data. Output operations are similarly controlled, so that output does not have to occur all at one time, but intermediate results may be obtained as the program progresses.

To obtain the detailed operating timing diagram of Figure 4-6, it is necessary to vastly expand certain portions of the processing time scale. Actually, since the input and output functions remain the same for the purposes of this discussion, they are not shown in detail on the expanded diagram but are assumed to be present.

4-2 THE CONTROL FUNCTION

As pointed out earlier, the computer operates in a sequential manner, as determined by a stored program of instructions. But the program by itself is not enough. For example, the computer must know where in memory to go to obtain the *first* instruction of the program. It must develop the control signals that obtain this instruction and then determine what action the instruction calls for. The STORAGE, INPUT/OUTPUT, and ARITHMETIC/LOGIC sections must be

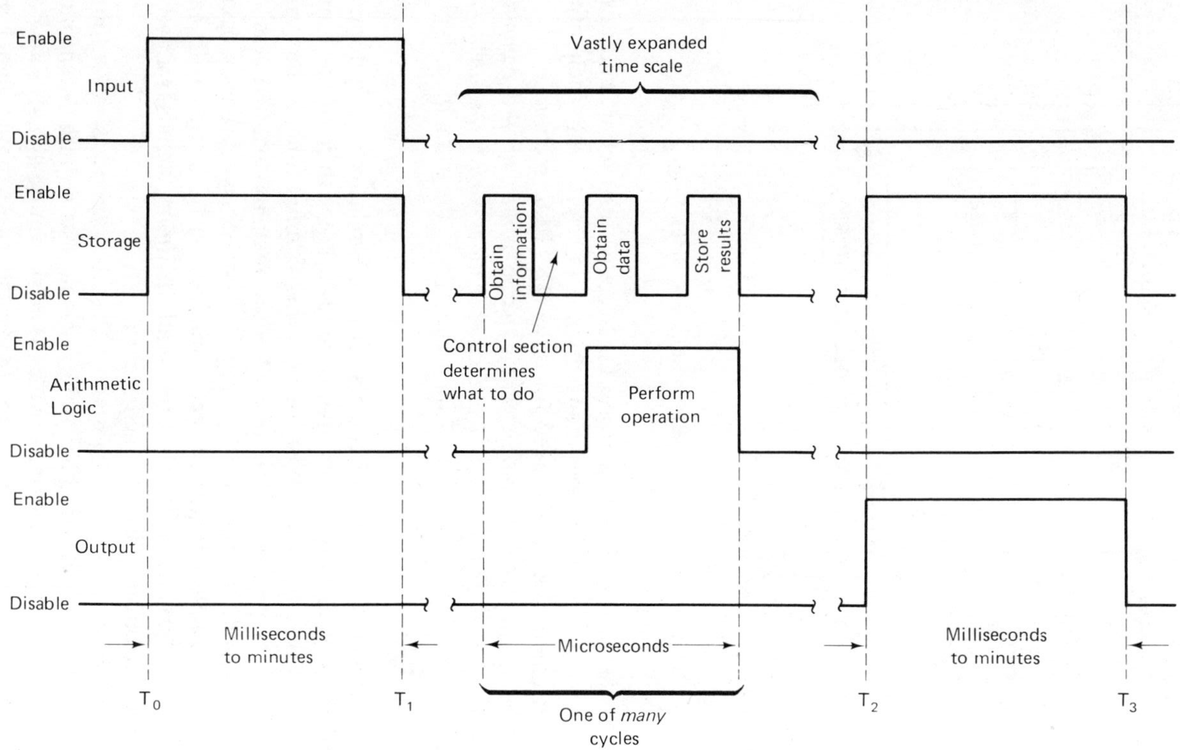

Figure 4-6 Computer Operation, Timing

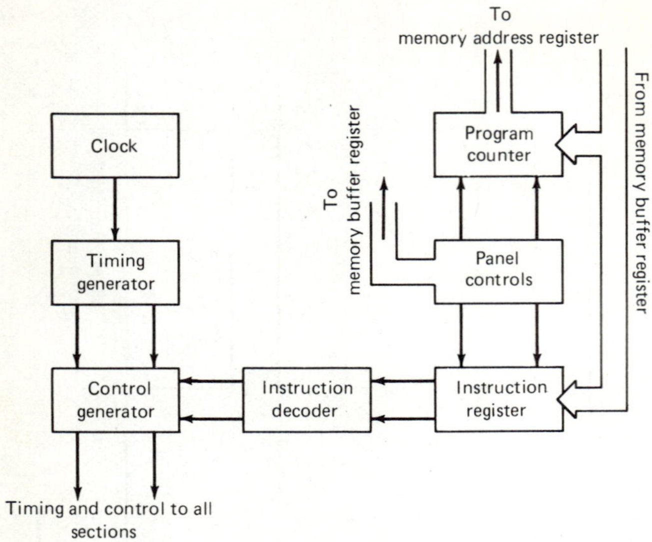

Figure 4-7 Control Section Functional Block Diagram

enabled at the proper time to perform the decoded instruction, and in the proper order so that their operations do not interfere with each other. These requirements define the need for the CONTROL section.

In short, the *control section is the nerve center of the computer, providing the timing, decoding, and enabling operations for all the other parts of the computer*. Figure 4-7 is a functional block diagram of a typical CONTROL section.

Timing. Timing requirements vary from relatively long duration (seconds) to very short duration (nanoseconds). The nature of the timing signals is such that they may exist for only the required time, or they may be repeated at a predetermined rate, or they may be required "on demand," and so on. In many cases a requirement for synchronous operations (where multiple operations occur simultaneously, in step with each other) may even exist. The basic source of all the timing signals is a precise and accurate electronic circuit called a *clock*.

The *clock* operates in such a manner that the *shortest-duration* timing signal is its basic output. All other timing signals that require synchronous characteristics are then developed from the basic timing signal. Flip-flops are used in the timing generator to divide the basic timing signal into longer-duration timing signals to be made available at the required times.

Decoding and Control. The *instruction decoder* determines what actions are required. In Section 4-1 it was stated that an instruction word was arranged in a format that separated the required operation and the address of the data to be operated upon. The operation portion of the instruction word (*op code*) is used by the instruction decoder in the CONTROL section to determine the required actions. As a result of a previous CONTROL section sequence, the op-code portion of the instruction word has been transferred to the *instruction register*. The instruction register is used to store the op code while it is being decoded so that the memory may be free to perform other duties during the decoding time.

The *instruction decoder* is made up of gates connected to recognize *only* specific combinations of binary symbols. A separate combination of gates may be used for recognition of each of the op codes, and the output of the instruction decoder is a separate enabling signal to the *control generator* for each of the op codes.

The *control generator* is also composed of gates that combine the output of the instruction decoder with timing divider outputs to supply all required timing signals in the proper sequence to the remainder of the functional units of the computer.

4-3 THE STORAGE (MEMORY) FUNCTION

Requirements. One merely needs to consider why the computer is capable of its fantastic operating speeds to understand the need for information storage. For example, if during the solution of a problem it were necessary to obtain problem data from the input function every time new data were required, the computer would be speed-limited by the input device. Storage of problem data allows *immediate* access (at least five or six orders of magnitude faster) to the data. Without data-storage capability, the computer would obviously be operating much slower. Furthermore, if it were necessary to physically record intermediate results each time a portion of the problem was solved, speed of operation would again suffer.

Consider what would happen to the speed of operation if the computer had to be told what to do each time an operation was to be performed (as with a hand calculator). The difference between the slow speed of the hand calculator and the high speed of the digital computer is the storage capability of the computer. Storage of instructions and data, whether it be initial problem information or intermediate results, defines the requirements for the storage function of the digital computer.

Concepts of Information Storage. How is it possible to store such a large volume of information in the computer? Because of the nature of the electronic circuits used, it has been necessary to employ binary symbols (1 and 0) throughout. Numbers, letters, punctuation marks, special symbols, and so on, are all assigned a specific combination of binary symbols for use with the computer. Information is supplied to the computer, in the example already discussed, in groups of 16 binary symbols, and may be either instructions or data. Such information, to be stored, must then have available a device with 16 separate storage *cells*, one cell for each of the binary symbols in the computer word.

Single-bit storage is relatively simple to visualize, since a device which is capable of only two separate conditions (states) is all that is required. Consider a simple switch such as is used to control a light in a room. It may be either OFF (binary 0) or ON (binary 1). In fact, the light to which it is connected provides an indication of its state. If the light is illuminated, the switch may be considered to be in the 1 (ON) state, while if the light is not illuminated, the switch is in the 0 (OFF) state. The switch, then, is storing either a binary 1 or a binary 0, and the light indicates which.

Flip-flops are stable in either one of two possible states, which may be identified by voltage/current relationships within the circuit. Magnetic devices may be used to store binary information, either by the presence or absence of a magnetic charge or the direction of a magnetic field. Electrical devices called capacitors may be either charged with an electric field or not charged, giving rise to a binary condition. Optical techniques, where a beam of light (or perhaps a laser beam) cause electrical circuits to respond or not to respond depending on the existence or nonexistence of an optical path, are also binary in nature. Many of these techniques are discussed in Chapter 9, but for now only the *concepts* of storing binary information are investigated.

Memory Operation. A practical memory, in addition to storing information, must have the means to identify the storage location (address) and circuitry to provide a means for moving the input data into the memory or the extracted data from the memory to external circuits (see Figure 4-8). The address signals are derived from the address portion of the instruction word, which contains the binary equivalent of the location of the data to be used in the present operation. The address information is temporarily stored in a *register (memory address register)* so that the rest of the computer may go on about its business while the memory is obtaining the data required.

Another register, called the *memory buffer register*, provides temporary storage for input and output information. If the information is to be written into storage cells, it is temporarily stored in the memory buffer register until it is time for the *WRITE* operation. Electronic circuits (driver amplifiers) that match the power requirements of the memory to the buffer register route information into the storage medium (*memory*). Similar action occurs when information is to be taken from storage. The information at selected cells is provided to a circuit called a *sense amplifier* and, when the *READ* operation is initiated, this information is stored in the buffer register for routing to other circuits in the computer.

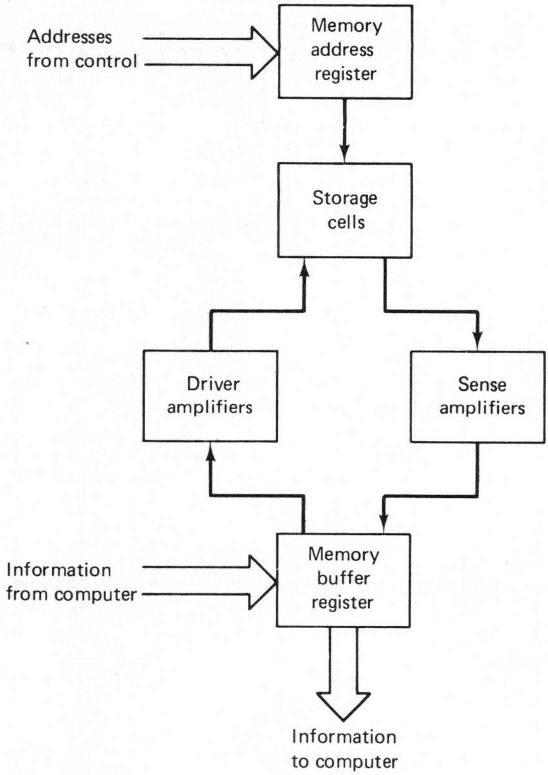

Figure 4-8 Storage Section Block Diagram

In one common memory configuration, storage cells are arranged in a *matrix* form as shown in Figure 4-9. Each horizontal row contains as many storage cells as there are bits in the computer word—that is, in a 16-bit word, 16 storage cells in each row. All cells in a row possess the same address, so that when the information stored at an address location is desired, all storage cells at that address are selected. There are as many rows as there are words storable in the memory. A READ/WRITE input is provided to all cells in the memory simultaneously. However, unless the cell is addressed, the READ/WRITE input is ineffective; therefore, only the addressed cells in the memory are activated. Each cell also has provision for feeding data in and reading data out. All cells in a given column are supplied the same data to be stored, for example. Only the addressed cell, however, will store the data. The data outputs of all cells in a column are also connected together, but once again only the addressed cell will provide any information on the data output line. Since each column represents one bit in the computer word, there must be n columns, where n represents the number of bits in the word.

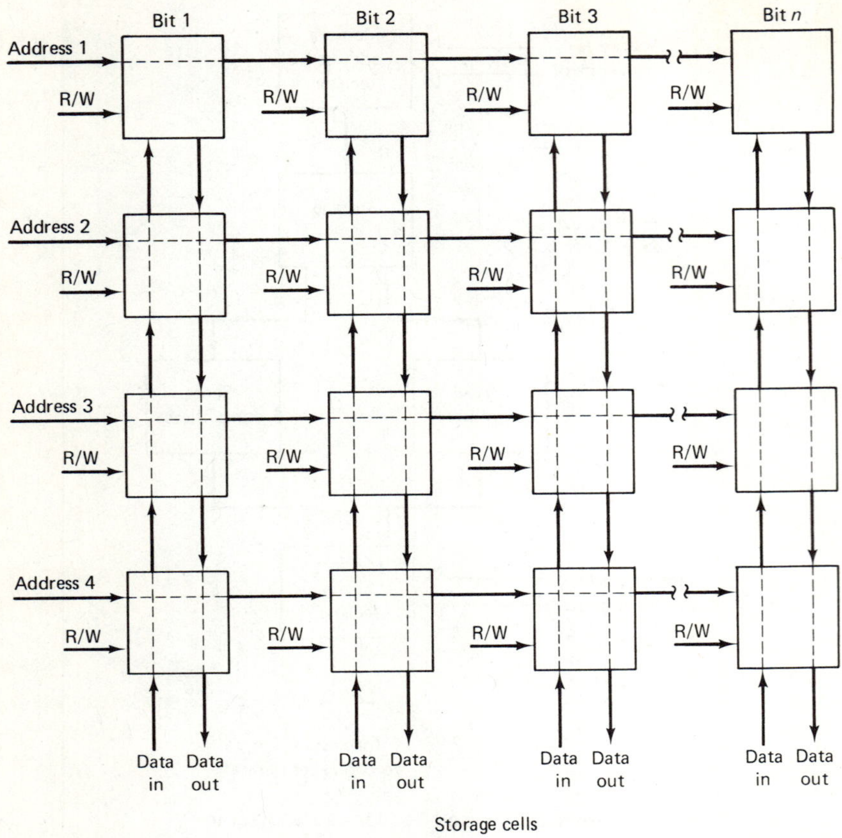

Figure 4-9 Multibit Memory Organization

A practical 16-bit memory, then, supplies a complete 16-bit computer word each time an address is selected and the READ command is valid. Similarly, the WRITE command stores a complete 16-bit word when properly addressed. Thus information is handled in groups of bits, or computer words. Complete lists of instructions, made up of many computer words, may be stored in the type of memory just described. It is necessary, of course, to make available thousands of words of storage, but it should be apparent by now that this is easily accomplished merely by increasing the number of cells and expanding address decoding capability.

4-4 THE ARITHMETIC/LOGIC FUNCTION

Requirements. The data manipulation, arithmetic, and logical operations required by the computer's instruction repertoire are performed within the part of the computer commonly called the ARITHMETIC, or

ARITHMETIC/LOGIC section. Implementation of these operations may be accomplished completely by built-in hardware or by a compromise between built-in hardware and programmed instructions. When it is considered that such operations as arithmetic addition, subtraction, multiplication, division, root finding, exponentiation, differentiation, integration, plus data manipulations that include movement of binary information, comparison of data, and logical operations must be performed, it should be apparent that the pure hardware implementation approach may be out of the question. Individual hardware circuits are possible for each of these operations, or they may be constructed from existing hardware, but considerable space and expense are required for this type of approach. The basic advantage, however, is that speed of operation is maximized with the pure hardware implementation of the operations.

When programmed instructions are used, the ARITHMETIC/LOGIC section contains, usually, devices capable of performing arithmetic addition and data storage and manipulation, and the other operations are performed by the use of a number of instructions. This method requires minimum hardware, but, since each instruction requires time to execute, speed of operation is reduced. The degree of compromise between speed of operation and hardware requirements is a choice to be made by the designer of the computer system.

An ever-increasing approach to arithmetic/logic operations is that of *microprogramming*, where a single instruction from the programmer results in performance of a number of computer operations at a much higher rate of speed. Separate *control memories* contain the *microprogram*, which responds to the programmer's general instructions and guides the ARITHMETIC/LOGIC section through a sequence of steps to perform the required operations.

ARITHMETIC/LOGIC Section Operation. Arithmetic/logic operations can be performed with the functional units shown in Figure 4-10. The data that the ARITHMETIC/LOGIC section must work with are stored in memory and are a result of program steps requiring arithmetic/logic operations. Reviewing the operation of the STORAGE and CONTROL sections aids in understanding the operation of the ARITHMETIC/LOGIC section. Recovery of a computer word containing the code defining the operation to be performed and the address of the information to be operated upon is the first step in the process. Decoding the operation to be performed results in a signal to the ARITHMETIC/LOGIC section from the CONTROL section that enables the required functional units. The data are then extracted from the memory address found in the computer word and supplied to the ARITHMETIC/LOGIC section. The operation to be performed is executed, and the CONTROL section obtains the next instruction, which will ultimately result in extraction of the next package of data to be operated upon. This cycle continues under control of the program until another type of operation is required.

Most ARITHMETIC/LOGIC sections contain the functional devices shown in Figure 4-10. The arrangement of devices and interconnections may vary from computer to computer, but fundamentally the following devices must exist:

1. An *adder* to perform the basic arithmetic operations.
2. An *operand register* to store one of the numbers to be operated upon.
3. An *accumulator/operand register* to store the other number to be operated upon and to accumulate the results of the operation.
4. An *accumulator extender* to accommodate the extra digits that result during multiplication and division, and so on.
5. A *comparator* to look for magnitude relationships between the numbers placed in the operand registers.
6. A *logic operators unit*, which performs logical operations on numbers placed in the operand registers.

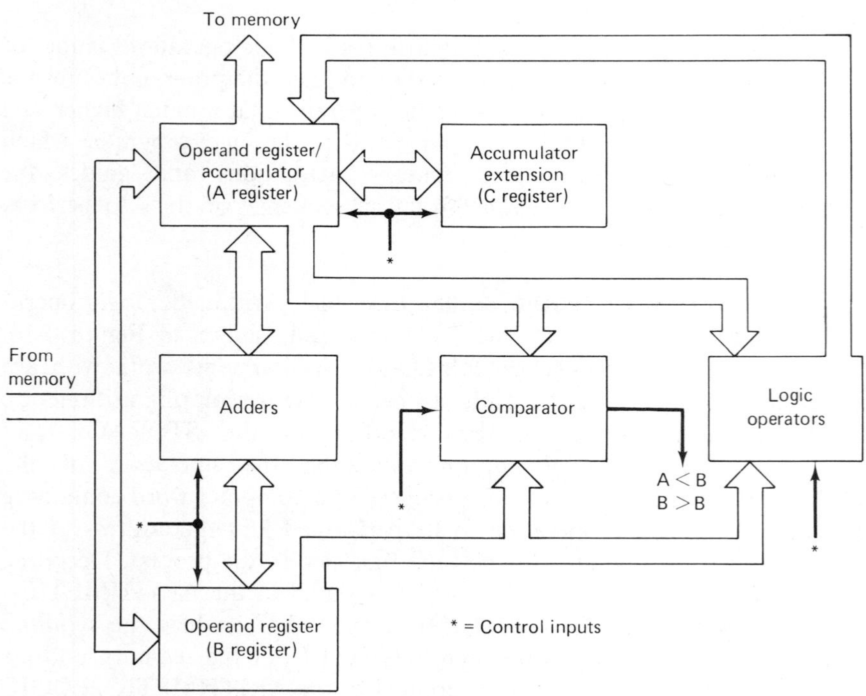

Figure 4-10 Arithmetic/Logic Section Block Diagram

For ease of explanation, the registers in the ARITHMETIC/LOGIC section are given arbitrary letter designation names. In Figure 4-10 the accumulator/operand register is designated the *A* register, the operand register is the *B* register, etc.

A simplified explanation of a simple arithmetic addition operation follows. All information in the registers is removed (the registers are "cleared") as the first step. A memory cycle is then initiated, and one of the numbers to be acted upon is placed in the accumulator. The second memory cycle loads the other number to be acted upon, and the ARITHMETIC/LOGIC section is ready to accomplish its task. The adder is enabled, and the number in the operand register is added to the number in the accumulator, with the sum of the two numbers replacing the number that was initially in the accumulator. A simplified waveform diagram describing these operations is shown in Figure 4-11. If more additions or arithmetic operations are to take place, the sum in the accumulator is retained, and the information to combine with the sum is entered into the operand register for processing. If not, the sum is routed from the accumulator to memory under orders from the CONTROL section. The addition operation is now complete and the next instruction is obtained and acted upon.

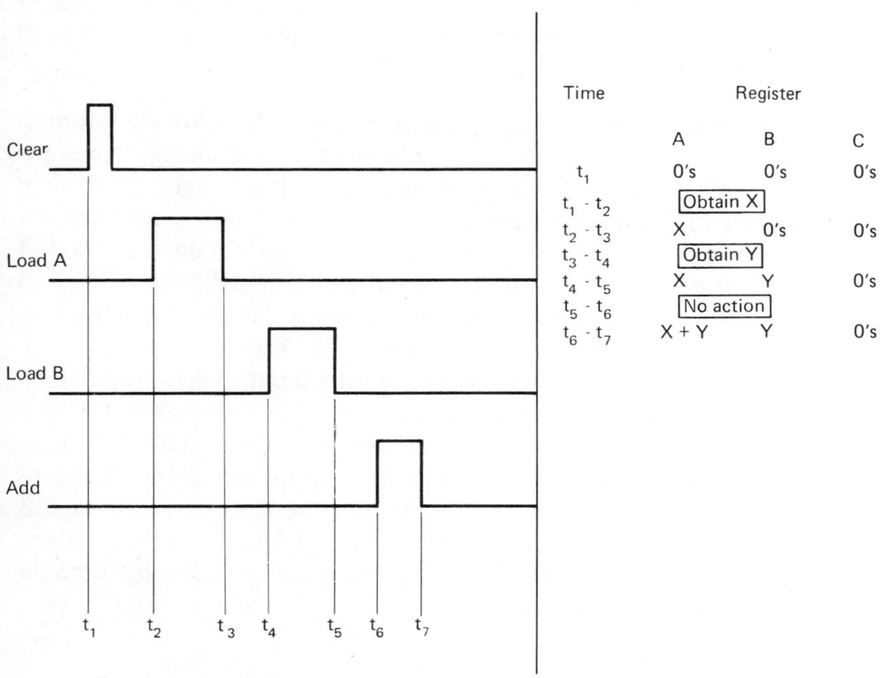

Figure 4-11 Arithmetic/Logic Section Waveforms

4-5 THE INPUT FUNCTION

Direct Input. From a strictly practical standpoint, the computer is *initially* incapable of *any* useful function. The memory has no information stored; and since all computer instructions and data must come from memory, those instructions and data must be *loaded* via the INPUT section.

Perhaps the simplest direct input device is available at most computer control panels. A group of switches, one for each bit in the computer word, is provided. Each switch can be positioned to generate a binary 0 or a binary 1. When all switches are properly positioned, the information may be loaded into memory (usually via an intervening storage register). Addresses, instructions, or data may thus be entered *directly* into the computer by means of panel controls. The programmer merely arranges the required information in the proper order, and word by word, at selected addresses, the instructions and data are loaded. Manual entry of information by use of console switches is, however, an inefficient means of loading the computer.

Another convenient, though still relatively slow, method of directly inputting information to the computer is via a keyboard that closely resembles the electric typewriter. Numerous keyboard devices are in use, but a common example is the Teletype.* When used as an input device, its speed is limited to that of the operator and is relatively slow therefore in terms of internal computer speeds.

Indirect Input. Any input device that does not provide information in direct binary computer word format provides a challenge. Attempts to make digital computers user-oriented have, unfortunately, resulted in input devices that are *not* binary-computer-word-formatted, and conversion methods must be provided to make them so. The common method of providing this capability is to write a computer program that will accept the input information in the format of the input device and convert it, within the computer, into binary-computer-word format.

The most common methods of information input to the computer are of an indirect nature. Direct input devices, such as the console switches and keyboards, are severely limited by operator speeds and can make computer use ineffective. Indirect methods of input are characterized by recording information on some medium that the computer can rapidly read. In addition, indirect input may be prepared *off-line*, that is, without the computer operating. Input may thus be prepared at the same time the computer is performing other tasks.

*Trademark registered by Teletype Corporation.

One of the most common indirect input media is *punched cards*. A *card punch* prepares precut cardboard cards by punching holes in specific locations and combinations for each character to be fed to the computer. Figure 4-12 shows a typical punched card and the code used with this type of card. Each card can contain up to 80 characters, and card readers may process up to 2000 cards per minute, although 500 cards per minute is more common. This equates to approximately 670 characters per second.

The Teletype keyboard often has a *paper tape punch* associated with it. Again, a specific combination of holes and spaces in a column corresponds to a character to be entered into the computer. The reader station on the Teletype processes the punched tape, entering the information into computer memory as the tape is read. Tape-punching capabilities exist at speeds up to 300 characters per second, while readers process punched tape up to 1000 characters per second.

Still higher input speeds may be achieved by the use of *magnetic tape recorded information*. Instead of combinations of holes and spaces as in the paper tape, combinations of magnetized and nonmagnetized locations in columns on the magnetic tape are used. A very high density of information is obtainable with this method, and up to 1600 characters per inch are not unusual. When it is considered that magnetic tape is read at speeds up to 75 inches per second, it can be seen that readings of many thousands of characters per second are easily obtainable. Magnetic tape may be directly recorded as an output of the computer, or direct key-to-tape systems are available.

Reel-to-reel magnetic tape systems are supplemented by cassette systems which operate on the same principle. However, the cassette is

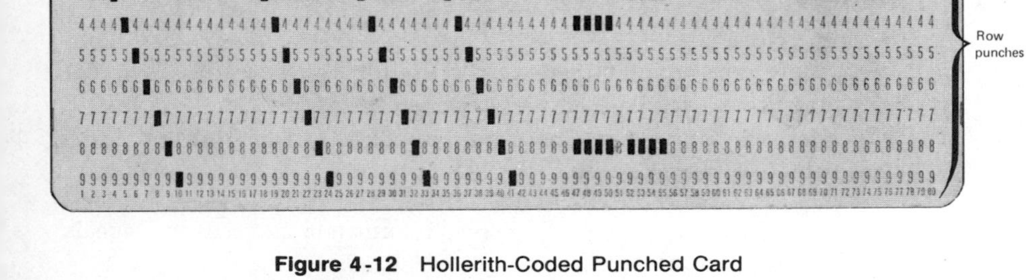

Figure 4-12 Hollerith-Coded Punched Card

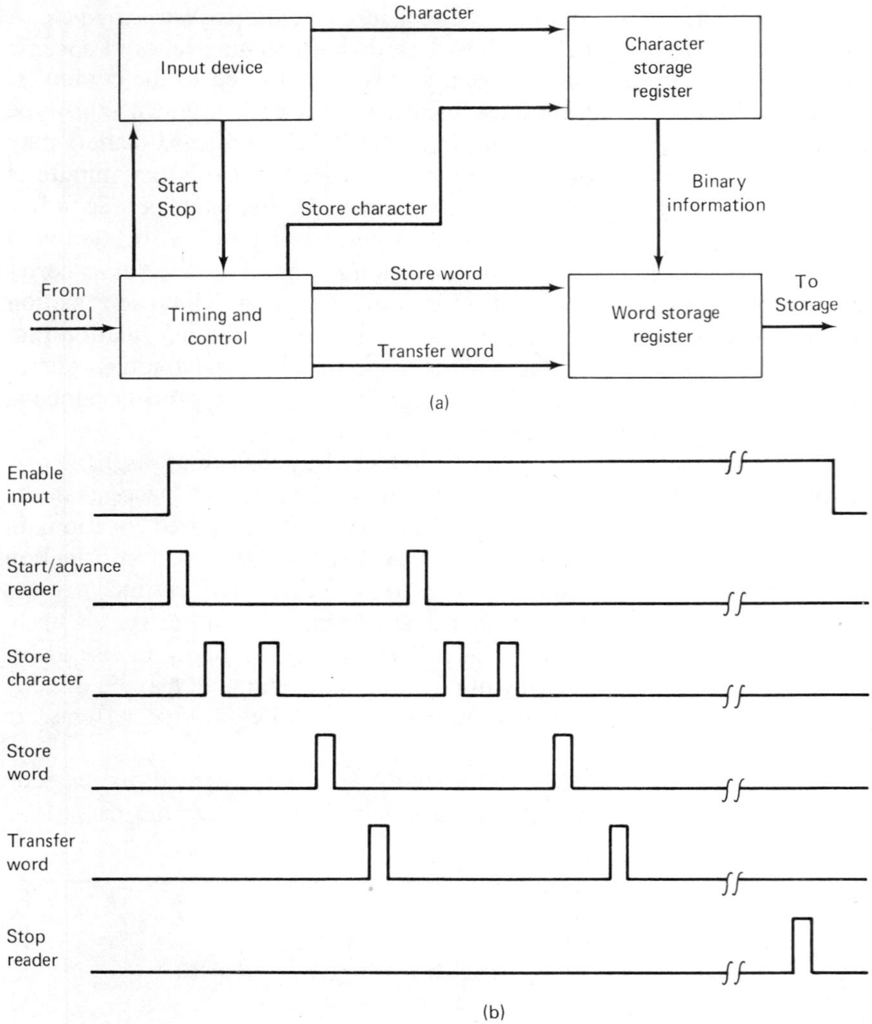

Figure 4-13 Input Section: (a) Functional Block Diagram; (b) Timing

easier and more convenient to use and is finding widespread use. Speeds are somewhat slower than reel-to-reel systems.

Consideration of only the bare minimum needs for an input device shows that the following elements are required:

1. A device to convert the recorded information into electrical signals, such as a reader.
2. A means of temporarily storing the information from the reader until a computer word is composed.

3. Timing and control to synchronize the input element with the rest of the computer.

A block diagram shows these requirements (Figure 4-13a), while the gross timing for a hypothetical INPUT section is shown in Figure 4-13b.

4-6 THE OUTPUT FUNCTION

Once the concept of input operations have been mastered, it is relatively simple to explain output operations. Whereas the INPUT section was the interface between the outside world and the computer, the OUTPUT section interfaces between the computer and the outside world. This requires conversion of the binary computer word to a format that is acceptable to the output device in use. A typical simplified OUTPUT section closely resembles the INPUT section except that the flow of data is in the reverse direction (Figure 4-14). Timing requirements are similar.

Output Devices. OUTPUT sections have devices similar in concept to INPUT sections. Direct information output is obtainable at the operating console of the computer in the form of indicator lights. Complete computer words may be examined, bit by bit. The electrical condition of important control operations may also be displayed. Individual register contents are accessible by the use of console display lights. In other words, the complete operation of the computer is available to the operator with the indicators provided on the console of most computers. Once again, however, this manual method of obtaining output is very slow and most often is used only for troubleshooting purposes.

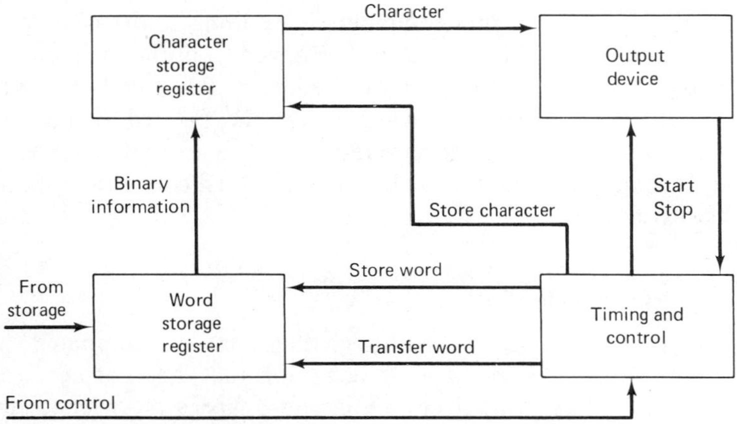

Figure 4-14 Output Section Block Diagram

Output may also be obtained by the use of printing devices which furnish a *hard copy* of the output information. The complexity of the printing mechanism also dictates the amount of information conversion that must take place. For example, use of the printing portion of the Teletype furnishes a printed record of the information and requires a minimum of conversion effort. Of course, the information in the computer word must be converted into the code that is recognized by the Teletype, but that is true of all output devices other than the indicator lights. Writing speeds on the order of 100 to 150 words per minute may be expected from the Teletype.

Many printing-type output devices do not print a single character at a time. The *line printer*, for example, includes intermediate storage capability (usually called *buffer storage*) which accumulates all of the information necessary to print a complete line before actually printing that information. Since information transfer between the computer memory and the buffer storage may take place at speeds much greater than the printing mechanism can use, actual information output speed is much greater. Line printers with speeds of up to 1400 lines per minute, 120 characters per line, are available.

Once again the emphasis on making the operation of the computer user-oriented is resulting in numerous innovative output devices. Equipment complexity usually increases in such instances, but the user of the computer realizes much greater results. The *cathode-ray tube* (*CRT*) is an example of a widely used information display device. Operating on the same principles as the television picture tube, the CRT makes available such information as letters, numbers, special symbols, graphical portrayal of information, and electrical schematics. The electronic circuitry required for operation of this type of display device is extremely complex, but once again the need usually justifies the cost.

Expanding use of digital computers is resulting in the need for both input and output capability at locations removed from the main computer. The Teletype and CRT display device represent this trend. A keyboard allows input to the computer, while either the printed output or the displayed output provides immediate response. The programming required, however, is somewhat complex to allow for the quasi-conversational operations that exist.

4-7 SUMMARY

It has been shown in this chapter that the computer can be approached on a *functional* basis. Recognition that it is not necessary to resort to *detailed* electrical/electronic circuit analysis in order to investigate computer operation should make study of the computer/microcomputer much more palatable. The *functional* level of detail is adequate to understand gross computer operation and should prepare the reader for subsequent excursions into the microprocessor field. Microprocessors are approached in the same functional manner in following chapters.

QUESTIONS

1. What is a computer program?
2. Describe the functions of a computer program.
3. Discuss the differences between an instruction word and a data word.
4. How does the computer know the difference between an instruction word and a data word?
5. What is a stored program?
6. Explain how stored programs are used in digital computers.
7. Discuss the flow of information and sequence of operation of a digital computer using the diagram shown in Figure 4-5.
8. List the functions of the CONTROL section.
9. What is the purpose of the instruction register; the instruction decoder; the control generator?
10. List the functions of the STORAGE section.
11. How is information stored in the STORAGE section?
12. What is the purpose of the memory address register; the memory buffer register; the sense amplifier?
13. List the functions of the ARITHMETIC/LOGIC section.
14. What is the purpose of the accumulator; the adder; the comparator?
15. List the functions of the INPUT section.
16. List the functions of the OUTPUT section.

CHAPTER 5

Microprocessor/ Microcomputer Concepts

As stated in Chapter 1, a microcomputer is a microprocessor-based computer. A microprocessor is, by generally accepted definitions, a large-scale-integrated (LSI) circuit assembly that contains much of the computing capability of a very small computer. It occupies an area on the integrated-circuit *chip* on the order of less than 1 square centimeter. When compared with discrete components, the microprocessor can be considered to be replacing thousands of transistors and diodes. Functionally it is a device that performs the arithmetic, control, and logical operations of a conventional digital computer.

A microprocessor actually contains, to some degree, all the functions shown inside the block labeled CPU in Figure 4-1. Thus, in most microprocessors one can expect to find the complete arithmetic/logic capabilities, means to generate the majority of the control signals required to operate a computer, plus a small portion of the input, output, and storage functions.

When compared with the conventional digital computer CPU, it will be seen that the *scope* of the microprocessor's capabilities is somewhat reduced. Its basic unit of information, the byte,* is 8 or less bits instead of the larger 16-, 24-, and 32/36-bit machines. (Some 12-bit and 16-bit microprocessors exist, but the trend is toward the shorter word, because of the space availability on the microprocessor chip. As integrated circuit technology advances, 12- and 16-bit oriented microprocessors are likely to become more popular.) Arithmetic operations as part of hardware are generally limited to those that can be performed by adder circuits (i.e., addition and subtraction). Other arithmetic operations are performed by software (computer programs). The microprocessor is much more dependent on storage capability external to the chip, once again because of space limitations. These are but a few of the reduced capabilities of the microprocessor. However, the versatility of the microprocessor, considering its physical size, is nothing short of phenomenal.

It should be pointed out, however, that there are about as many different ways to implement these functions as there are engineers to design them. Therefore, in this book a "composite" microprocessor will be used to develop the readers' functional knowledge.

5-1 THE BUS

Concepts. Although the microcomputer could be represented by the same block diagram as the general-purpose digital computer, enough differences exist to make it advisable to develop a new functional diagram. Most microcomputers tend to be organized along *bus*† lines rather than the more classic von Neumann approach. A bus-organized computer contains all the functions of the classically organized machine. Instead of information flowing from function to function, however, all functions are connected to all other functions in the bus-organized microcomputer. The control section of the CPU decides which of the functional units is to originate the information to be used and which of the functional units is to receive that information. Figure 5-1 shows the general concepts of bus-organized devices.

All the microcomputer assemblies are connected together by the bus. Some of the wires in the bus are committed to handling memory location information; others carry only data; still others transmit only control signals. Each of the assemblies uses only those address, data, and control signals related to the specific operation being performed.

*A *byte* is a unit of information consisting of a fixed number of bits. It is treated as an entity. "Byte" and "computer word" are used interchangeably.

†A bus is a group of interconnections between all sections of the microcomputer. It connects all control, data, address, and power signals of one section to all other sections. Its function is to provide a communication path between two or more of the computer's sections.

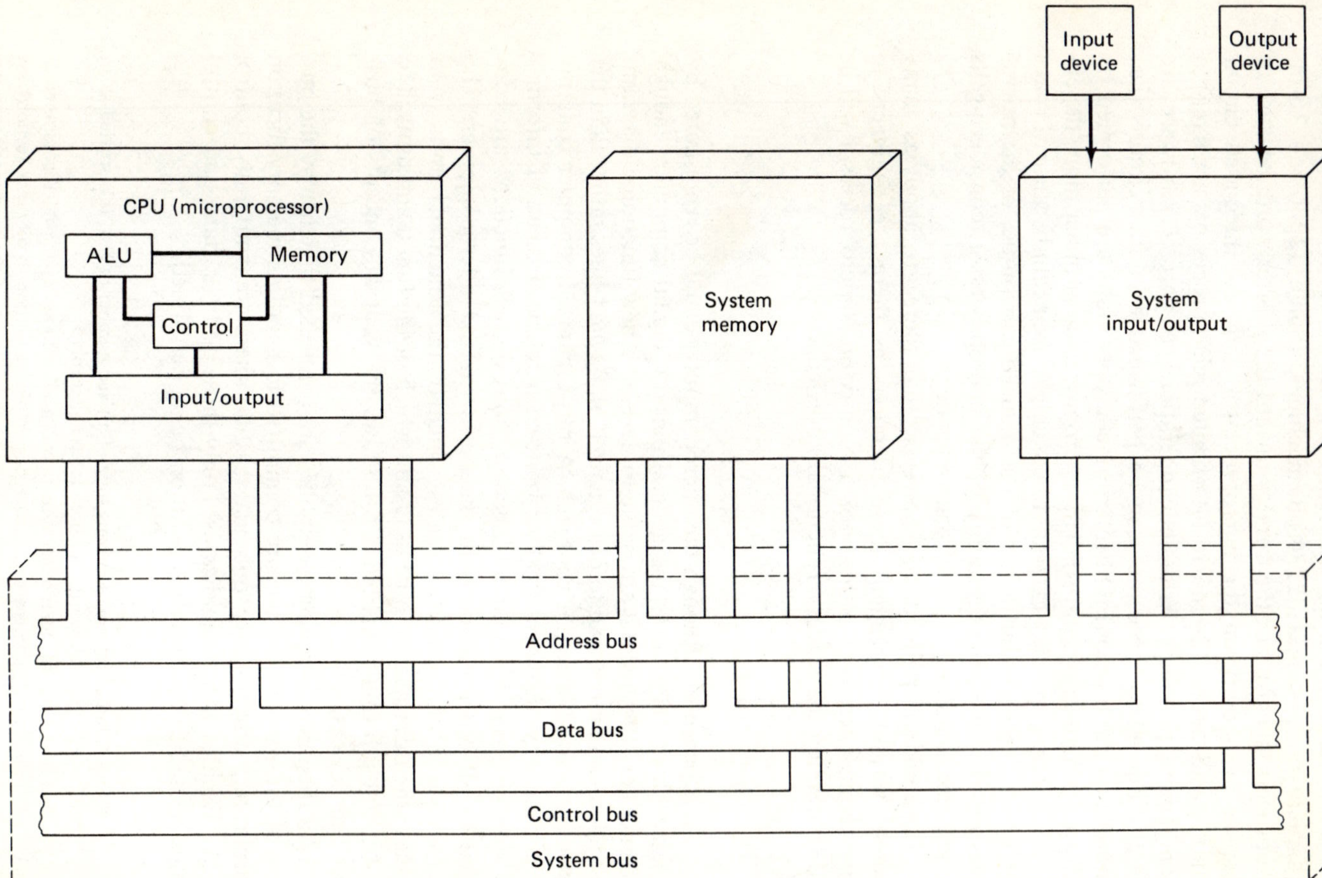

Figure 5-1 Concepts of Bus-Organized Computers

Bus Functions. Microcomputers generally have a number of buses. For ease of understanding, it will be considered that a data bus, an address bus, and a control bus are adequate to interconnect all the functional units of a microcomputer.

The *data bus* carries either instructions to be decoded or information to be manipulated. It is bidirectional; that is, it carries information from the microprocessor to other units of the microcomputer at one time, and at other times information may be carried from another unit of the microcomputer to the microprocessor. Direction of data flow is determined by the CONTROL section of the microprocessor. Microcomputer word size (8-bit, 12-bit, etc.) determines the number of connecting lines in the data bus. Thus, an 8-bit microcomputer has 8 connecting lines in the data bus; a 12-bit machine has 12 lines, and so on.

A unidirectional *address bus* contains a binary code representing the location of the next data to be used or the location where data are to be stored. Many microprocessors also use the address bus to identify input/output devices that are to be used. Thus, although information flows in only one direction, the address bus performs a dual function. The CONTROL section of the microprocessor supplies a signal to indicate whether an input/output device or the computer memory is to act on the binary code on the address bus. Total memory size (i.e., number of memory locations) fixes the number of interconnections provided by the address bus. A memory with 65,536 (64 K) storage locations, for example, requires 16 address lines ($2^{16}=65{,}536$) in the address bus.

The unidirectional *control* bus carries information describing what operation is to be performed, which functional units are to respond, and so on. Control bus size (number of interconnecting lines) is determined by the number of control signals required for the microprocessor in use, and is determined by the designer. Typical control bus signals include system timing, memory read/write, input read/write, and output read/write.

Control, data, and address buses also extend into each of the separate microcomputer assemblies. For example, the CONTROL, ALU, MEMORY, and INPUT/OUTPUT functions of the microprocessor itself are all connected, either directly or indirectly, to each of the buses. Thus, the bus signals not only control all other assemblies of the microcomputer, but allow the functional operations of each section to gain access to the bus signals.

Bus Gating. How all the sections of the computer can be connected together without causing massive interference problems may be confusing. Simple application of gates, plus precise timing, solves such problems. Figure 5-2 shows a simplified logic diagram of the busing concept. Each of the sections of the microprocessor is connected to each of the wires of the bus through a gating network. In its simplest form, the

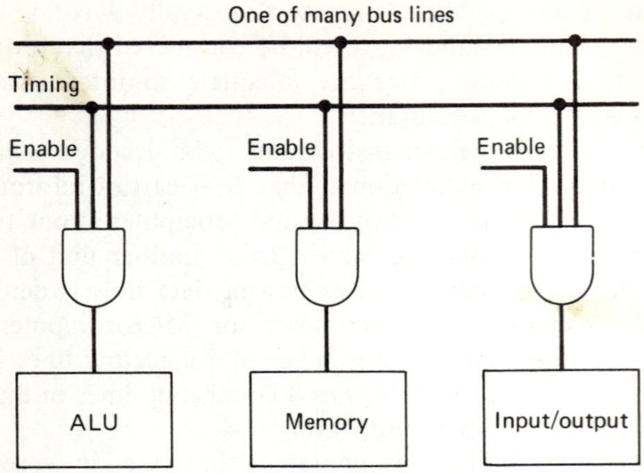

Figure 5-2 Bus Logic Diagram

gating network consists of nothing more than hundreds of AND gates. One input of each gate connects to a specific bus line. Another input connects to the source of timing in the microprocessor, while the third input connects to an output of the device performing the microprocessor control operations. The information on the bus line is passed to a functional section of the microcomputer only when its AND gate is enabled, and at the time when the computer timing is correct.

It is very important that the reader recognize the microprocessor as merely a collection of digital circuits, most of which are not "connected" in the conventional manner. The connection of the circuits is a function of the software (computer programs). Actually, most of the circuits are connected by gates that are enabled only to perform the desired functions. Thus, the person working with the microprocessor must learn a new "wiring" skill—that of computer programming. "Wiring with software" is the subject of Chapter 7.

5-2 MICROPROCESSOR DATA FLOW

Functional Organization. Since most microprocessors contain all the functional sections found in a conventional general-purpose digital computer it becomes relatively easy to discuss microprocessor operation. The bus organization of the microprocessor may be seen in Figure 5-3. The control bus for example, which originates in the CONTROL section, supplies inputs to both the ALU and the on-chip MEMORY sections as well as to external units via the system control bus. Address bus routing is to both on-chip MEMORY and external units. The address bus also originates in the CONTROL section of the microprocessor.

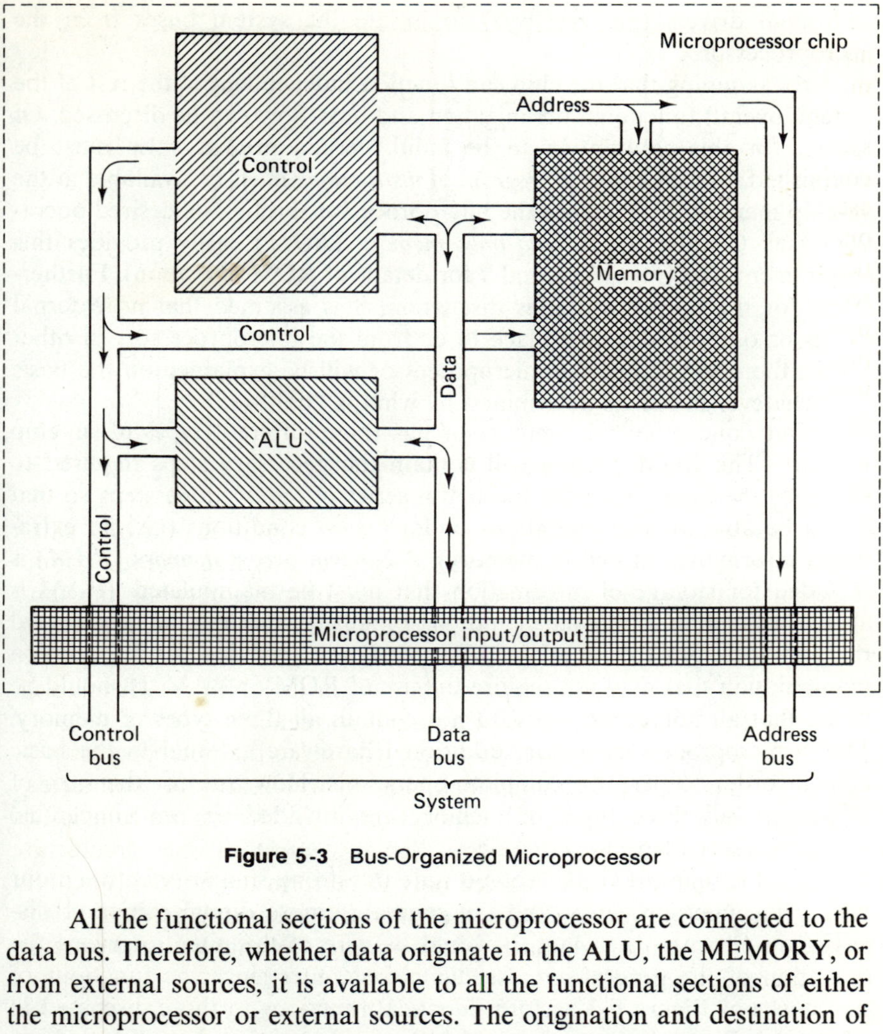

Figure 5-3 Bus-Organized Microprocessor

All the functional sections of the microprocessor are connected to the data bus. Therefore, whether data originate in the ALU, the MEMORY, or from external sources, it is available to all the functional sections of either the microprocessor or external sources. The origination and destination of the data on the data bus are determined by the CONTROL section of the microprocessor.

Microprocessor input/output is shown as a single functional block with all buses supplying either control, address, or data information. While not strictly accurate, it will serve the purpose of this preliminary discussion. Since many external devices may be connected to the system buses, it is important that such connections do not change the characteristics of signals on the buses. Unidirectional buses interface with the system bus via devices known as *bus drivers* (*bus transmitters*) which isolate the microprocessor from the external world. Similarly, bidirectional buses contain bus transmitters to isolate the microprocessor from the system buses and additional drivers (*bus receivers*) to isolate the system buses from the microprocessor.

By assuming that the chip can function independent of the rest of the system, overall functional organization and data flow can be discussed. Of course, for this assumption to be valid, certain ground rules must be established. For example, a *program of instructions* must be available in the on-chip memory to sequence the microprocessor through its desired operations (see Chapter 7). A *read-only memory* (*ROM*) easily provides this requirement (see Chapters 6 and 9 for details of ROM operation). Furthermore, for the purposes of this discussion, it is assumed that no external inputs or outputs will take place to or from the microprocessor. In other words, the operation of the microprocessor will be explained on the basis that whatever it does is determined by what is "built-in."

Our conceptual microprocessor has three different types of on-chip memory. The ROM portion will contain the necessary steps required to sequence the microprocessor through a series of *initialization* steps so that it will be able to start operations under *known* conditions (i.e., no extraneous information stored in memory). A *random-access memory* (*RAM*) is included for storage of information that must be manipulated. (RAM is also discussed in Chapters 6 and 9). Finally, a number of single-word memory locations (registers; see Chapter 2) are provided for information manipulation that may not require the use of ROM or RAM. It should be noted that all microprocessors do not contain all three types of memory. Many microprocessors require additional hardware external to the basic chip in order to perform complete operations. However, for the sake of discussion, all three types of memory are included on our conceptual microprocessor chip.

At this time the reader should note the strong similarity between our conceptual microprocessor and the general-purpose digital computer discussed in Chapter 4. In fact, it would be wise at this time to review the operation of the general-purpose digital computer since the functions of the blocks of Figure 5-3 perform identical functions to those discussed in Chapter 4. Particular attention should be paid to the timing relationships, as they are extremely important in understanding computer operation.

A Typical Operation. Now let us follow the operation of our conceptual microprocessor as it performs its *initialization* program. Initialization is commonly performed as a result of initial power application. The program stored in ROM consists of the steps necessary to clear RAM of all data so that new data may be inserted. ROM data are permanently stored and cannot be changed. If the CONTROL section is forced to start at the first-step address of the initialization program when power is applied, then the first operation performed by the microprocessor is one that will begin initialization.

A typical method used to clear memory of extraneous data is to place all zeros in each location. Thus, our initialization program must go to the first location to be cleared, place all zeros in that location, and go to the next location to perform the same operation. This sequence must be continued until the required number of locations have been cleared. The program must recognize when all locations are cleared and stop to await further orders.

One possible sequence of operations (program) to perform initialization appears below:

1. Load the starting address in general-purpose register 1. This is the first location in RAM that is to be cleared.
2. Load the stopping address in general-purpose register 2. This is the last location in RAM that is to be cleared.
3. Move all zeros to the location in memory addressed by general-purpose register 1.
4. Compare the current address with the address stored in general-purpose register 2.
5. If both addresses are the same, stop and wait for further instructions. All memory locations have been cleared.
6. If both addresses are not the same, increment the address in general-purpose register 1.
7. Move all zeros to new address.
8. Go back to step 4.

This program is by no means an optimum solution to initialization, but it does demonstrate the many different kinds of information manipulation and decisions that must be made in even the simplest of programs.

As noted in Chapter 4, a specific cycle of events occurs for each of the steps in the program. The CONTROL section supplies the timing to sequence each cycle of the microprocessor. When power is initially applied, the CONTROL section is forced to furnish the ROM address of the first step in the initialization program. The information present at that location in memory is supplied to the data bus and is routed to the CONTROL section for examination. In the case of step 1 in the initialization program, the CONTROL section determines that it must return to memory for information before the step can be completed. Therefore, it enables general-purpose register 1 and returns to the next location in memory to obtain the starting address. When the memory is enabled the second time, the data at the second location are routed to register 1, and the first step in the program has been completed. Note that it has taken *two* trips to memory in order to accomplish the first step. This is a common occurrence in microprocessors due to the relatively short (8-bit) word length. Many operations require up to *three* trips to memory to handle both instructions and associated addresses.

Step 2 also requires two trips to memory for accomplishment. The first trip obtains the instruction, while the second trip obtains the address to be loaded into register 2. Throughout the complete execution of the program the CONTROL section is keeping track of which memory location is being accessed, and also determining where in memory to go in case the present location is not in the sequential order. Step 3 initializes the starting address location in memory, while step 4 is used to determine whether the stopping address has been reached. If both addresses (the present address and the stopping address) are the same, the microprocessor stops and awaits further instructions. If, however, the addresses are not the same, the address in register 1 is increased by one and zeros are moved into that location in memory. The program now "loops" back to step 4 to compare addresses. The loop is repeated until both the present and the stopping address are the same. Step 5 then stops the program.

The ALU also came into play during execution of step 4. Comparison of two numbers is performed by subtracting one from the other and determining if the difference is zero. If the numbers are not the same, the result will not be zero, and the microprocessor proceeds in its normal sequence. If, however, the difference *is* zero, the program stops at step 5.

Thus, all the sections of the microprocessor come into play even in the simplest of programs. Although the microprocessor INPUT/OUTPUT section was not used in the initialization program, any information coming into or going out of the microprocessor is routed through this section so that external parts of the computer will be properly interfaced with the microprocessor.

5-3 SUMMARY

The important differences between the general-purpose digital computer and the microprocessor/microcomputer have been discussed in this chapter. Functional organization, based on control, address, and data buses, should be familiar to the reader as a result of reading this chapter. The reader should also be able to follow the execution of a simple program at the level of the functional diagram in Figure 5-3. A more detailed functional diagram is developed in Chapter 6, where the internal functions of each of the major sections of the microprocessor are explained.

QUESTIONS

1. Define the term microprocessor.
2. Define the term microcomputer.
3. What is a byte?
4. What is a bus?
5. How does a bus-organized computer differ from most conventional digital computers?
6. Explain the purpose of the data

bus; the control bus; the address bus.

7. What is the purpose of read-only memory (ROM) in a microprocessor?
8. What is the purpose of random-access memory (RAM) in a microprocessor?
9. How many times must memory be accessed to perform step 3 in the initialization program of Section 5-2?
10. How many times must memory be accessed to perform step 4 in the initialization program of Section 5-2?
11. How many times must memory be accessed to perform step 6 in the initialization program of Section 5-2?

CHAPTER 6

The Microprocessor as Hardware

Chapter 5 established that the microprocessor was in actuality a very small scale digital computer on a single integrated-circuit chip. Furthermore, the microprocessor was shown to be organized in a bus-connected configuration rather than in the more conventional von Neumann organization. Gross flow of data during the execution of a simple program of instructions was shown using a simplified block diagram.

More detailed operation of the microprocessor can be shown by expanding the functional block diagram to that shown in Figure 6-1, which is the manner in which a hardware-oriented person would view the microprocessor chip. Each of the major sections are still uniquely identified, but it should be noted that portions of the MEMORY section are shown attached to other sections, and that the INPUT/OUTPUT section is shown in three different locations. These changes are necessary to more adequately explain detailed hardware operation of the microprocessor, as will be seen shortly.

Each of the functional sections, with attached portions from the MEMORY section, will be discussed separately to show what is required to allow the sections to perform their operations.

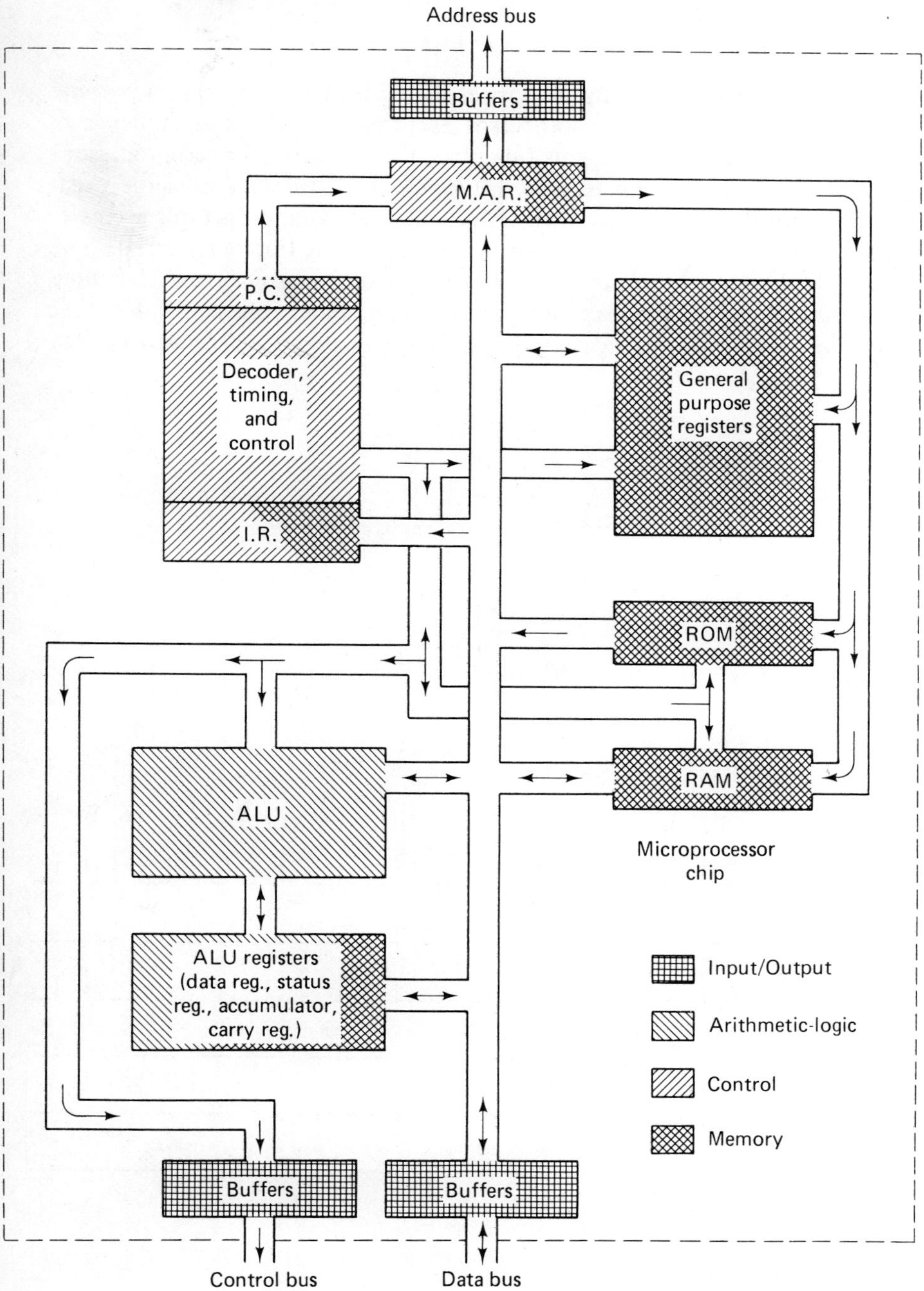

Figure 6-1 Microprocessor Detailed Functional Diagram

6-1 THE CONTROL FUNCTION

The Processor Cycle. Just as its big brother (the general-purpose digital computer), the microprocessor operates in a repetitive sequence of operations when performing its functions. Our conceptual microprocessor's sequence of operations is seen in Figure 6-2. Much confusion exists concerning the name of this sequence, since each manufacturer seems to use a different meaning for each descriptive term. For our purposes, we will use the term *processor cycle*, which will be *the sequence of operations required by the microprocessor to obtain information from either a storage location on-chip or off-chip, and to perform the operations required by that*

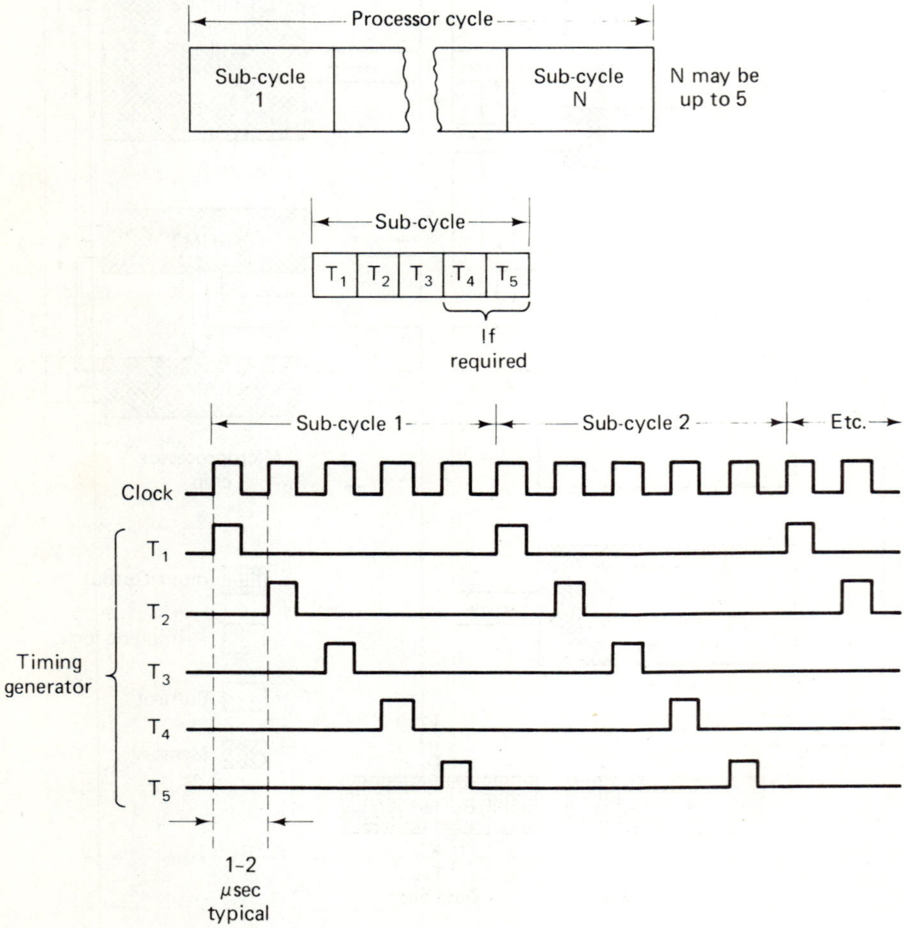

Figure 6-2 Processor Cycle and Timing

information. Each processor cycle is made up of one or more subcycles. Furthermore, each subcycle consists of up to five timing periods, which are the smallest increment of operation available to the microprocessor.

One subcycle is used for each byte of information to be obtained, or for each basic operation to be performed. The first subcycle of each processor cycle is always a "fetch" subcycle, and consists of at least four timing periods. The first period moves the contents of the program counter onto the address bus for the purpose of obtaining information from either input/output or memory. During the second timing period the program counter is incremented so that the next location to be accessed is readily available for use. The third timing period is used to move the information at the address noted in the first time period to a temporary storage register on the microprocessor chip. Depending on the operation to be performed, the fourth and fifth timing periods are used to move information between registers and other portions of the chip or perhaps to perform simple arithmetic operations.

The first subcycle always considers the information obtained as an instruction, and routes that information to the instruction register via a temporary storage register. All of the remaining subcycles consider information obtained to be data to be operated upon. Once again, the first time period of the remaining subcycles places an address on the address bus. If the address came from the program counter, the second timing period increments the program counter, and the third (and fourth and fifth, if used) timing period(s) perform operations on the information obtained. If the address came from an internal register, then information manipulation begins immediately with the second timing period.

Most processor cycles require only three subcycles to perform their operations. However, some of the more complex operations may require up to five subcycles, and some even more. Chapter 7 contains a more detailed study of processor cycles.

Timing. A functional diagram of a typical CONTROL section is shown in Figure 6-3. The length of each timing period in the processor cycle is determined by the *clock*. Our conceptual microprocessor, like so many of the modern chips, will contain the basic timing capability within the chip, requiring only an external frequency determining element (usually a piezoelectric crystal). Clock output is a continuous series of equally spaced timing pulses occurring at the rate determined by the external frequency-determining element. The timing pulses are fed to other functional sections within the microprocessor, and determine the basic operating speed of the microprocessor. Timing for each subcycle of the processor cycle is developed by the *timing generator*. Figure 6-2 showed

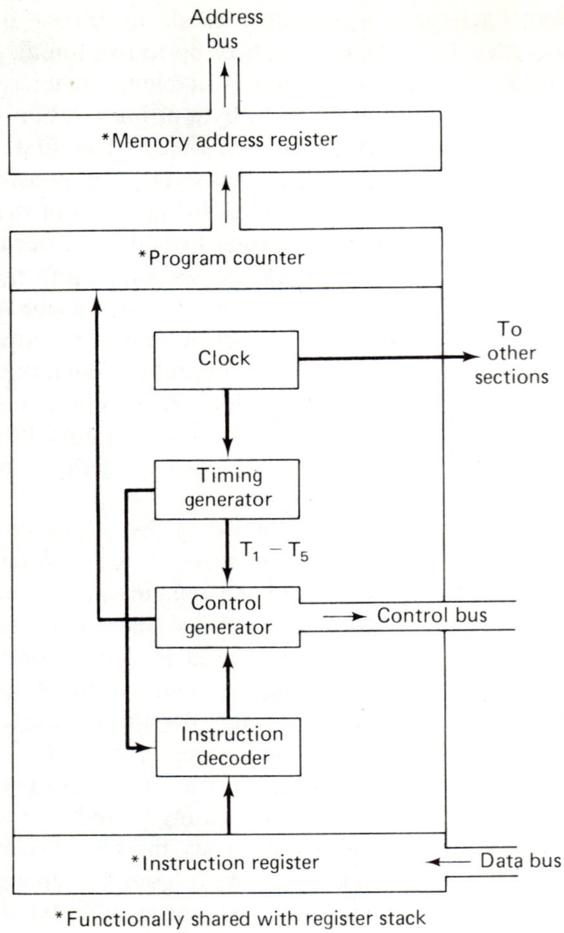

Figure 6-3 Microprocessor Control Section

typical clock and timing generator signals that could occur in a typical microprocessor. Both the clock and timing generator contain the integrated-circuit equivalents of counters, registers, and gates (see Chapter 2 for digital logic circuit discussion).

Decode and Control. Although physically a part of the register stack, the *instruction register* is functionally a part of the CONTROL section of the microprocessor. It receives a byte of information from the data bus and temporarily stores that information while the *instruction decoder* determines what action is represented in the specific arrangement of 1s and 0s in that byte. In our conceptual microprocessor the instruction decoder will be enabled (allowed to function) only during the subcycle in which it is known that information on the data bus represents an actual instruction. The timing generator is responsible for enabling the instruction decoder. Many other schemes are possible, but this method of assuring that only actual instructions will result in control signals represents a typical hardware-oriented solution. The instruction decoder may be implemented either by the integrated-circuit equivalent of networks of gates, or by a Read-Only Memory whose addressed inputs are the outputs of the instruction register and whose contents at each addressed location are those signals which will eventually represent the control signals that appear on the control bus.

The control signals on the control bus originate in the *control generator*. Here the outputs of both the timing generator and the instruction decoder are combined to develop the various control signals used throughout the microprocessor, and also throughout the microcomputer when the microprocessor is used in control and computing applications. Additionally, the control generator in our microprocessor oversees the operation of the *program counter*. Since the program counter must contain the address of the next location in memory to be used, it must be loaded with a given address, incremented, or decremented according to program instructions for each memory access. The control generator is well suited for these operations, since it is aware of each operation to be performed. Integrated-circuit equivalents of gating networks usually make up microprocessor control generators, while the program counter's implementation is usually as a physical part of the register stack even though functionally part of the CONTROL section.

Thus, within the CONTROL section, the important functions of timing, decoding, and control signal generation are accomplished. Integrated-circuit technology has allowed the packaging of these three functions in the incredibly small space of less than 0.1 square centimeter.

Many microprocessors also have a dedicated register, called the *memory address register* (*MAR*), that performs functions identical to a similar register in a general-purpose digital computer. The address of the location currently being processed is usually in the MAR, which obtains its inputs from either the program counter or one of the general-purpose registers in the register stack.

6-2 THE STORAGE (MEMORY) FUNCTION

Those sections of our conceptual microprocessor that are coded as "memory" include the *register stack*, *random-access memory* (*RAM*), and *read-only memory* (*ROM*). As noted in Chapter 5, all microprocessors do not contain this complete capability. The trend, however, seems to be toward including a register stack, RAM, and ROM on the same micro-

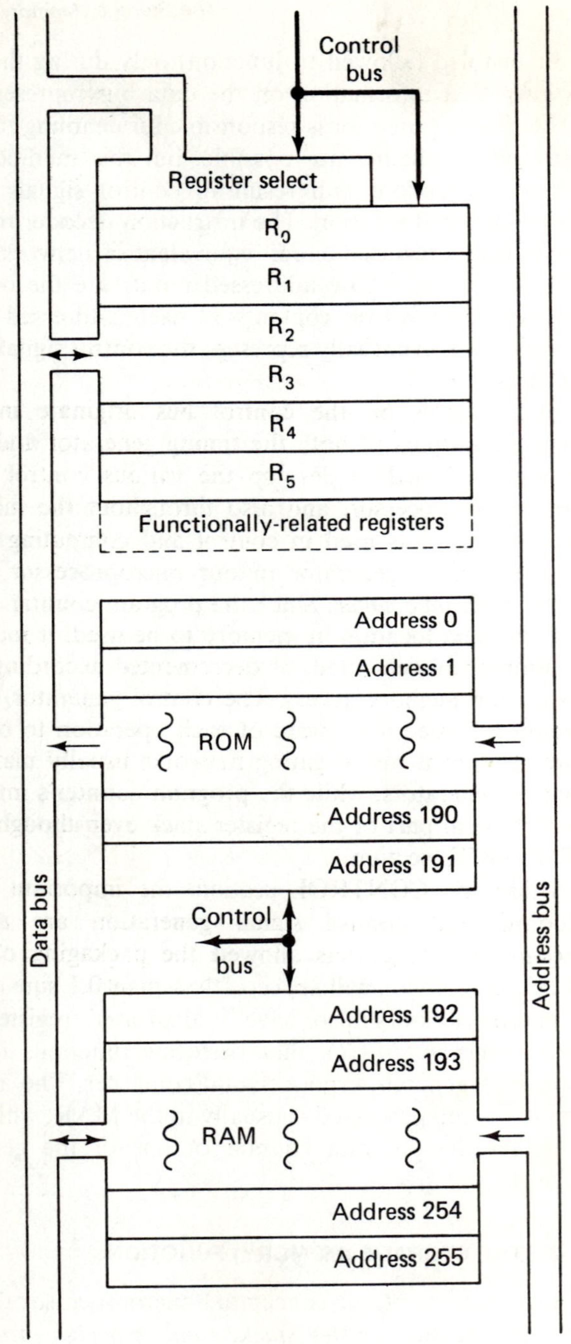

Figure 6-4 Microprocessor On-Chip Memory Organization

processor chip. Therefore, our conceptual microprocessor represents modern approaches to microprocessor design and operation. Figure 6-4 shows the general organization of microprocessor on-chip memories.

The Register Stack. The *register stack* is merely a collection of individual temporary storage registers that may be classified as *special-purpose registers* and *general-purpose registers*. The special-purpose registers, except for the *stack pointer*, are discussed as they are functionally used in other sections of the microprocessor, even though they may physically be a part of the register stack. For example, the instruction register and the program counter, both physically a part of the register stack, were discussed functionally as they were used with the CONTROL section.

Another use for registers appears when it becomes necessary to stop computer operations to respond to a higher-priority operation. Rather than giving up all the work previously performed, the microprocessor temporarily stores the information necessary to resume the original operation following completion of priority operation. Some microprocessors provide on-chip register storage for this function, while others use a dedicated register in the stack to store the address in RAM where the "present program" information is to be stored. The register used to "point" to the RAM address is called the *stack pointer*.

The general-purpose registers are used principally for the manipulation and temporary storage of data. Since they are used somewhat like a person would use a pad of paper to write down intermediate results of an arithmetic operation, they are often called "scratch-pad" registers. From a hardware viewpoint, most register stacks possess the following capabilities:

1. Each register in the stack may be loaded with any 8-bit byte desired. This requires gating circuits and a storage register with parallel-load configuration (see Chapter 2).
2. The information in any register in the stack may be moved to any other register in the stack, or the information in any register in the stack may be exchanged with the information in any other register in the stack. Once again, gating circuits and storage registers with parallel-load capability are required.
3. Registers in the stack may be treated in pairs to "trick" the microprocessor into thinking that 16-bit bytes are available. Only simple gating circuits are required to accomplish the pairing of registers.

Many other register operations exist with most microprocessors, but such operations are not hardware-oriented. Use of the general-purpose registers is most adequately discussed from a software viewpoint, as will be seen in

Chapter 7. It should be noted, however, that the number, organization, and judicious use of the register stack is one of the primary determining factors in obtaining maximum efficiency from a microprocessor. Our conceptual microprocessor will contain six general-purpose registers. The actual number of general-purpose registers available on commercial microprocessor chips is a design decision, and may vary from as few as 2 to as many as more than 48.

Selection of the general-purpose register to be used and the operation to be performed is accomplished by the CONTROL section. The byte in the instruction register is decoded by the instruction decoder, and if a register operation is indicated, certain of the bits in the instruction byte are selected and decoded by the *register select* function of the register stack. The register select function is enabled *only* when the register operations are indicated. Thus, as a result of an instruction byte on the data bus, any of the registers in the register stack and any of the register operations in the microprocessor's repertoire may be selected.

RAM and ROM. RAM and ROM used in a microprocessor differ little from the general concepts of storage devices discussed in Chapter 4. Both consist of arrays of storage cells that may be addressed in groups of eight cells at a time (considering that 8-bit bytes are the unit of information in the microprocessor). Furthermore, access to the information stored in both RAM and ROM may be considered to be of a random nature. That is, *any* of the locations may be "accessed" by merely placing the binary equivalent of that address location on the address bus. It is *not* necessary to start at location 0 and sequentially pass through all other locations to reach the desired address. However, each location that contains or can contain information must have a unique address, whether it be RAM or ROM.

Thus, when an address appears on the address bus, that location in memory is accessed. The CONTROL section determines what is to be done at that location. If an instruction that requires information from memory has been decoded (a READ instruction), the information at that location appears on the data bus. An instruction that requires information to be stored in memory (a WRITE instruction) will allow the information on the data bus to be written into memory at the addressed location.

The primary difference between RAM and ROM is that ROM information is permanently stored and is not subject to change by actions that go on within the microprocessor. RAM locations may be "erased" or changed under program direction, and new information stored. As noted in Chapter 5, ROM normally is used to store programs of instruction that will be used repetitively, while RAM stores information needed for short periods of time, such as intermediate results of arithmetic operations, or one-time use programs.

Design philosophy determines the RAM and ROM capabilities of commercial microprocessors. Upward of 2048 8-bit ROM locations and 128 8-bit RAM locations are common. Our conceptual microprocessor will contain 192 8-bit ROM locations and 64 8-bit RAM locations. This amount of ROM/RAM has been selected so that explanation of our microprocessor can be kept in simple terms. As the microprocessor is mated with external memory (Chapter 9) it will be seen that some modifications must be made to address more than 256 memory locations. Detailed discussion of RAM and ROM operation and characteristics appears in Chapter 9.

6-3 THE ARITHMETIC/LOGIC FUNCTION

The ARITHMETIC/LOGIC section of our microprocessor (Figure 6-5) performs the data modifications required to accomplish both logical and arithmetic operations. A collection of gates called the *arithmetic/logic unit (ALU)* does the actual data manipulations, while dedicated registers from the register stack temporarily store operands* during ALU operations.

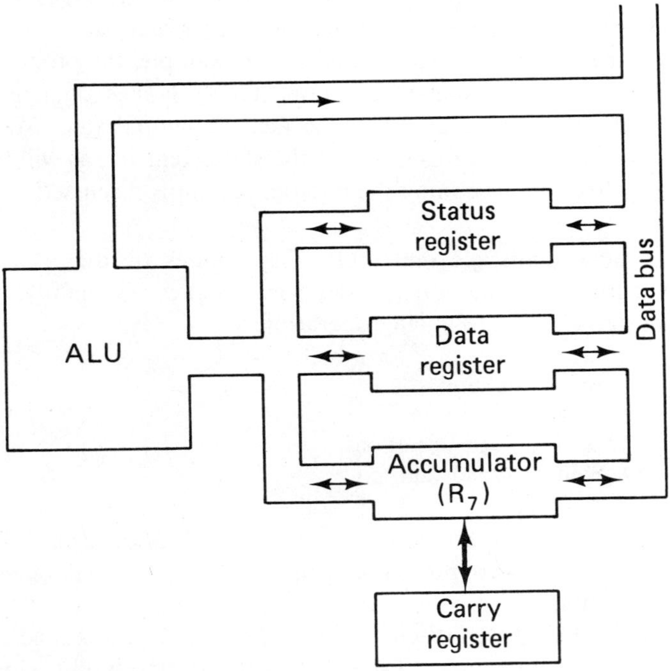

Figure 6-5 Microprocessor Arithmetic/Logic Section

*An operand, as used here, is a quantity on which a mathematical operation is performed.

Registers. Our microprocessor uses four registers during ALU functions. Since most arithmetic and logical operations require two operands, a register for each of the operands must be provided. One of the operands is temporarily stored in the *data register*; that is, the information on the data bus is moved into the data register as instructed by the CONTROL section. A second register, called the *accumulator*, similarly stores the second operand. In most microprocessors, the accumulator also stores the *results* of the mathematical operation, which replaces the operand that was originally stored there. Inputs to the data register and the accumulator may be from either the data bus or from the ALU. Therefore, the registers are bidirectional, under program control.

Certain arithmetic operations provide results that contain more bits than the regular 8-bit accumulator can hold. This is called an *overflow*, and means must be provided to retain this information. A separate single-bit register, called the *carry register*, is used to store any overflow information so that it may be included in subsequent calculations.

Arithmetic and data manipulation operations in a microprocessor generate various *flags** to define the results of the operation. For example, an arithmetic operation may generate a *carry*; the result may be −, 0, or +; and so on. One register, often called the *status register*, may be dedicated to storing the *status* information. The stored status may then be used to control *conditional* operations. If, for example, the program detects a zero result in an arithmetic operation, it may *branch* or *jump* to some portion of the program other than the next sequential step. Many variations are possible using information in the status register, as will be seen in Chapter 7, where programming the microprocessor is discussed.

Arithmetic/Logic Unit (ALU). Capabilities of the ALU are determined entirely by the microprocessor designer. Generally, however, ALUs can perform the following operations:

1. Binary addition.
2. Binary subtraction.
3. Boolean logic operations.
4. Shift right and shift left.

Other mathematical operations, such as multiplication and division, are implemented by software programs. Our microprocessor will operate under these ground rules.

A typical ALU was discussed in Chapter 3, and a simple arithmetic operation was demonstrated. From a functional standpoint, which is the

*A flag is some kind of indicator, usually a single bit in the status register, that indicates a specific condition has occurred.

preferable approach to microprocessors, only the *operations* that are performed by the ALU are important. One should understand *how* the processes of addition, subtraction, and so on, are performed, but it is not necessary to delve into detailed flow of data in a logic diagram. When the software methods of Chapter 7 are integrated with the hardware orientation of this chapter, the reader should possess enough knowledge of microprocessors to proceed with actual applications.

6-4 THE INPUT/OUTPUT FUNCTION

Concepts. From the hardware viewpoint, little can be said concerning our microprocessor INPUT/OUTPUT section. As mentioned in Chapter 5, chip input/output is generally limited to the hardware required to isolate the chip from outside influences. Except in some specific types of microprocessors, input/output capability is determined by software. Some microprocessors do possess the hardware necessary to input and output information in serial (one bit at a time) form, but generally such operations are under program control.

In most microprocessors, input/output is treated in the same manner as memory read/write. An address is placed on the address bus to represent the input/output device to be used, data are placed on the data bus, and the CONTROL section generates discrete control signals that define whether an input or an output operation is to be performed.

Buffers. The address and control buses in our microprocessor are one-way buses, transferring information to the external world from within the microprocessor. A bus transmitter (buffer) is required for each of the address or control lines leaving the chip. Each of the buffers is supplied with an enable input (Figure 6-6a) from the CONTROL

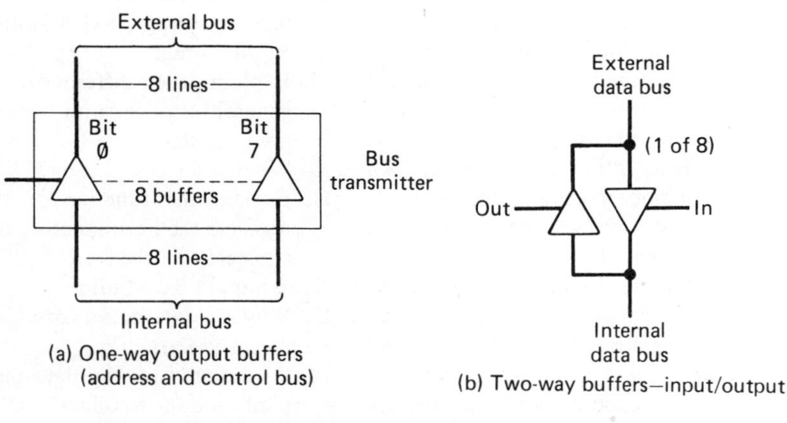

Figure 6-6 Microprocessor Input/Output Buffers

section so that the external lines may be effectively disconnected from the microprocessor if no activity is taking place. Considerable saving in power consumption and reduction in spurious response of the microprocessor is accomplished by disabling the bus transmitters when not in use.

Two-way buffers (Figure 6-6b) are necessary on the data bus outputs, since information must be both received and transmitted by the microprocessor. Enabling inputs are again provided from the CONTROL section, and the appropriate buffers (transmitters or receivers) are enabled to select the input or output function.

Input/output will be discussed in greater detail during the software investigation of the microprocessor's operation in Chapter 7. Specific input/output examples appear in Chapter 9.

6-5 SUMMARY

Each of the functional sections of the microprocessor has been discussed in detail. The major internal operations of each section have been shown, and the reader should have sufficient knowledge to follow the simple program of Chapter 5 through the detailed diagram of Figure 6-1. Complete understanding, however, still hinges on how the software causes each of the operations to exist. Chapter 7 provides this information, and upon completion of the software discussions, detailed portions of the simple program will be explained. The reader is encouraged, however, to correlate the initialization program of Chapter 5 with Figure 6-1 to review the hardware operation of the microprocessor.

QUESTIONS

1. What is a processor cycle?
2. What is a subcycle?
3. What is a timing period?
4. Why must the first subcycle always consider the information obtained as an instruction?
5. Explain the sequence of events occurring in the first subcycle of an instruction-processing operation.
6. What determines the length of a timing period?
7. Explain the function of the clock; the timing generator; the instruction register; the instruction decoder; the control generator; the program counter; and the memory address register.
8. Discuss the difference between register storage, ROM storage, and RAM storage.
9. Explain the difference between general-purpose and special-purpose registers.
10. What is a stack pointer?
11. Explain the function of the data register; the accumulator; the carry register; and the ALU.
12. What are data buffers?
13. Why are data buffers disabled when not in use?
14. Why must two-way data buffers be used on data bus lines?

CHAPTER 7

The Microprocessor as Software

All our effort to this point has been toward understanding the microprocessor from a hardware viewpoint. This has been effort well spent, since efficient application of the microprocessor requires a good background in the functional operation of the device. Now, however, the time has come to approach the other half of the microprocessor. As has been stated before, the microprocessor may also be considered almost entirely from a software standpoint. It is the function of this chapter to explore the software aspects of the microprocessor, and to integrate with those capabilities the hardware knowledge gained in previous chapters.

7-1 PRELIMINARY BACKGROUND

Instruction Format. Large computers work with units of information consisting of 16 or more bits, and both instructions to the processor (op codes) and the location (addresses) of the information to be processed *or* the actual information to be processed can be contained in the same computer word. In most microprocessors, however, information is stored and manipulated in groups of 8 binary digits called *bytes*. Microprocessors compensate for this apparent deficiency by using as many bytes as necessary to adequately define the instruction and other information. For example, most microprocessors can operate with single-byte, two-byte, or three-byte words.

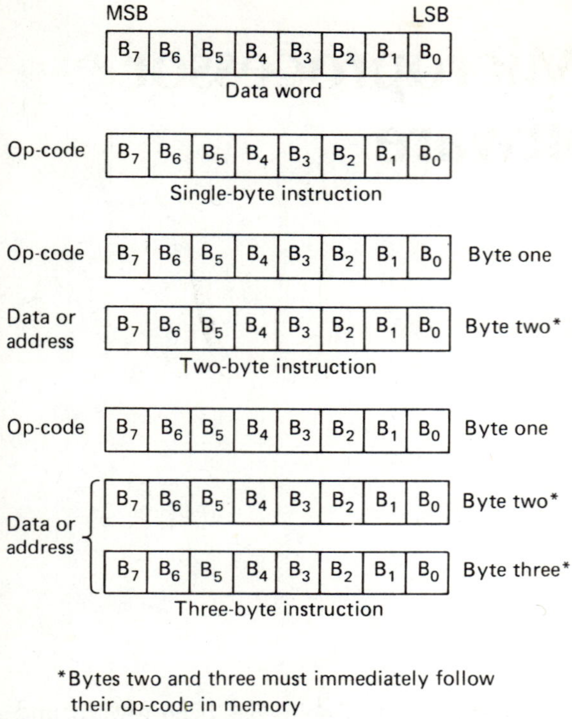

Figure 7-1 Instruction Formats

Fortunately, a large number of instructions can be represented by 8 bits (256 to be exact), and that is usually more than adequate for most microprocessors. However, many of the instructions require reference to memory locations for information to be processed, loading of a specific numerical value, identification of a particular input/output device, and so on. Two-byte instructions are generally adequate for references to memory locations in a small memory (less than 256 bytes), loading of numerical values less than 256, or identification of an input/output device. When memories with a capacity greater than 256 bytes are referenced, or when large numerical values must be loaded, three-byte instructions are required. Some instructions require three bytes *regardless* of the memory size so that a small-scale memory can be easily expanded without modification to the instruction set. Examples of all three types of instruction words are shown in Figure 7-1.

Addressing Modes. To execute any program, the microprocessor must constantly refer to storage locations in either the register stack, ROM, or RAM. Reference to storage locations is called *addressing*, and

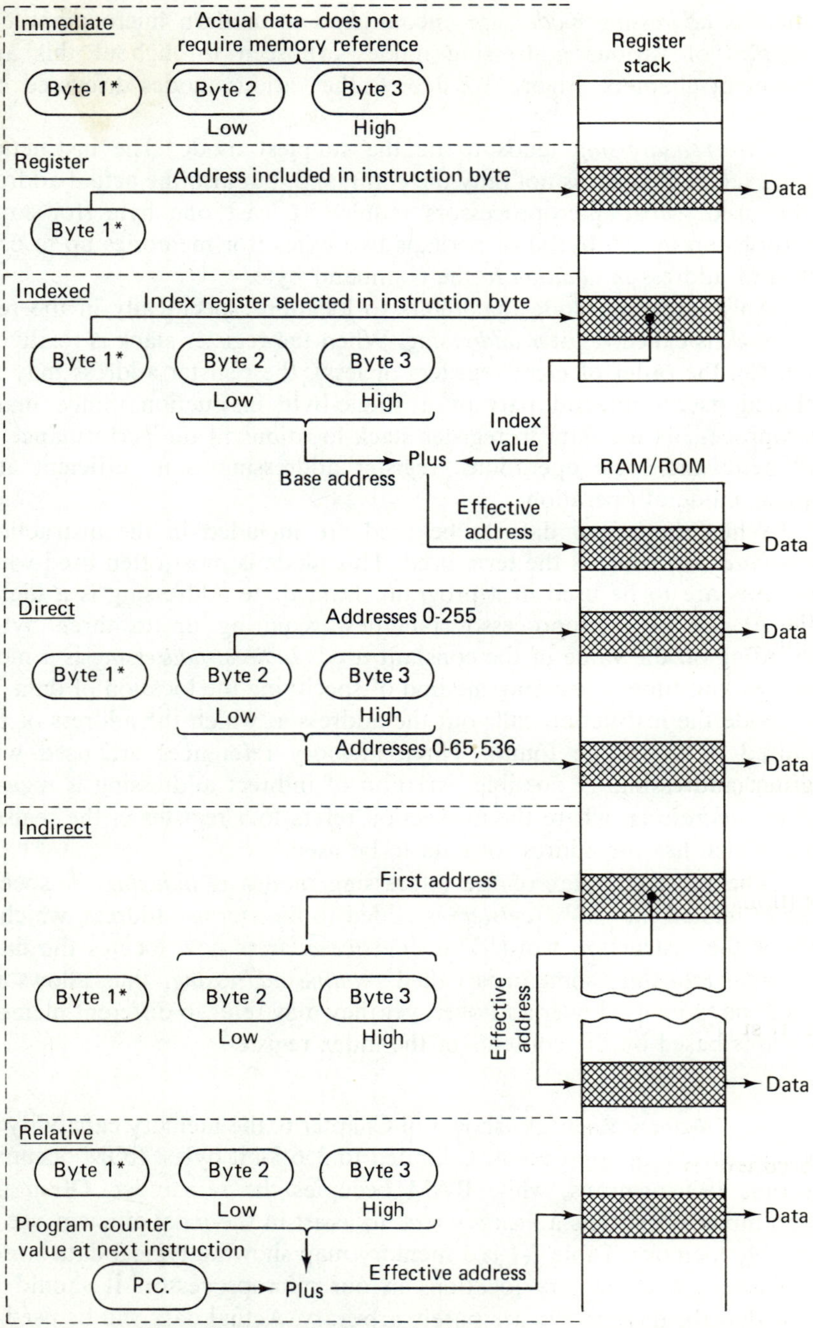

*Byte 1 is always the instruction byte

Figure 7-2 Addressing Modes

numerous *addressing modes* are encountered in modern microcomputers. Examples of various addressing modes are seen throughout this and subsequent chapters. Figure 7-2 depicts the various modes discussed below.

Direct addressing tends to be the simplest mode. The instruction being executed contains not only the command, but also the actual address to be used. Most microprocessors require at least one byte (for small memories up to 256 bytes) or perhaps two bytes (for memories up to 65K bytes) of address *in addition to* the command byte.

An addressing mode that refers to locations specifically in the register stack is called *register addressing*. When the register stack is relatively small (on the order of eight registers or less), the register address may be included as an integral part of a single-byte instruction. Since many microprocessors use data in register stack locations in the performance of arithmetic and logic operations, register addressing is an efficient and popular mode of operation.

When the actual data to be used are included in the instruction, *immediate addressing* is the term used. This mode is most often used when constants are to be used in a program. Immediate addressing is a highly efficient use of microprocessor facilities, requiring up to three bytes, depending on the value of the constant used. *Indirect addressing* is a more complex and time-consuming method of specifying the location of data. In this mode the instruction calls out the address at which the address of the required data will be found. Three memory references are used with indirect addressing. A possible variation of indirect addressing is *register indirect addressing*, where the instruction refers to a register in the register stack which has the address of data to be used.

The most complex of the addressing modes is *indexing*. A special register (called the *index register*) is added to the storage address, which is part of the instruction word. This *effective address* now locates the data. *Indexed addressing*, sometimes called *relative addressing*, thus allows instructions to be used over and over, yet they may refer to different memory locations based on the contents of the index register.

Memory Maps. As shown in Chapter 6, the memory capability of our conceptual microprocessor is limited to 256 8-bit bytes. ROM occupies the first 192 locations, while RAM occupies the remainder. Often the programmer constructs a *memory map* to assist in the most efficient use of available memory. Table 7-1 is a memory map showing typical data stored at representative memory locations in our microprocessor. It should be noted that the data shown are purely arbitrary. Actual data will be used in subsequent examples.

Table 7-1 Composite Memory Map

DATA				LOCATION			
(d) Decimal	(c) Hexadecimal	(b) Octal	(a) Binary	(a) Binary	(b) Octal	(c) Hexadecimal	(d) Decimal
85	55	125	01010101	00000000	000	0	0
109	6D	155	01101101	00000001	001	1	1
170	AA	252	10101010	00000010	002	2	2
204	CC	314	11001100	10111111	277	BF	191
51	33	063	00110011	11000000	300	CO	192
146	92	222	10010010	11111111	377	FF	255

Column (a) is the most primitive of memory maps, since it is shown in the form that the microprocessor recognizes (binary number system). Such a map is extremely difficult for the programmer to use, and other forms are often seen. Octal (base 8), hexadecimal (base 16), and decimal (base 10) notations may be used, depending on the manufacturer's technical data and the programmer's background. Octal memory maps are common because of the former widespread use of the octal number system and its close correlation to the binary number system. Representation of the basic digits of the octal number system (0 through 7) is easily accomplished with 3 binary digits (bits). In fact, if any binary number is separated into groups of 3 bits, the octal number is immediately obtained. Limiting each group to 3 bits provides equivalent octal digits from 0 to 7, which is the range of the octal system. The relationship between the binary and octal number systems is diagrammed in Figure 7-3. The octal equivalent of the binary map of column (a) may be seen in column (b).

For the very common 8-bit byte-oriented microprocessor, the hexadecimal number system is quite prevalent. Many newcomers to the computer field are unfamiliar with the hexadecimal number system, but it

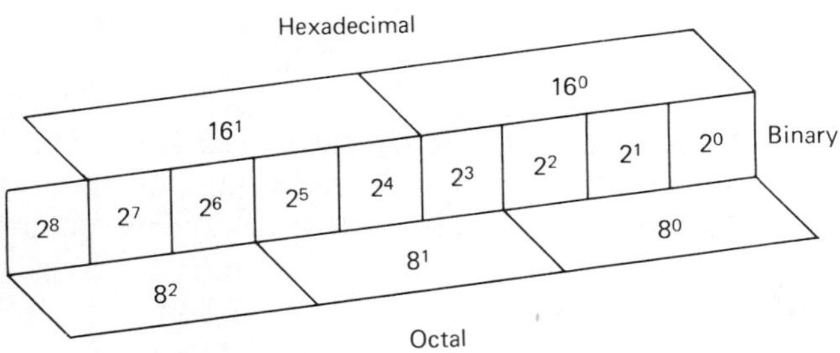

Figure 7-3 Binary/Octal/Hexadecimal Relationships

Table 7-2 Binary and Related Radix Codes

Decimal	*Radix 2* *Binary*	*Radix 8* *Octal*	*Radix 16* *Hexadecimal*
0	0	0	0
1	1	1	1
2	10	2	2
3	11	3	3
4	100	4	4
5	101	5	5
6	110	6	6
7	111	7	7
8	1000	10	8
9	1001	11	9
10	1010	12	A
11	1011	13	B
12	1100	14	C
13	1101	15	D
14	1110	16	E
15	1111	17	F
16	10000	20	10

is a natural selection. The hexadecimal number system employs 16 symbols, which can be represented easily with 4 bits. Thus, *two* hexadecimal symbols may be used to represent an 8-bit binary byte. The relationships among the binary, octal, and hexadecimal number systems are further diagrammed in Figure 7-3, and Table 7-2 shows the symbolic relationships. Column (c) of Table 7-1 is the hexadecimal-oriented memory map, equivalent to the binary map of column (a). Appendix B is a table of decimal-to-hexadecimal conversions to aid the reader during the use of hexadecimal notation in this book.

Decimal maps [column (d)] are the easiest for the newcomer to recognize, but do present difficulty in application as a result of the requirement to convert each decimal location or representation of data to the computer-compatible binary system. Binary/decimal relationships were shown in Chapter 2.

The Programmer's View of the Microprocessor. The *programmer's* primary concern with the microprocessor is that of application of its capabilities. Since the microprocessor is merely a data-manipulation device, the programmer is most interested in those parts of the microprocessor that perform that function *and* are available for use. Figure 7-4 is a possible programmer's viewpoint of the makeup of our conceptual

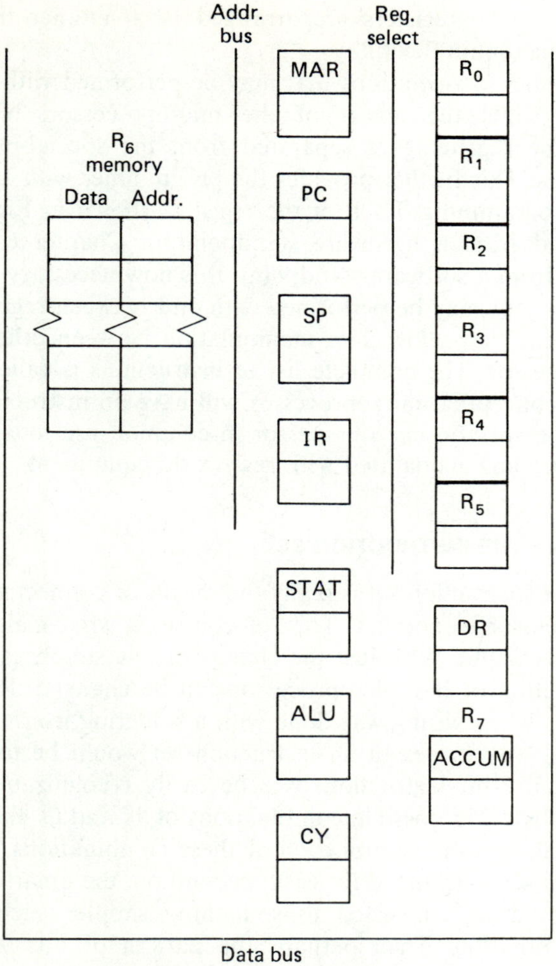

Legend:

MAR	– Memory address register
PC	– Program counter
SP	– Stack pointer
IR	– Instruction register
STAT	– Status register
ALU	– Arithmetic/logic unit
CY	– Carry register
DR	– Data register
R_0–R_5	– General purpose registers
R_6	– Memory designator
R_7	– Accumulator

Figure 7-4 Programmer's View of the Microprocessor

microprocessor. The registers are arranged in a manner that gives the programmer maximum flexibility.

The number of operations that may be performed with and between registers represents the power of the microprocessor. Note that the general-purpose registers are separated from the special-purpose (dedicated) registers. This further provides the programmer with an operational aid during programming. Each of the registers shown in Figure 7-4 have been discussed from a hardware standpoint in Chapter 6. In order to discuss them from a software standpoint, it is now necessary to investigate the operations that may be performed with and between registers, plus the other operations that allow data manipulation between other portions of the microprocessor. The complete list of instructions is called the *instruction set*. Our conceptual microprocessor will have an instruction set that is derived from a specific microprocessor in common use today, so that the reader may become acquainted with real-world equipment.

7-2 THE INSTRUCTION SET

The instruction set is really the means of connecting one part of the microprocessor to another. This, of course, is what makes the microprocessor so versatile. With just the change of one simple instruction, the complete function of the microprocessor can be changed. Remember the good old days when wiring was done with a soldering iron?

An ideal way to design an instruction set would be to have the bit pattern in each 8-bit instruction byte be easily recognizable. Of course, with an 8-bit byte 256 possible combinations of 1s and 0s exist, and it is a major undertaking to recognize each of these combinations.* If, however, the instruction set is arranged for easy recognition, the binary nature of the 8-bit byte could be converted to something simpler such as octal or hexadecimal notation. Then perhaps the pattern of bits would still be recognizable, and the task of using the instruction set is greatly simplified.

There are numerous methods of discussing the instruction set of a microprocessor. Each manufacturer seems to have a favorite method of organization, and little correlation is seen in the technical information available. A classification method that fits not only our conceptual microprocessor, but most commercial devices, is developed below.

*The instruction set used in our conceptual microprocessor will not be complete. A full complement of 256 instructions is just too much to work with this early in the learning process. A full instruction set is used in subsequent chapters where commercially available microprocessors are discussed and applied.

INSTRUCTION CLASSIFICATION

Information Movement Instructions

***Group 0*—General information operations.** Instructions in this group allow incrementing and decrementing (increasing and decreasing value of contents) of registers, loading and storing registers, shifting contents of registers, and so on.

***Group 1*—Information transfer operations.** Group 1 instructions move information from one location (called the *source*) to another location (called the *destination*) without modifying the information. Typical sources are on-chip registers, on-chip memory, and input devices. Typical destinations are on-chip registers, on-chip memory, and output devices. Actually, these instructions perform the function of connecting one location to another, just as though they were actually wired together.

Information Modification Instructions

***Group 2*—Arithmetic/logic operations.** The contents of certain on-chip registers may be modified by these instructions. Modification is accomplished by performing either *arithmetic* or *logic* operations on the contents of the registers.

Control Instructions

***Group 3a*—Program control operations.** These instructions transfer execution of a program from its present location to some other location in memory. Transfer may be conditional based on contents of the status register, or unconditional with no restrictions. Furthermore, each transfer may be a returning transfer, in which the present address of execution is saved so that control may be returned, or nonreturning, where the transfer does not require saving the present address.

***Group 3b*—Processor control operations.** Group 3b instructions *perform operations directly on the microprocessor* and are relatively few in number. Typical are processor halt, enable/disable interrupts, and no operation.

Assigning specific meaning to certain bits in the instruction byte allows easy determination of the major operations (groups) to be per-

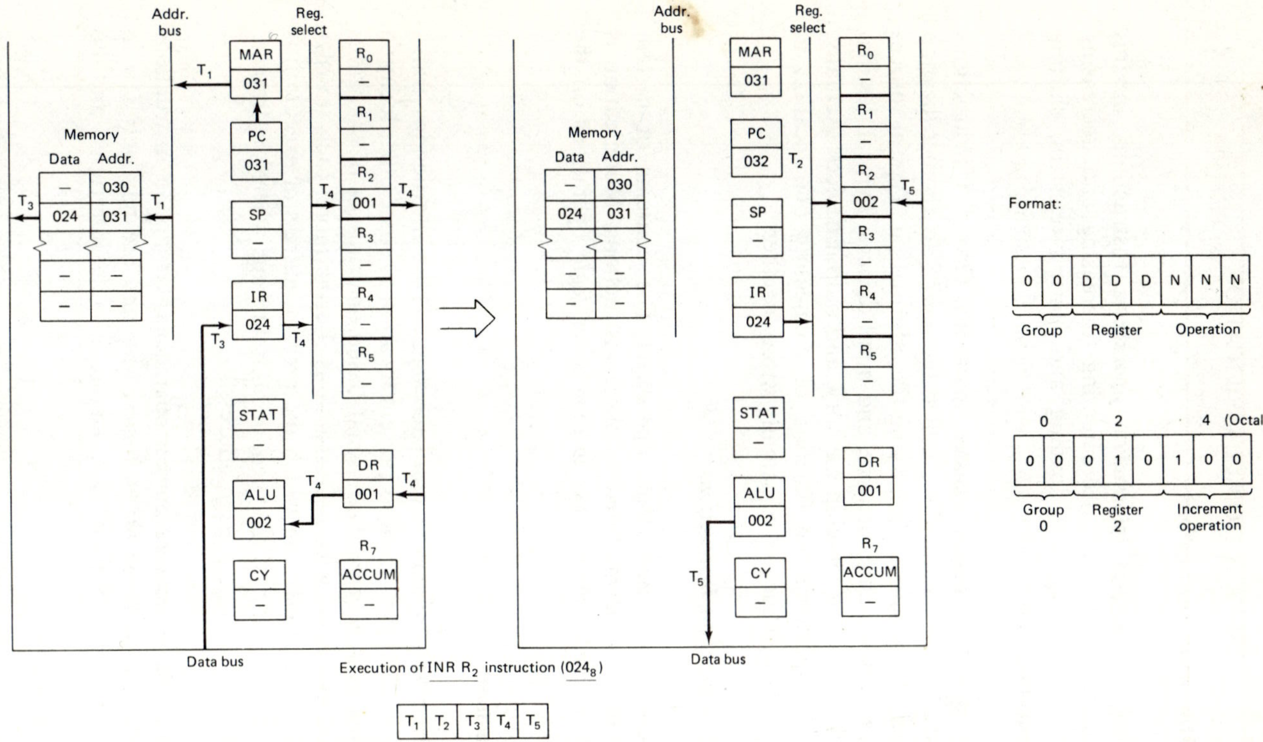

T_1 – PC output to MAR and addr. bus
T_2 – Increment PC (PC = PC + 1)
T_3 – Data at memory location in MAR moved to IR
T_4 – Data in IR decoded, contents of R_2 moved to ALU via DR, and ALU adds 1
T_5 – Contents of ALU returned to R_2

Figure 7-5 Group 0 Instruction Example

Table 7-3 Instruction Set Organization

	Group classification		*Register Pair (RP)*
0	General information operations	0	$R_0 - R_1$
1	Information transfer operations	1	$R_2 - R_3$
2	Information modification operations	2	$R_4 - R_5$
3	Control operations	3	Stack Pointer (SP)

	Register Designations		*Arithmetic/Logical designations (XXX)*
0	R_0	0	Add
1	R_1	1	Add with carry
2	R_2 DDD-Destination Registers	2	Subtract
3	R_3 SSS-Source Registers	3	Subtract with borrow
4	R_4	4	Logical AND
5	R_5	5	Logical EXCLUSIVE-OR
6	R_6 (Memory treated as register and requires address)	6	Logical OR
		7	Compare
7	R_7 (Accumulator)		

	General Information Operations (NNN)		*Branch Operations (YYY)*
0	Not used	0	Conditional RETURN
1	Immediate operations (Register pair)	1	Miscellaneous
2	Load & Store (Register pair)	2	Conditional JUMP
3	Increment/Decrement (Register pair)	3	Miscellaneous
4	Increment (Register)	4	Conditional CALL
5	Decrement (Register)	5	Miscellaneous
6	Immediate operations (Register)	6	Special A/L operations (Immediate)
7	Register shifting/miscellaneous	7	RESTART (Special Unconditional JUMPs)

Conditions of Branch (CCC): Result of last arithmetic/logic operation

0	was not zero	(NZ)
1	was zero	(Z)
2	did not generate a carry	(NC)
3	generated a carry	(C)
4	generated odd parity	(PO)
5	generated even parity	(PE)
6	was positive	(P)
7	was minus	(M)

formed. Our instruction set will use the two most significant bits to define each operational group as shown in Figure 7-5. Use of two bit positions results in four different combinations which are adequate for representation of Groups 0 through 3. It should also be noted that octal notation is more convenient than pure binary notation when discussing complete instruction format. Therefore, unless otherwise noted, the octal form of instructions will be used for explanatory purposes. Hexadecimal notation is employed in later chapters.

Table 7-3 is used to allow generalization of various groups of instructions. Register designations, arithmetic/logic codes, and general information codes are included so that the reader may "decode" any of the instructions used in this chapter.

Group 0 Instructions. As seen in Figure 7-5, the Group 0 instructions consist of the 2-bit group classification, a 3-bit register classification (DDD), and a 3-bit general operation classification (NNN). Referring to Table 7-3, it can be seen that NNN indicates what operation is to be performed, while DDD indicates the register to be operated upon. Once again, the complete instruction set will not be investigated in this chapter in the interest of developing a solid foundation without the confusion that normally attends evaluation of a complete instruction set.

A simple Group 0 instruction will demonstrate the use of Table 7-3. Assume that the instruction word 024_8 is stored at location 031_8 in ROM. The operation to be performed is decoded using the information in Table 7-3. Octal digit 0 first defines the word as a General Information Operation. The second octal digit (2) shows that the operation is to be performed on register 2, while the third octal digit is an indication that the contents of the register are to be incremented (increased in value by 1). Figure 7-5 shows the register operations that take place, correlated with subcycle timing periods. It should further be noted in Figure 7-5 that a mnemonic aid to remembering the instruction being performed is shown. *A mnemonic aid is a group of letters and or numbers that is remindful of actual English-language words*. For example, INR R_2 is much easier to remember than 024_8. Most microprocessor specification sheets include the mnemonic aids that correlate with the actual numerical designations of each operation. Group 0 instructions are included in the instruction set listing for our conceptual microprocessor (Table 7-4).

Group 1 Instructions. Group 1 instructions take the general form of Figure 7-6a. The conventional group designation appears in the 2 most significant bits (first octal digit) while the next 6 bits (next two octal digits) define a destination register and a source register, respectively. Thus, the Group 1 instructions take on the task of moving information from one

Table 7-4 Typical Instruction Set

Mnemonic	*Description*	*MSD*			*Instruction Code*				*LSD*
$MOV_{r1,r2}$	Move register to register	0	1	D	D	D	S	S	S
MOV M, r	Move register to memory	0	1	1	1	0	S	S	S
MOV r, M	Move memory to register	0	1	D	D	D	1	1	0
HLT	Halt	0	1	1	1	0	1	1	0
MVI r	Move immediate register	0	0	D	D	D	1	1	0
MVI M	Move immediate memory	0	0	1	1	0	1	1	0
INR r	Increment register	0	0	D	D	D	1	0	0
DCR r	Decrement register	0	0	D	D	D	1	0	1
INR M	Increment memory	0	0	1	1	0	1	0	0
DCR M	Decrement memory	0	0	1	1	0	1	0	1
ADD r	Add register to A	1	0	0	0	0	S	S	S
ADC r	Add register to A with carry	1	0	0	0	1	S	S	S
SUB r	Subtract register from A	1	0	0	1	0	S	S	S
SBB r	Subtract register from A with borrow	1	0	0	1	1	S	S	S
ANA r	And register with A	1	0	1	0	0	S	S	S
XRA r	Exclusive Or register with A	1	0	1	0	1	S	S	S
ORA r	Or register with A	1	0	1	1	0	S	S	S
CMP r	Compare register with A	1	0	1	1	1	S	S	S
Add M	Add memory to A	1	0	0	0	0	1	1	0
ADC M	Add memory to A with carry	1	0	0	0	1	1	1	0
SUB M	Subtract memory from A	1	0	0	1	0	1	1	0
SBB M	Subtract memory from A with borrow	1	0	0	1	1	1	1	0
ANA M	And memory with A	1	0	1	0	0	1	1	0
XRA M	Exclusive Or memory with A	1	0	1	0	1	1	1	0
ORA M	Or memory with A	1	0	1	1	0	1	1	0
CMP M	Compare memory with A	1	0	1	1	1	1	1	0
ADI	Add immediate to A	1	1	0	0	0	1	1	0
ACI	Add immediate to A with carry	1	1	0	0	1	1	1	0
SUI	Subtract immediate from A	1	1	0	1	0	1	1	0
SBI	Subtract immediate from A with borrow	1	1	0	1	1	1	1	0
ANI	And immediate with A	1	1	1	0	0	1	1	0
XRI	Exclusive Or immediate with A	1	1	1	0	1	1	1	0
ORI	Or immediate with A	1	1	1	1	0	1	1	0
CPI	Compare immediate with A	1	1	1	1	1	1	1	0
RLC	Rotate A left	0	0	0	0	0	1	1	1
RRC	Rotate A right	0	0	0	0	1	1	1	1
RAL	Rotate A left through carry	0	0	0	1	0	1	1	1
RAR	Rotate A right through carry	0	0	0	1	1	1	1	1
JMP	Jump unconditional	1	1	0	0	0	0	1	1
JC	Jump on carry	1	1	0	1	1	0	1	0
JNC	Jump on no carry	1	1	0	1	0	0	1	0

TABLE 7-4 (Continued)

Mnemonic	*Description*	*MSD*			*Instruction Code*				*LSD*
JZ	Jump on zero	1	1	0	0	1	0	1	0
JNZ	Jump on no zero	1	1	0	0	0	0	1	0
JP	Jump on positive	1	1	1	1	0	0	1	0
JM	Jump on minus	1	1	1	1	1	0	1	0
JPE	Jump on parity even	1	1	1	0	1	0	1	0
JPO	Jump on parity odd	1	1	1	0	0	0	1	0
CALL	Call unconditional	1	1	0	0	1	1	0	1
CC	Call on carry	1	1	0	1	1	1	0	0
CNC	Call on no carry	1	1	0	1	0	1	0	0
CZ	Call on zero	1	1	0	0	1	1	0	0
CNZ	Call on no zero	1	1	0	0	0	1	0	0
CP	Call on positive	1	1	1	1	0	1	0	0
CM	Call on minus	1	1	1	1	1	1	0	0
CPE	Call on parity even	1	1	1	0	1	1	0	0
CPO	Call on parity odd	1	1	1	0	0	1	0	0
RET	Return	1	1	0	0	1	0	0	1
RC	Return on carry	1	1	0	1	1	0	0	0
RNC	Return on no carry	1	1	0	1	0	0	0	0
RZ	Return on zero	1	1	0	0	1	0	0	0
RNZ	Return on no zero	1	1	0	0	0	0	0	0
RP	Return on positive	1	1	1	1	0	0	0	0
RM	Return on minus	1	1	1	1	1	0	0	0
RPE	Return on parity even	1	1	1	0	1	0	0	0
RPO	Return on parity odd	1	1	1	0	0	0	0	0
IN	Input	1	1	0	1	1	0	1	1
OUT	Output	1	1	0	1	0	0	1	1
LXI H	Load immediate register Pair H & L	0	0	1	0	0	0	0	1
LXI SP	Load immediate stack pointer	0	0	1	1	0	0	0	1
PUSH B	Push register Pair B & C on stack	1	1	0	0	0	1	0	1
PUSH D	Push register Pair D & E on stack	1	1	0	1	0	1	0	1
PUSH H	Push register Pair H & L on stack	1	1	1	0	0	1	0	1
PUSH PSW	Push A and Flags on stack	1	1	1	1	0	1	0	1
POP B	Pop register pair B & C off stack	1	1	0	0	0	0	0	1
POP D	Pop register pair D & E off stack	1	1	0	1	0	0	0	1
POP H	Pop register pair H & L off stack	1	1	1	0	0	0	0	1
POP PSW	Pop A and Flags off stack	1	1	1	1	0	0	0	1

TABLE 7-4 (Continued)

Mnemonic	*Description*	*MSD*			*Instruction Code*				*LSD*
STA	Store A direct	0	0	1	1	0	0	1	0
LDA	Load A direct	0	0	1	1	1	0	1	0
STAX B	Store A indirect	0	0	0	0	0	0	1	0
STAX D	Store A indirect	0	0	0	1	0	0	1	0
LDAX B	Load A indirect	0	0	0	0	1	0	1	0
LDAX D	Load A indirect	0	0	0	1	1	0	1	0
INX H	Increment H & L registers	0	0	1	0	0	0	1	1
CMA	Complement A	0	0	1	0	1	1	1	1
STC	Set carry	0	0	1	1	0	1	1	1
CMC	Complement carry	0	0	1	1	1	1	1	1
SHLD	Store H & L direct	0	0	1	0	0	0	1	0
LHLD	Load H & L direct	0	0	1	0	1	0	1	0
NOP	No-operation	0	0	0	0	0	0	0	0

Note 1: Not *all* instructions are discussed in this text; only those necessary to an introductory understanding are covered.
Note 2: Abbreviations are discussed in the text.
Note 3: Register designations

R_0-(B)
R_1-(C)
R_2-(D)
R_3-(E)
R_4-(H)
R_5-(L)
R_6-MEMORY
R_7-ACCUMULATOR (A)

register (or storage location) to another. Group 1 instructions lend themselves well to a matrix representation, such as that seen in Figure 7-6b. Reference to the matrix and to Table 7-3 will show that a definite pattern exists in Group 1 instructions, except for R_6 to R_6 transfers. (Since R_6 is defined as a memory location that must have an associated address, it would be meaningless to transfer information from that location to itself.)

If it was necessary to move the contents of R_2 to R_7, the octal instruction 172_8 could be used. The instruction can be verified by referring to Table 7-3 and Figure 7-6. The first octal digit (1) represents an information movement instruction; the second octal digit (7) identifies the destination register as R_7; and the third octal digit (2) lists R_2 as the source of the information to be transferred. MOV R_7, R_2 is the mnemonic

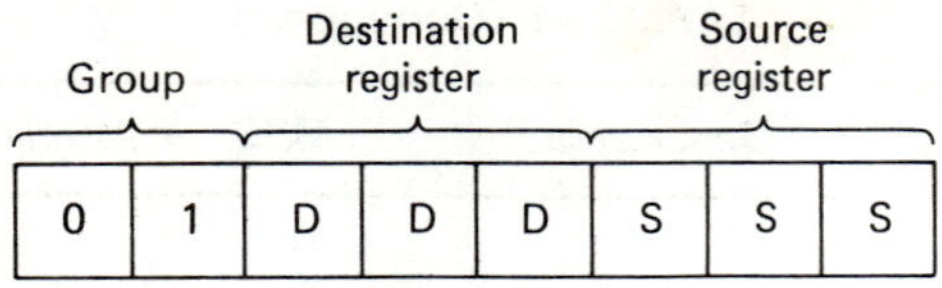

(a) General format

From → / To ↓	R_0	R_1	R_2	R_3	R_4	R_5	▻ R_6	R_7
R_0	100	101	102	103	104	105	106	107
R_1	110	111	112	113	114	115	116	117
R_2	120	121	122	123	124	125	126	127
R_3	130	131	132	133	134	135	136	137
R_4	140	141	142	143	144	145	146	147
R_5	150	151	152	153	154	155	156	157
▻ R_6	160	161	162	163	164	165	*	167
R_7	170	171	172	173	174	175	176	177

▻ Requires address

(b) Instruction matrix

Figure 7-6 Group 1 Instructions

designation for the instruction 172_8. Subcycle timing periods and register operations for MOV R_7, R_2 (stored at location 120_8) are shown in Figure 7-7.

Group 2 Instructions. Arithmetic and logical operations are performed by Group 2 instructions. Once again, the 2 most-significant bits (first octal digit) define the overall group of instructions. The format for Group 2 instructions varies from other formats, however. An arithmetic or logical operation (see Table 7-3) is coded in the second octal digit (next 3 bits), while the register that is to supply the data is identified by the last octal digit (last 3 bits). The matrix of Figure 7-8 shows the simple arithmetic/logical operations available in the instruction set of our conceptual microprocessor.

The reader should be aware that the Group 2 instructions are somewhat unique, in that they *all* modify the contents of the accumulator

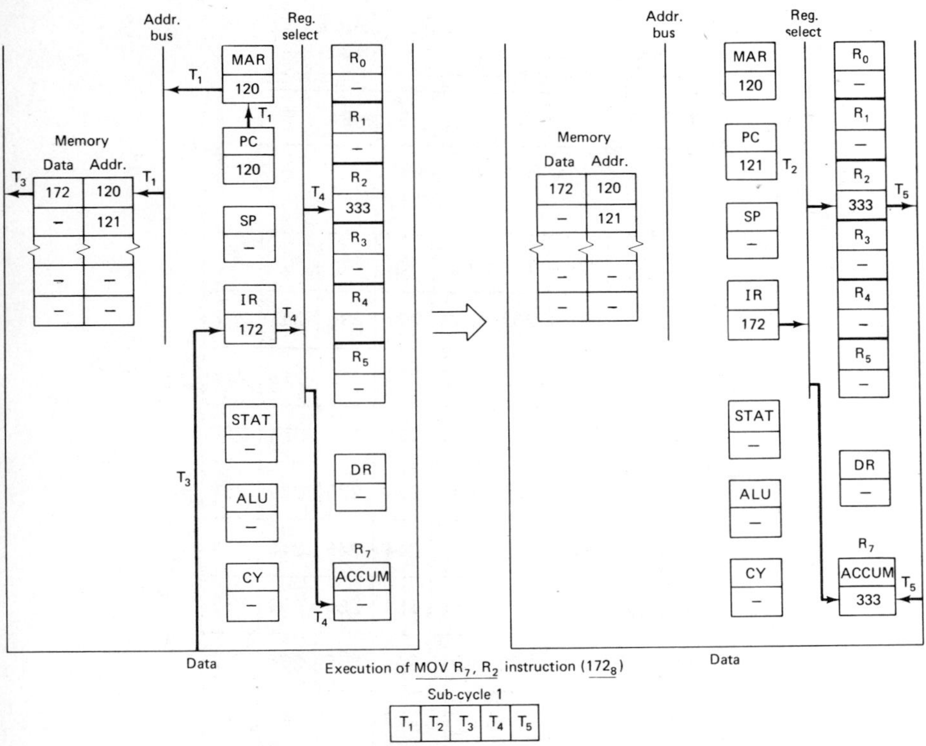

T_1 – PC output to MAR and address bus
T_2 – Increment PC (PC = PC + 1)
T_3 – Data at location in MAR moved to IR
T_4 – Data in IR decoded and registers selected
T_5 – Data exchanged

Figure 7-7 Group 1 Instruction Example

(R_7). That is, an instruction such as ADD R_4 (204_8) stored, for example, at location 226_8 obtains the contents of R_4, moves the information to the Data Register (DR), adds the contents of DR to the accumulator contents via the ALU, and then places the result in the accumulator. Register manipulations keyed to subcycle timing periods are depicted in Figure 7-9.

Group 3 Instructions. Instructions in Group 3 do not fit as neatly into specific categories as do those in previous groups. However, enough of the instructions may be classified as *branching*-type instructions to make it worthwhile discussing their organization. For those instructions that do classify as branching instructions, the configuration of the instruction byte may be used to determine the actual function to be accomplished. The most-significant octal digit once again determines the major group of

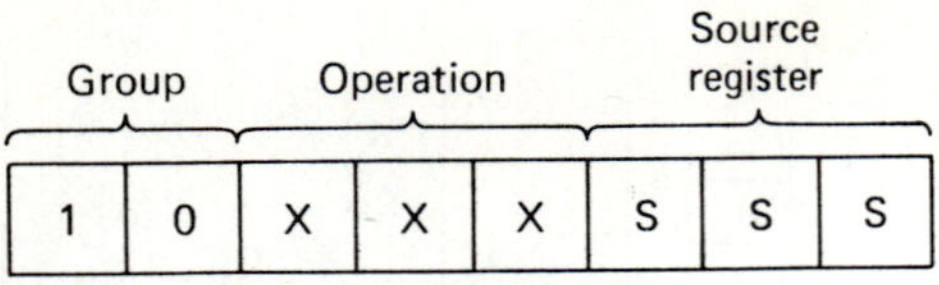

(a) General format

	R_0	R_1	R_2	R_3	R_4	R_5	R_6	R_7
ADD	200	201	202	203	204	205	206	207
ADC	210	211	212	213	214	215	216	217
SUB	220	221	222	223	224	225	226	227
SBB	230	231	232	233	234	235	236	237
AND	240	241	242	243	244	245	246	247
XOR	250	251	252	253	254	255	256	257
OR	260	261	262	263	264	265	266	267
CMP	270	271	272	273	274	275	276	277

(b) Instruction matrix

Figure 7-8 Group 2 Instructions

instructions. The least-significant digit is used to further define the instruction function as shown in Figure 7-10 and Table 7-3.

A *branching* instruction allows deviation from the normal sequential execution of the steps of a computer program. That is, as a result of either an instruction included in the program, or the results of executing a step in the program, it may be necessary to perform program steps that do not follow sequentially. Those deviations that result from the execution of a normal program step are called *conditional branches*. They are based on examination of the status word, which collects information concerning results of arithmetic and/or logic operations in the ALU.

The status word contains information concerning whether the result of the arithmetic/logic operation:

1. Was zero (Z) or not zero (NZ).
2. Generated a carry (C) or did not generate a carry (NC).
3. Generated odd parity (PO) or even parity (PE).
4. Was plus (P) or minus (M).

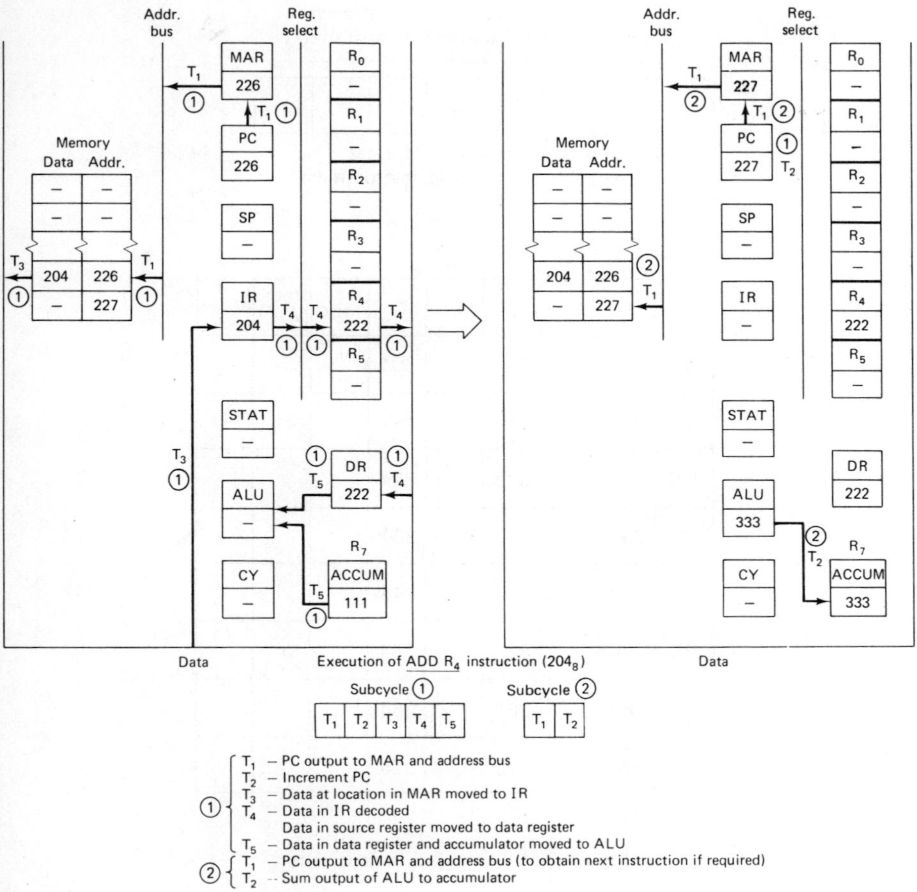

Figure 7-9 Group 2 Instruction Example

This status information is tested by the instruction byte, using the information appearing in the second octal digit of the byte (see Table 7-3).

Three types of conditional branches are allowed by our instruction set. They are selected by the third octal digit of the instruction byte, and may be either RETURN (0), JUMP (2), or CALL (4). All three conditional branch instructions require the existence of one of the conditions for branching, which results in 24 different instructions. Conditional JUMP instructions direct the program to a new step based on the tested condition. For example, the octal instruction 332_8 (mnemonic abbreviation JC) causes the program to transfer to the location directed by the byte(s) immediately following the JC step *if* the result of the previous arithmetic/logic operation caused a carry to be generated. The program then con-

Group		Conditional branch (status)			Branch operation		
1	1	C	C	C	Y	Y	Y

(a) Conditional instruction format
(see table 7-4 for misc. instructions)

	R	Misc*	J	Misc*	C	Misc*	Imm. (A/L)*	Restart*
NZ	300		302		304			
Z	310		312		314			
NC	320		322		324			
C	330		332		334			
PO	340		342		344			
PE	350		352		354			
P	360		362		364			
M	370		372		374			

*See table 7-4

Figure 7-10 Group 3 Instructions

tinues sequentially from the new location. If a carry had not been generated, the program would have continued with the step immediately following the conditional JUMP instruction.

The CALL conditional branch instruction is used to transfer control to a subroutine* upon recognition of the desired condition. Each CALL instruction must be accompanied by the location of the subroutine in the immediately following byte(s). If, for example, a subroutine was located at 000 123_8, and it was required whenever the result of a logical operation was zero, it could be accessed by the instruction CZ followed by the subroutine location $(314\ 000\ 000)_8$. The program would then continue at location 123_8

*A subroutine is a small program (subprogram) which performs operations that may be required more than once during the execution of the main program. It may be called upon when required and control transferred back to the main program (or elsewhere) upon completion of the subroutine steps.

until informed of other actions. If the result of the logical operation had not been zero, the program would have continued with the step immediately following the conditional CALL instruction and its attached address.

Finally, the conditional RETURN instruction is used to get the computer program back to its original sequence upon recognition of a specific condition, or to a different sequence. Conditional RETURN instructions are used in conjunction with the CALL instructions to assure that the main program continues normally following completion of the subroutine. The instruction byte 360_8 (RP) returns the control of the program to the step immediately following the CALL instruction that accessed the subroutine if the result of the previous arithmetic/logic operation was positive. If the result was negative or zero, the program would continue to the next step, whether a part of the subroutine or not. Therefore, it is necessary to have a RETURN instruction, whether it be conditional or unconditional, in order to return the program to its normal sequence. An unconditional RETURN instruction does just exactly that. It returns to the main program whenever the instruction appears; no conditions are tested.

Unconditional JUMP and CALL instructions are also used, and they perform the same basic functions as their conditional counterparts without the requirement to test for status conditions. Since all unconditional branch instructions (JUMP, CALL, and RETURN) are not restricted by status conditions, they may be used when appropriate. They do not fit into an overall grouping scheme. Their mnemonic codes are shown in Table 7-4.

The third-digit octal codes 1, 3, 5, and 7 do not follow any of the previously mentioned organizations. Stack pointer operations, input/output codes, special unconditional jumps (RESTARTS), processor controls, and so on, are included in this group and are available in Table 7-4. When the third-digit octal code is 6, however, special *immediate* arithmetic/logical operations are performed. The second-digit octal code used under these circumstances is identical to that used with normal arithmetic/logical operations. The *immediate* arithmetic/logical operations combine the information contained in the byte immediately following the immediate instruction byte with the contents of the accumulator.

Instruction Set Summary. By using Table 7-3 it is possible to determine practically the complete instruction set for our conceptual microprocessor. As an example, the complete Group 1 register operations were summarized by the matrix of Figure 7-6. It is only necessary to decide which registers are to be used, and the instruction may be constructed. More commonly, however, a list of instructions is provided for the use of the programmer, requiring only the use of the register designations of Table 7-3. A partial instruction set summary is provided in Table 7-4.

A complete programmer's manual for a microprocessor also often includes a detailed description of *every* instruction. Such information as number of subcycles used during execution of the instruction, addressing modes, status flags affected, and so on, is commonly included. Since this type of information differs for every microprocessor, only the details necessary to explain the specific instructions are shown in this chapter.

7-3 USING THE INSTRUCTION SET

Basic Concepts of Programming. Now that we have a set of instructions to interconnect all the electronic circuits in the microprocessor, let us see what it takes to actually provide those interconnections. First, a list of instructions (called a *program*) must be developed so that the microprocessor will know what to do. Writing a computer program is a demanding task. The computer programmer must understand the problem being programmed with great clarity, problem analysis must be complete, and no errors can be tolerated. The programmer's job can be separated into five separate tasks: (1) *analysis*, (2) *organization*, (3) *coding*, (4) *testing*, and (5) *documentation*.

During the *analysis* phase, the programmer must develop a concise statement of the problem, analyze the facts, and write the mathematical expressions for problem solving. The more efficiently a computer is used, the better planned must be the program. The key to a good program is an *algorithm*, which is a foolproof, step-by-step procedure that takes into account *all* possible situations, including exceptions. The algorithm, then, is an unambiguous procedure that always leads to an answer (or at least an indication that an answer cannot be determined) in a finite number of steps. An algorithm can be likened to a recipe in a cookbook or the choreography of a ballet. All combine a number of individual small steps into a composite action.

The *organization* phase of the programmer's job is closely tied to the analysis phase. The algorithm constructed during analysis is now organized and sequenced into the operations required to perform the problem solution. Most programmers use *flowcharts*, which are block diagrams of the algorithm, for the organization task. Flowcharts use standard symbols (see Figure 7-11) and serve as a very good communications link between programmers, much as a blueprint does among engineers. The use of both algorithms and flowcharts is demonstrated in subsequent sections of this chapter.

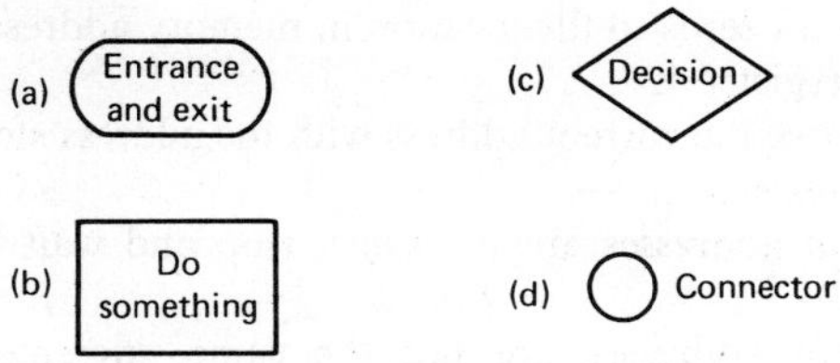

Figure 7-11 Flowchart Symbols

With the help of the flowchart, the computer program is *coded.* Coding is simply the process of converting the flowchart actions into instructions that will result in computer operations. It is here that the instruction set developed earlier is applied. After the coded program is prepared, it is first *tested* by comparing each step with the appropriate flowchart action. Often the coded program is flowcharted by another programmer so that any errors may be detected. When these preliminary tests are completed, the program is prepared for computer input and a solution is attempted. It may be necessary to run the program a number of times to remove all the errors (*debug* the program). *Documentation* is the final, and perhaps the most important, step. The corrected flowchart, a copy of the final program, sample input/output data (if appropriate), and so on, are included so that future changes to the program may take advantage of the experience gained during the original program evolution. All these steps will be demonstrated as the memory clear operation discussed in earlier chapters is programmed.

The Analysis Phase. As stated earlier, during the analysis phase the programmer develops the step-by-step procedure for solution of the problem at hand. Much of this has been performed for the memory clear problem as it was discussed in Chapter 5. The basic requirement was to assure that no extraneous information was stored in memory prior to beginning normal computer operations. This was to be accomplished by placing all zeros in all memory locations, and it was decided that the following general steps were required:

1. Place all zeros in the first location to be cleared.
2. Repeat until the required number of locations were cleared.
3. Recognize completion and stop to await further orders.

The algorithm (by no means an optimum solution) developed to perform initialization was:

1. Load the starting address in general-purpose register 1.
2. Load the stopping address in general-purpose register 2.

3. Move all zeros to the location in memory addressed by general-purpose register 1.
4. Compare the current address with the address stored in general-purpose register 2.
5. If both addresses are the same, stop and wait for further instructions.
6. If both addresses are not the same, increment the address in general-purpose register 1.
7. Move all zeros to the new address.
8. Go back to step 4.

Modifications to the algorithm often occur as the organization, coding, and testing phases of the programming task are accomplished. The programmer should be alert to necessary changes as the task progresses.

The Organization Phase. Each of the steps in the algorithm is investigated during the organization phase, and a flowchart is developed. Figure 7-12 is one way that a programmer might initially view the memory clear algorithm. Each of the blocks on the flowchart is numbered to correspond with the steps in the original algorithm. However, when the information is put in flowchart form, it immediately becomes apparent that steps 3 and 7 are performing the same function. Therefore, step 7 can be deleted and operations can go directly back to step 3 following step 6. The same operations are being performed, but note that the use of the flowchart allowed simplification of the task.

One of the goals of programming is to perform the task at hand with the minimum number of program steps so that maximum utilization of available memory space may be achieved. It should be noted that as much simplification as possible should be performed before actually beginning the coding phase of programming. Knowledge of the instruction set is not required, and only minimal microprocessor background is needed. Often the analysis and organization phase of programming is accomplished by a *systems analyst*, whose skills are more of an analytic nature. However, with the microprocessor, the user is becoming more and more involved, and the analysis and organization skills must be incorporated with hardware knowledge to most effectively use the capabilities of the microprocessor.

The Coding Phase. Now it becomes necessary to apply knowledge of the instruction set of the microprocessor as the flowchart is converted to an actual list of instructions (the program). Although the function of the program is to place zeros in all memory locations bounded by the starting and ending addresses, the operation of the program revolves around the *comparison* of current and ending addresses. An instruction (or instructions) is required that compare the two addresses and

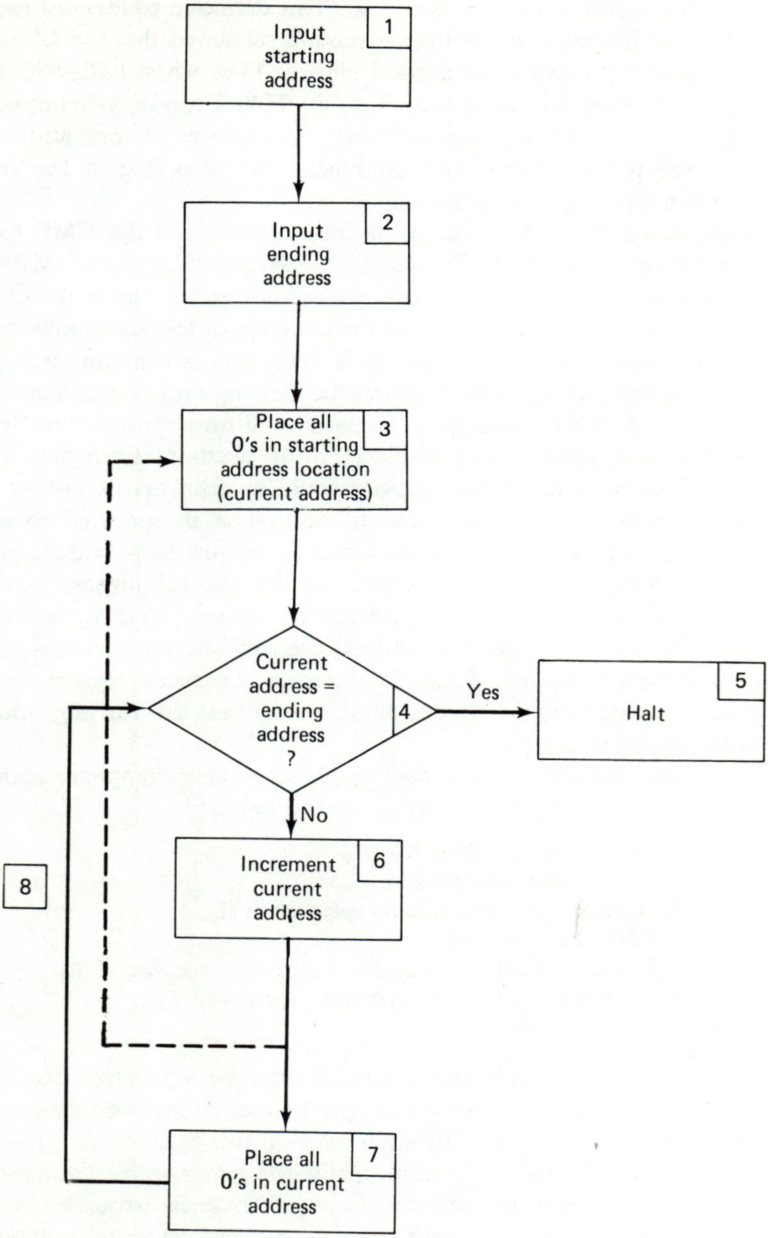

Figure 7-12 Memory Clear Flowchart

takes specific action when they are equal. This defines a *conditional branch* operation, and investigation of the instruction set shows that the JZ (jump on zero) instruction would be a good choice. Thus when both addresses were equal, the program could branch to the HALT mode, as required by the algorithm. It now becomes necessary to perform an operation that compares the two addresses and establishes the zero flag in the status register when equality is determined.

Once again the instruction set is investigated, and the CMP (compare) instruction is selected. The detailed description of the COMPARE instruction states that the contents of the register called out in the COMPARE instruction are compared with the contents of the accumulator and appropriate status word flags established. With this information it is now possible to select the registers in which the starting and ending addresses are to be loaded. Since comparison is made each time through the "loop" with the ending address, it may be placed in the accumulator register (R_7). Selection of the starting address register could be arbitrary except for one additional requirement. If zeros are to be placed in specified memory locations, some means of defining those locations must be provided. In our conceptual microprocessor, one register in the general-purpose stack is always considered as a memory reference register. That is, when an instruction refers to memory, the address used will be the contents of the memory reference register. R_4 is the memory reference register in our conceptual microprocessor. Under these guidelines, the starting address should be loaded into R_4.

The basic analysis of the coding phase is now complete, and the preliminary program can be written as shown below:

1. Load starting address into R_4.
2. Load ending address into R_7.
3. Load all zeros into address specified in R_4.
4. Compare R_4 and R_7.
5. Branch to HALT if contents of R_4 equals contents of R_7.
6. If contents of R_4 not equal to R_7, increment R_4.
7. Branch to step 3.

Each of the general steps above should now be converted to actual program steps. Remember, only 8 bits may be used at any one time, and if the instruction requires more than 8 bits, a two-byte or a three-byte instruction may be used.* Conventionally the program is arranged on paper just as it appears in memory (i.e., each byte is associated with a memory location). In keeping with previous convention, octal notation is used and correlated with both the binary notation and mnemonic codes of Table 7-4.

*The sample program is limited to two-byte instructions for the sake of simplicity.

Memory Address			Memory Contents	
Binary	*Octal*	*Mnemonic*	*Binary*	*Octal*
00000000	0	MVI R_4	00100110	046
00000001	1	300	11000000	300
00000010	2	MVI R_7	00111110	076
00000011	3	377	11111111	377
00000100	4	MVI *M*	00110110	066
00000101	5	000	00000000	000
00000110	6	CMP R_4	10111100	274
00000111	7	JZ	11001010	312
00001000	10	014	00001100	014
00001001	11	INR R_4	00100100	044
00001010	12	JMP	11000011	303
00001011	13	004	00000100	004
00001100	14	HLT	01110110	166

Assume starting address 192_{10} (300_8) ending address 255_{10} (377_8)

The Testing Phase. The program just written is now compared with the flowchart of Figure 7-12 to assure that all operations are accounted for and in the correct order. Block 1 in the flowchart is implemented by the instruction word consisting of the contents of memory addresses 0 and 1. Addresses 2 and 3 accomplish flowchart block 2. The contents of location 4, according to the detailed description of the MVI *M* instruction, places the contents of the next location in memory (location 5, all zeros) in memory at the location that appears in R_4. R_4 has been loaded with the starting address, and upon accomplishment of the MVI *M* instruction, location 300_8 should be all zeros. Block 3 has now been accomplished. During the CMP R_4 operation (location 6), the present location (300_8) is compared with the value in R_7 (377_8). Since the two values are not equal, the program continues with the next instruction following the JZ instruction in locations 7 and 10 (block 4). R_4 is incremented by the instruction in location 11 (block 6), and locations 12 and 13 automatically cause the program to return to location 4, which once again moves zeros into the memory location contained in R_4 (now 301_8). The "loop" continues until equality is detected, and the JZ instruction forces the program to memory location 14 (HLT), completing blocks 4 and 5. Note that block 7 was not used in the final program. Thus, it would appear that the program does accomplish the requirements of the flowchart.

In a real-life situation, the next action would be to load the program into the microprocessor and verify that it actually does clear all locations from 300_8 to and including 377_8.

The Documentation Phase. Without the actual hard-copy print-out verifying the operation of the Memory Clear program, it is a bit difficult to perform the documentation phase. However, the reader should realize that every program should be thoroughly documented with as much background information as possible. The rationale behind the algorithm is extremely useful when changes must be made to the program, as are the various levels of flowcharts that were developed as the program was being organized. If intermediate programs preceded the final program, these intermediate programs should be a part of the documentation, since they are a record of the thinking of the programmer. Finally, the actual results of executing the program should be an integral part of the documentation package.

7-4 SUMMARY

The reader should be fully aware that it is highly improbable that the material in this chapter will create a proficient programmer. Software use and creation is a matter of much studying and hard work, preferably with a microprocessor at hand to make the inevitable errors most obvious. Chapter 7 has merely introduced the subject of programming and the instruction set. Future chapters will explore the subject more fully, but the reader should now be able to appreciate the magnitude of the work entailed in programming the microprocessor. Additional and very valuable information may be obtained by acquiring the user's manuals for the specific microprocessor that is to be used. Therein will be located such data as the microprocessor's basic architecture, its instruction set, methods of interfacing to peripheral devices, and hints on programming. Many of these subjects are covered in subsequent chapters in this book, and the reader is invited to continue.

QUESTIONS

1. Why do many microprocessors require multibyte words?
2. Define direct addressing; register addressing; immediate addressing; indirect addressing; and index addressing.
3. What is a memory map and how is it used?
4. Define an instruction set.
5. Classify an instruction set and explain the purpose of each general type of instruction.
6. Explain how each bit position in an instruction byte can be used.
7. What is conditional branching; unconditional branching?
8. How does a microprocessor perform conditional branching?
9. What is a subroutine?
10. How is a subroutine used and how is it obtained for use by the microprocessor?
11. List the five steps of programming and explain what takes place at each step.
12. What is a flowchart and how is it used?

PROBLEMS

Using the instruction set for our conceptual microprocessor, develop the following partial programs following the five steps of programming:

1. A program to place all 1s in a selected number of memory locations.
2. A program to place all 1s and all 0s alternately in a selected number of memory locations.
3. A program that adds the contents of R_3 and R_4 and places the sum in a selected memory location.
4. A program that subtracts the contents of R_4 from R_3 and places the difference in a selected memory location.
5. A program that loads the number 377_8 in R_2, decrements the contents of R_2, and stops when the contents of R_2 reaches zero.

CHAPTER 8

Microcomputer Input/Output

The microprocessor discussed in Chapters 6 and 7 is, by itself, almost useless. It does perform the operations required by the program stored in on-chip ROM, but without a means to convey the results to the outside world, it is of little use. Furthermore, without a means to acquire real-world information, the microprocessor is greatly limited in its applications. Chapter 8 discusses the devices, hardware, and software required to acquire real-world information (INPUT) and to convey program results to the outside world (OUTPUT).

8-1 PARALLEL INPUT

Concepts. The primary purpose of *any* input function is to convert a human-oriented representation of information into a form that the microprocessor can use. An *input device* is used to change the human-oriented representation of information into an electrical signal. Since most input devices tend to be incompatible with microprocessors, an *interface* must be provided to convert the input device's electrical output into electrical signals that are acceptable to the microprocessor.

Input devices serving a microcomputer must be tailored to the input capabilities of the specific microprocessor used in the microcomputer. Information at the output of input devices is presented in many different forms, but the microprocessor is capable of accepting that information in only a limited number of formats. Therefore, in most cases, it becomes necessary to change the magnitude, timing, and data formats of input devices before they may be used by the microprocessor.

The simplest type of input device to consider is one that provides representation of information in parallel (i.e., one computer word or byte at a time). If it is assumed that the input device supplies 8 bits at a time, each on separate lines, the information is in a form that is compatible with the input bus (or data bus) for most microprocessors. However, many incompatibilities exist between the information supplied by the input device and the requirements of the microprocessor. Such incompatibilities are corrected by the use of the *interface* function. It is assumed that the 8-bit representation of information is properly coded so as to provide an information representation that is usable by the program instructions provided by the microprocessor.

Since most input devices operate on a different time scale than the microprocessor, the interface must be able to store the information presented by the input device. The microprocessor may not be ready to accept the information at the time it is presented by the input device, and a storage function must be provided to prevent loss of data. Furthermore, the voltage and/or current level from the input device may not be compatible with the microprocessor. The interface must accept the input device levels and convert them to microprocessor levels.

When the interface has stored the information from the input device, it must inform the microprocessor so that the information may be transferred. The microprocessor must then inform the interface that it has accepted the information, and that the stored data in the interface may be destroyed or that new information may be accepted. Microprocessor acceptance of the interface-stored information removes the indication of data availability, and the microprocessor is free to return to its previous operations.

The description above, although simplified, is typical of the operations performed by the interface function of a typical microcomputer.

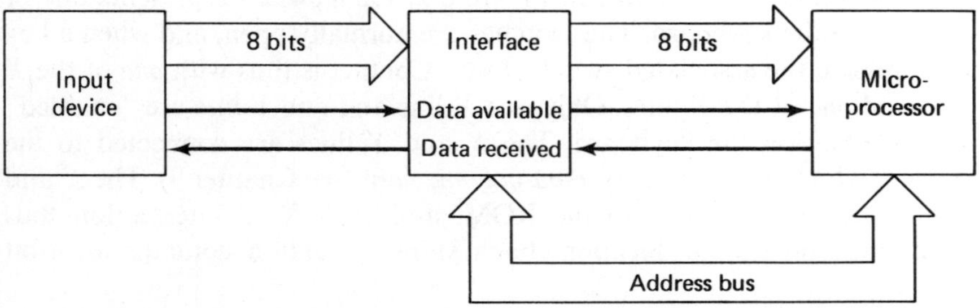

Figure 8-1 Typical Microprocessor Parallel Input Block Diagram

Figure 8-1 shows how a parallel-input device may be interfaced to a microcomputer.

Devices. Since the microprocessor is basically a device that accepts information in parallel form (all word bits simultaneously), the most appropriate input device to use is one that supplies information in the same form. Typical parallel input devices are keyboards (except for the Teletype*), panel switches, and paper tape readers. Each of these devices has one thing in common—parallel output, with all word bits being presented to the microprocessor at one time. Therefore, the microprocessor can use the information without special hardware and software, resulting in simpler and less expensive operation.

When panel switches are used for microprocessor input they supply either a HIGH or a LOW (1 or 0), depending on switch position. In most cases the switches do not connect directly to the output of the panel that feeds the microprocessor input. Switches tend to "bounce" when actuated, and multiple changes in level occur with obvious results. An output that changes numerous times for each switch closure can easily be considered as numerous switch changes, and the microprocessor becomes confused. Each switch that supplies input to the microprocessor may be "debounced" with hardware and the result stored in a register stage external to the microprocessor. The register output then becomes the input to the microprocessor, not the actual switch output. In a typical microprocessor, information is received in groups of 8 bits which originate in an 8-bit register in the input device.

Keyboards are more complex than panel switches. Whereas 8 switches must be set in a particular 1–0 pattern to cause a character to be represented and input to the microprocessor, only one key must be depressed on the keyboard to perform the same function. As usual, though, when an operation becomes easier for the operator, more complex hardware or software must be provided. Such is the case with the typical keyboard inputs to a microprocessor.

Most computer input keyboards merely contain many simple switches connected as shown in Figure 8-2. Each switch represents one of the keys on the keyboard. The switches are normally open, and when a key is depressed, the associated switch closes. Contact is thus with *one* of the X lines and *one* of the Y lines. Only one X line and one Y line are "enabled" for each key on the keyboard. The X and Y lines are connected to the *encoder*, which is effectively a *read-only memory* (see Chapter 9). The X and Y lines act as addresses to the ROM, and each $X-Y$ intersection thus represents one storage location. Each storage location contains an 8-bit

*Trademark registered by Teletype Corporation.

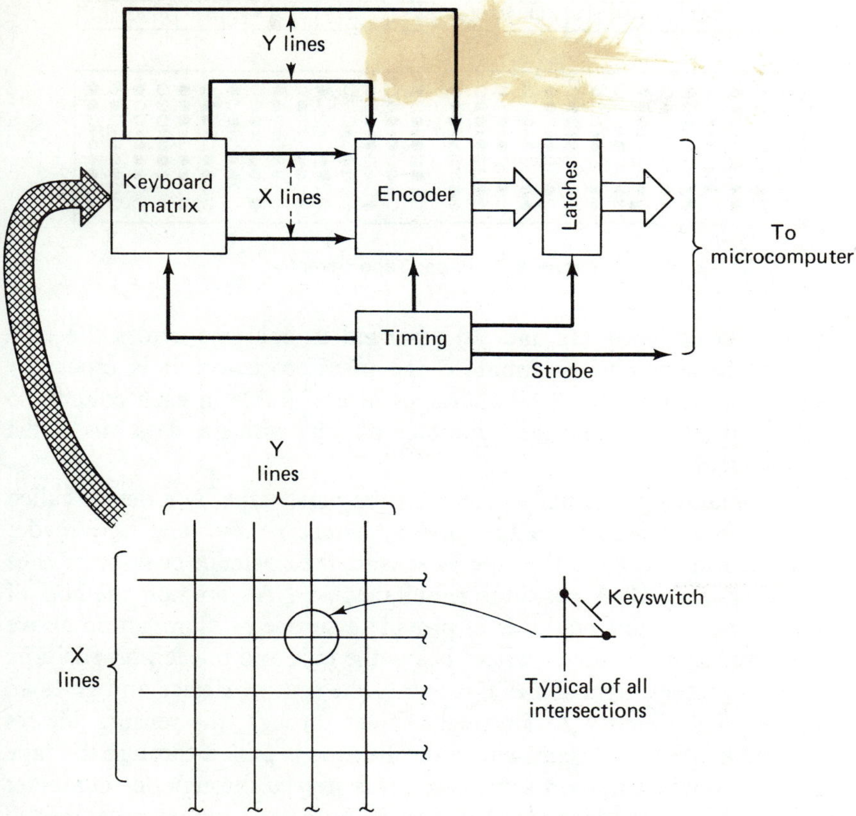

Figure 8-2 Keyboard Encoding

binary word which represents the character whose key has been depressed. Therefore, as each key is depressed, the output of the encoder is an 8-bit binary word. Latches are provided for each bit in the word to retain the information until the next character is selected. The output of the keyboard–encoder combination, then, is an 8-bit parallel binary word which may be fed directly (or perhaps via line drivers) to the data bus of the microprocessor.

Timing is also required to synchronize the operation of the keyboard, encoder, and the latches. A *strobe* is used to inform the microprocessor that the keyboard assembly has information ready to be transmitted.

When relatively large amounts of information must be input to the microprocessor, paper tape is often the medium used. Typically, paper tape is 1 inch wide and contains discrete locations on the tape where holes may or may not occur (see Figure 8-3). A logical 1 is represented by a hole, and a logical 0 by the lack of a hole where one could occur. The combinations

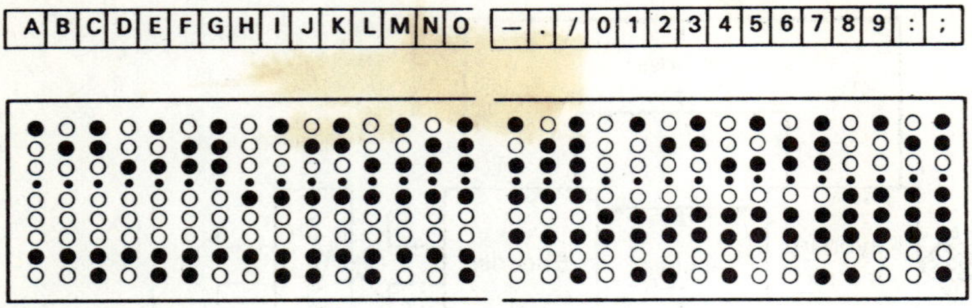

Figure 8-3 Paper Tape Sample

of holes and no holes (1s and 0s) arranged in columns across the tape represent characters to be input to the microprocessor. It is especially convenient to have a full 8-bit character representable in each column so that the paper tape reader may interface directly with the data bus of the microprocessor.

Information is originally placed on the paper tape by a device called a *tape punch* and read from the tape by a *tape reader*. The tape reader extracts information from the tape by sensing the existence or nonexistence of holes, reading all 8 positions simultaneously. A common method of implementing the tape reader is to provide a source of illumination above the tape, and light-sensitive devices below the tape and in each bit position. If a hole is present, light shines through to the sensing device and gives an indication of the binary 1. If no light shines through, the sensing devices gives no indication and represents a 0. The tape is pulled through the tape reader at a constant rate (in some cases, it is *stepped* through one character at a time), storing the information at each character location in latches so that a relatively stable input may be provided to the microprocessor. Figure 8-4 is a simplified block diagram of a tape reader.

The common point with the all input devices discussed has been that their outputs are in parallel form. As stated before, parallel input devices tend to be more practical and less expensive than other types when they are to be directly interfaced with the microprocessor. The basic concepts of parallel input, as far as the microprocessor is concerned, are shown in the following portions of this chapter.

Fundamental Hardware. The minimum hardware required to implement the parallel-input interface function consists of latches (flip-flops) to store input device information, other flip-flops to provide control signals, buffer amplifiers to isolate the interface from the microprocessor, and gates to determine microprocessor and input device states. If the logic signals from the input device are not compatible with the interface circuits, additional buffer amplifiers may be required at the input to the interface.

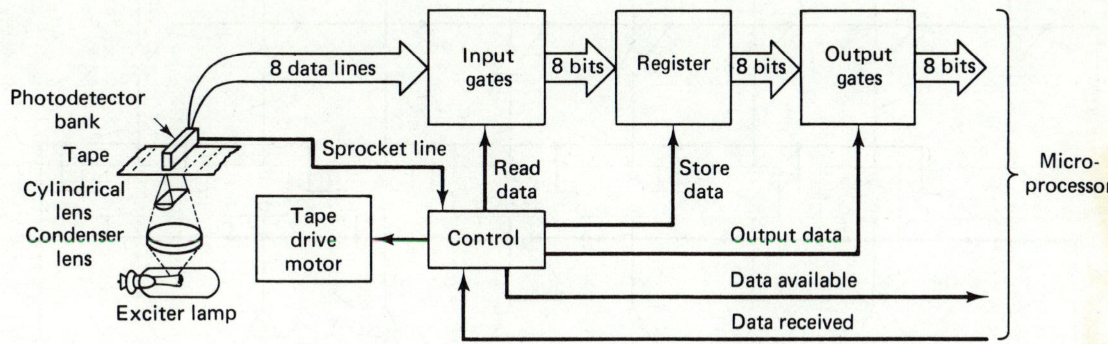

Figure 8-4 Paper Tape Reader Block Diagram

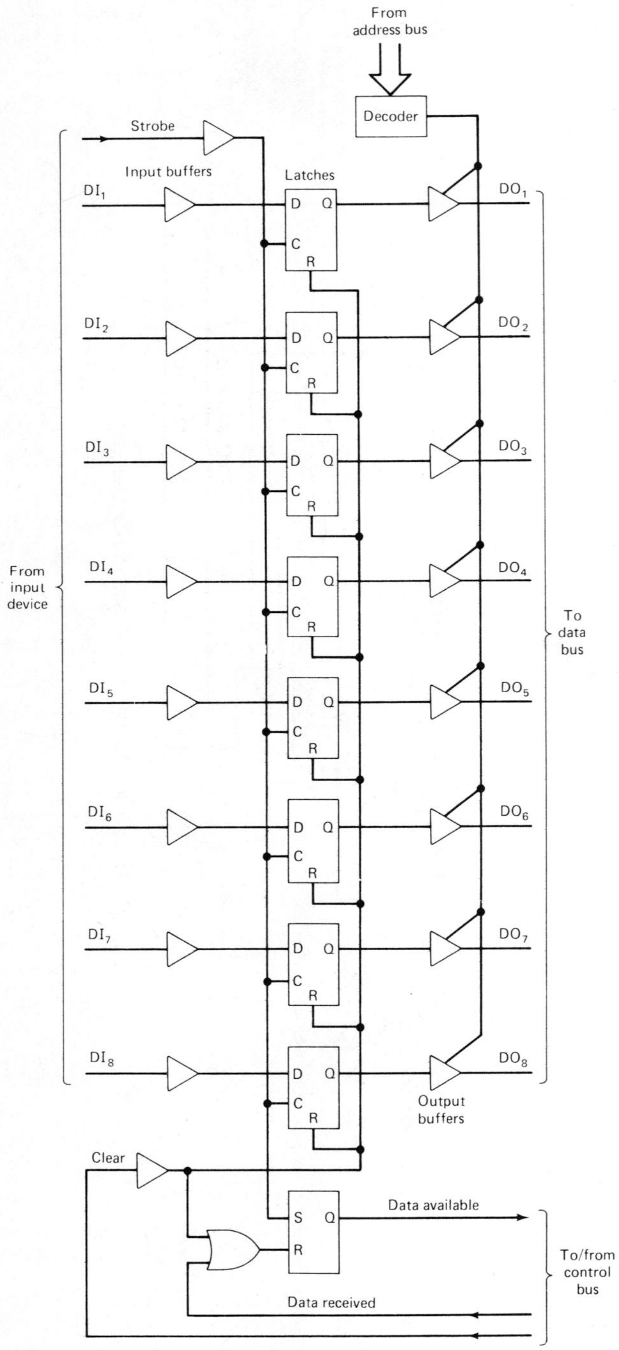

Figure 8-5 Input Interface

Figure 8-5 is a hypothetical input interface circuit that could be used to route information from, for example, a keyboard to a microprocessor.

When a key on the keyboard is depressed, a number of events take place almost simultaneously. An 8-bit byte is generated as a result of depressing the key, and this byte appears on Data Input lines DI_1 through DI_8. At the same time a *strobe* signal is generated, causing the information on the D inputs of each of the latches to be stored. In addition, the data available flip-flop is set, informing the microprocessor that information is ready for transfer. Thus, as a result of depressing a key on the keyboard, the electrical representation of the character depressed is stored in the latches of the interface. In the minimum hardware approach to processing parallel-input information, the data available flag is often treated as one bit of a *separate* input channel, commonly called the *status channel.* Since in the typical microprocessor an 8-bit byte is used, the other 7 bits are available for such functions as output device ready indications and additional input/output device conditions. It will be seen shortly that the interface register used for storing status information is shared by both the input and the output functions.

The remainder of the operations depend on action from the microprocessor. Usually, depending on the type of input program being used, the microprocessor is started into action by recognition of the data available signal from the interface. An INPUT command to the appropriate input interface is generated, which is decoded by the decoder gates in the interface, causing the output buffers to be enabled. The output buffers are commonly of the *tristate* variety, which appear as very high impedance at their outputs until enabled. This allows the output buffers to be connected directly to the microprocessor data bus without adversely affecting other devices connected thereto. As soon as the output buffers are enabled, the data stored in the latches are seen on output lines DO_1 through DO_8, and that data are accepted by the microprocessor. Upon acceptance, a data received signal is generated, resetting the data available flip-flop and removing the data available signal. The INPUT command is also removed and the output buffers are disabled. The microprocessor is now free to go on about its business, and will not query the interface again until another data available signal is generated.

One other signal is noted on Figure 8-5. A CLEAR input is provided to reset all flip-flops and latches to the no-information state. Such an operation is commonly performed upon power turn-on, or when it becomes necessary to perform a complete master reset of the microprocessor. Other means of recovering information from the interface are in use and will be discussed subsequently as the software aspects of information input are investigated.

Fundamental Software. The software required to process input information depends on the method selected by the programmer and the capabilities of the microprocessor. Minimum-hardware implementation of the input operation uses a simple sequence of program steps that constantly looks for the existence of the *data available indication (flag)* from the interface function. Upon detection of the data available *flag* the program branches to a series of steps that places the input information in microprocessor storage. The data received signal is generated by the microprocessor (resetting or clearing the data available flag) and the program resumes its search for the next data available flag.

Figure 8-6 is a flowchart that could be used to implement the minimum-hardware input operation. Table 8-1 is the simple program that may be written from the flowchart of Figure 8-6. The flowchart shows the basic functions that must be performed to obtain and store data from a parallel-input interface. Note that the program is located within the RAM area defined for our conceptual microprocessor in Chapter 6. The power of the microprocessor and its instruction set begins to appear in this short program.

Two of the instructions (LXI R_4 and INX R_4) are unique. They perform operations on *pairs* of registers as defined in Table 7-3. LXI R_4 will place the information in the next two bytes of the program into R_4 and R_5. The byte immediately following the instruction is the least significant number, while the third byte of the instruction is the most significant

Table 8-1 Input Program

Address		*Mnemonic*	*Memory Contents*		
Binary	*Octal*		*Binary*	*Octal*	*Comments*
11000000	300	LXI R_4	00100001	041	Load address where first input
11000001	301	000	00000000	000	data word will be stored. An
11000010	302	001	00000001	001	off-chip location is used.
11000011	303	IN	11011011	333	Input status channel to the
11000100	304	000	00000000	000	accumulator.
11000101	305	RRC	00001111	017	Rotate accum. 1 bit right.
11000110	306	JNC	11010010	322	Jump back to "Input Status" if
11000111	307	303	11000011	303	contents of Carry FF was not
11001000	310	000	00000000	000	set to 1.
11001001	311	IN	11011011	333	Input data channel to the
11001010	312	001	00000001	001	accumulator.
11001011	313	MOV *M, A*	01110111	167	Store accumulator contents
11001100	314	INX R_4	00100011	043	Increment R_4
11001101	315	JMP	11000011	303	Return to "Input Status"
11001110	316	303	11000011	303	
11001111	317	000	00000000	000	

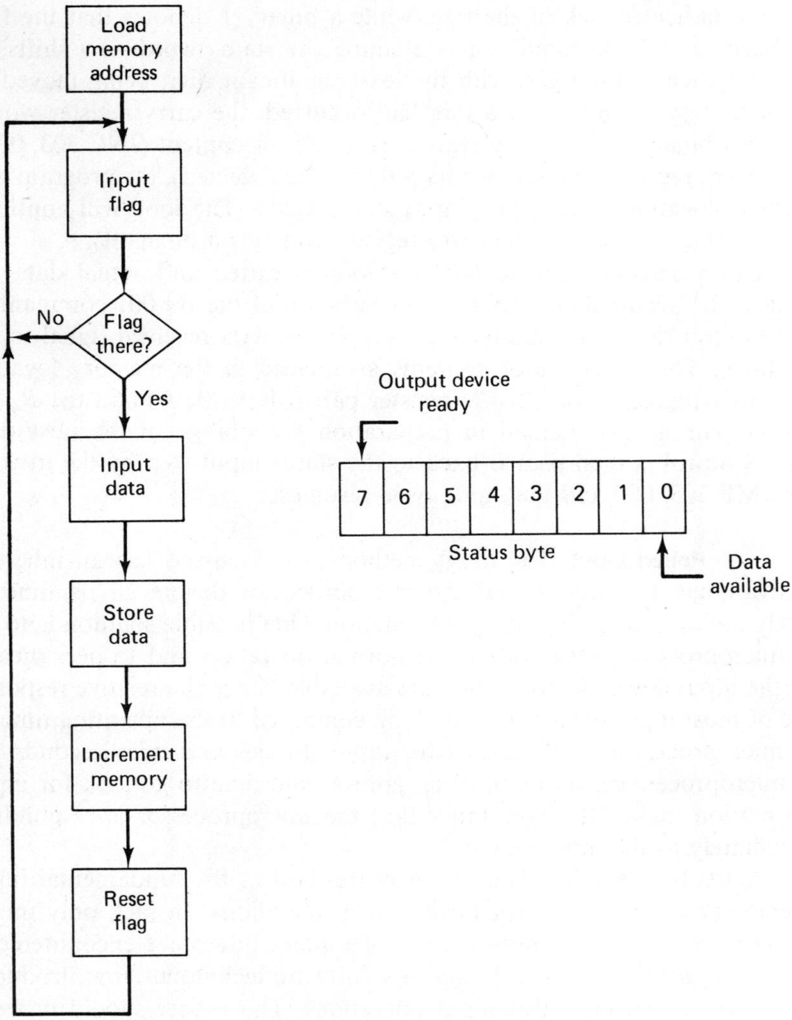

Figure 8-6 Input Program Flowchart

number. As a result of executing the instruction located at bytes 0, 1, and 2, register R_4 will contain 000_8 and R_5 will contain 001_8. These two instructions are used to establish the first location in memory where input data will be stored (LXI R_4) and to increment the R_4–R_5 register pair each time the instruction is executed (INX R_4) so that a fresh storage location will be available for each input byte.

After loading the R_4–R_5 register pair, the status information is input (IN 000) and placed in the accumulator. In our example, bit location 0 in the *status word* is used to identify the existence of the data available flag. A binary 0 indicates lack of the flag, while a binary 1 denotes that the flag has been set. The accumulator containing the status word now shifts all bits one place to the right, with the least-significant digit being moved to the *carry register* (*RRC*). If a flag had occurred, the carry register would contain a binary 1. The carry register is tested for content (JNC 303, 000). If the carry register contains a binary 0 (no flag detected), the program will return to location 000303 and input status again. The loop will continue until the flag is detected (the carry register contains a binary 1).

Upon detection of the flag, the loop is exited and actual data are input to the accumulator (IN 001). Generation of the IN 001 command is used to reset the data available flag (supply the data received signal to the interface). The accumulator contents are placed in the memory location previously placed in the R_4–R_5 register pair (MOV M, A) and the R_4–R_5 register pair is incremented in preparation for storage of the next data word. Control is then passed back to the status input step by the instruction JMP 303, 000 and the input cycle resumes.

Polled Input. The input method just discussed has an inherent disadvantage. It requires that the microprocessor devote all its time to merely looking for or inputting information. One possible solution is to let the microprocessor go on about its normal operation and to periodically *poll* the input device to see if data are available. Since the relative response time of most input devices is quite long compared to the operating time of the microprocessor (milliseconds for input devices and microseconds for the microprocessor), selection of an appropriate time to go look for input information makes the user think that the microprocessor is responding immediately to the input action.

Actually, the same type of hardware used in the fundamental input scheme can be used in *single-input* polled operations. In fact, only minor changes are required in the software. The major difference encountered is in *how* the software is used. Many new software techniques are introduced during discussion of polled input operations. The reader should proceed slowly as these new ideas are discussed so that their concepts are well understood. Much of the additional work to be done depends on the ideas of *subroutines* and *stack operations*, for it is here that the power of the microprocessor is put to use.

A *subroutine* or *subprogram* is a sequence of instructions that may be used many different times in the main program. The subroutine may be *CALL*ed each time the group of instructions is needed rather than repeating each instruction every time it is required. Economy in memory utilization results when subroutines are used, as may be seen in the conceptual memory map of Figure 8-7.

(a)

100	B
101	D
102	E
103	A
104	E
105	I
106	O
107	U
108	F
109	C
110	E
111	B
112	A
113	E
114	I
115	O
116	U
117	A
118	G
119	B
120	B
121	A
122	E
123	I
124	O
125	U
126	D
127	E
128	B

(b)

	100	B	
	101	D	
	102	E	
	103	CALL(X)	
	104	F	
	105	C	
	106	E	
	107	B	
	108	CALL(X)	
	109	A	
	110	G	
	111	B	
	112	B	
	113	CALL(X)	
	114	D	
	115	E	
	116	B	
	117		6 locations not used
	118		
	119		
	120		
	121		
	122		
(X)	123	A	
	124	E	
	125	I	
	126	O	
	127	U	
	128	RETURN	

Figure 8-7 Simple Memory Map

The section of memory between locations 100 and 128 is used to demonstrate subroutine concepts. Each location's contents is represented by a letter rather than actual instructions so that subroutines may better be visualized and understood. Figure 8-7a shows a segment of a hypothetical program, and it will be noted that certain steps are repetitively used (A, E, I, O, U).

These repetitively used steps (subroutine) may be stored in another location in memory, and retrieved when needed by using a CALL instruction. The location of the *first* step of the subroutine must accompany the CALL instruction. When the CALL instruction is recognized, the following steps occur:

1. The address of the next instruction to be executed is saved.
2. The instruction at the address accompanying the CALL instruction is retrieved and executed.

3. The subprogram continues, step by step, at the new location in memory.
4. Execution of the RETURN instruction sends the program back to the address saved in step 1 above and the main program continues normal execution.

Note that six memory locations were not used in the simple example of Figure 8-7b, making them available for additional program steps. Actually, when detailed investigation of subroutine processing is made, it is found that more memory space is required than is apparent. The use of subroutines may be economical only if the subroutine is CALLed many times in the main program.

In step 1 it was stated that the next instruction to be executed was saved. When an actual subroutine is investigated, it is noted that perhaps *more* than just the next instruction must be retained. If any of the registers in the microprocessor are in use in *both* the main program and the subroutine, the contents of those registers must also be stored when the subroutine is entered and retrieved when the subroutine is exited.

All the "saving" and "retrieving" operations take place in a portion of memory called the *stack*. Some microprocessors provide on-chip stack capabilities, with specific locations in memory permanently assigned. Others allow *any* portion of memory, both on- or off-chip, to be selected as the stack. The latter method tends to be more versatile and is used in all future examples.

The stack may be used as soon as its location is defined in the special-purpose register known as the *stack pointer*. Most microprocessors have both direct and indirect methods of placing a location in the stack pointer. Direct loading of the stack pointer is accomplished by supplying the memory location of the first byte of information in the stack as a part of the loading instruction. As an example, LXI SP 032 000 loads the stack pointer with the address 000032. The stack is then defined as starting at memory location 32. Indirect loading of the stack pointer is also provided so that the contents of certain microprocessor register pairs may be placed in the stack pointer. The location of the stack may thus be established by performing calculations in the microprocessor and placing the results in a specified register pair. Relocation of the stack is easily performed in this manner. Incrementing and decrementing of the stack pointer is also provided.

A subprogram that inputs information by polling the input device can now be examined. Since it takes such a short time to determine if an input device has data available, the microprocessor can "look" at the input device quite often. Periodic checking to see if the data available flag is active can be implemented by an internal timing device, or the check may be made anytime a convenient breaking point exists in the main program.

Before the subroutine can check data availability, though, certain "housekeeping" tasks must be accomplished. The present main program step location in memory must be saved so that normal sequencing of program steps can resume following input operations. Furthermore, all registers that contain information pertinent to main program execution must have their contents saved. Following execution of the subroutine, all registers and the program counter must be restored.

Figure 8-8 shows the parts of a memory map necessary to explain the use of the subroutine CALL and RETURN instructions. Only those addresses and instructions required to trace instruction execution are included. For example, it is not necessary to know what the main program is designed to do in order to visualize the operation of the subroutine. Only the instruction that loads an address into the stack pointer, plus a CALL instruction, are needed in the main program to implement subroutine operations.

As the main program executes it will encounter the LXI SP (load stack pointer) instruction, followed by two bytes containing the low portion of the stack's address followed by the high portion. The LXI SP instruction reserves a starting point in memory which defines the location of the stack. A CALL instruction at location 000201 is the next subroutine-related instruction to be encountered. Upon execution, the next main program address is PUSHed onto the stack at locations 010376 (high address) and 010375 (low address). Note that the stack pointer is actually pointing to a location *one greater than* the actual location of the data in the stack and must be established in that manner when the stack pointer is initially loaded.

After the address of the next main program step is saved in the stack, control of the microprocessor transfers to the address included as part of the CALL instruction. Subprogram execution then continues sequentially at the new location. A final step of all subroutines is a RETURN instruction. The return address (to the main program) is POPed from the stack in reverse order (high address first followed by low address) and placed in the microprocessor's program counter. Main program execution thus continues at the program step immediately following the CALL instruction. The stack pointer is also returned to its original value so that the empty stack is ready for the next CALL instruction.

The stack may also be used to temporarily save information from other registers in the microprocessor by use of the PUSH and POP instructions. Data entered in this manner behave exactly as the program counter contents behaved when it was placed in the stack. The stack stores all information in a "last in, first out" format so that any information is removed in reverse order from its input. Examples of the PUSH and POP operations are seen in Figure 8-9 and Table 8-2.

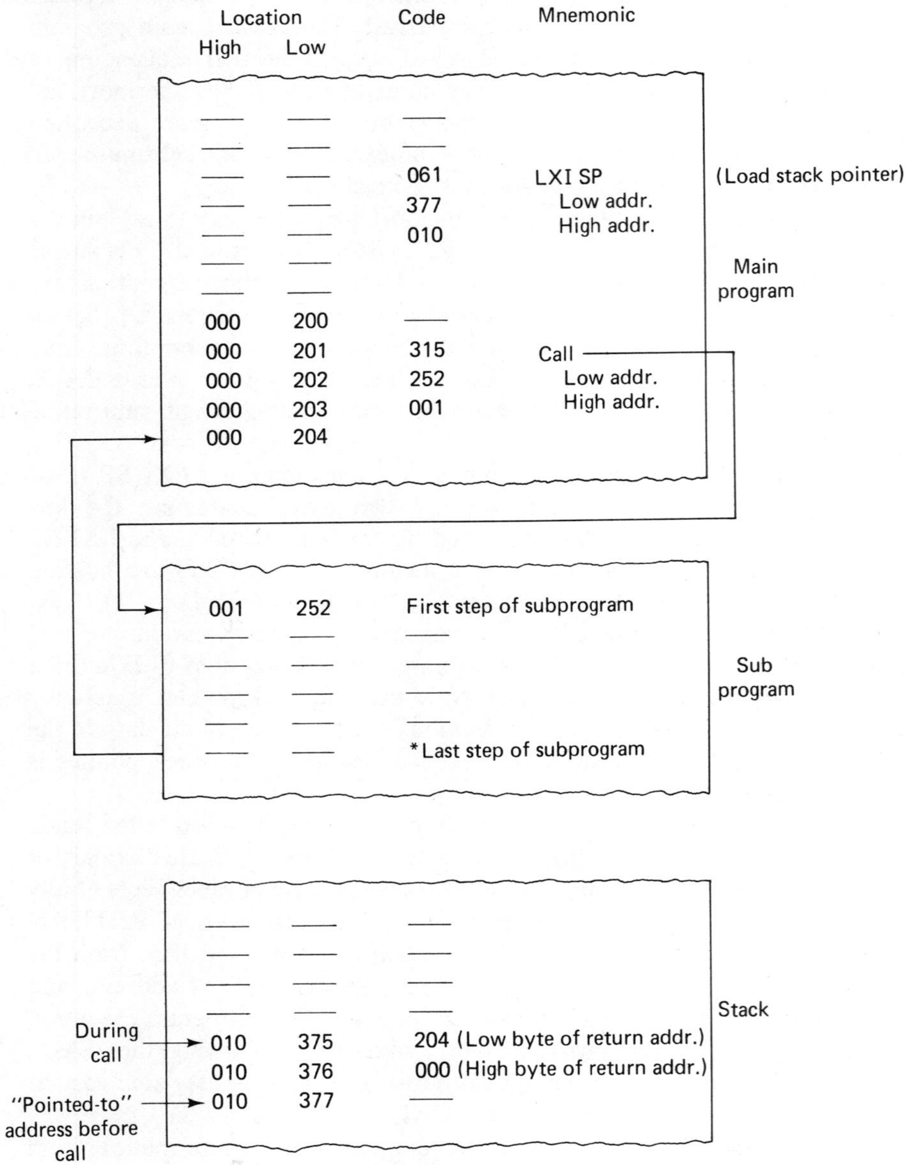

*Must be a RETURN instruction

Figure 8-8 Memory Map Showing Stack Operations

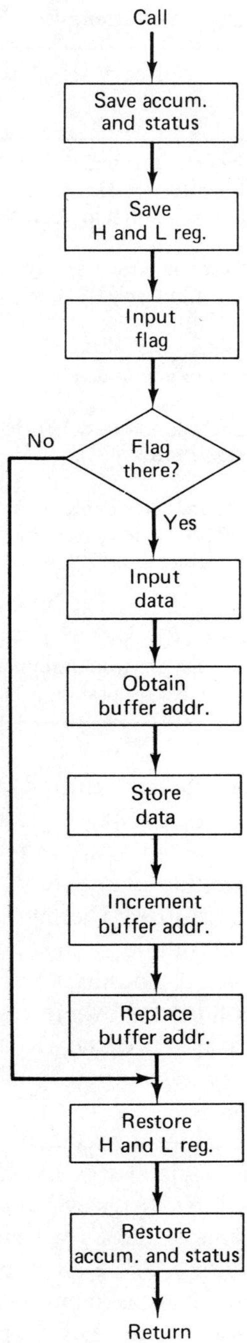

Figure 8-9 Subroutine Operations (Flowchart)

Table 8-2 Subroutine Operations Program

Location Radix 8	*Operation*	*Comments*
300	PUSH PSW	Save Accumulator and Status Register
301	PUSH H	Save H and L Register Main Program Data
302	IN	Input status channel to accumulator
303	000	
304	RRC	Rotate Accumulator 1 bit right (LSB into Carry FF)
305	JNC	Check for 1 in Carry FF—if not, Jump to Exit
306	322	
307	000	
310	IN	Input data to accumulator
311	001	
312	LHLD	Insert address where data is to be stored in H
313	xxx	and L Registers
314	xxx	
315	MOV *M, A*	Store accumulator contents at address in H & L Registers
316	INX H	Increment H & L Registers by 1
317	SHLD	Store incremented H & L register value where it was
320	xxx	before
321	xxx	
322	POP H	Restore main program H & L data
323	POP PSW	Restore main program Accumulator and Status Register
324	RET	Return to main program

It can be inferred from the background information provided on subroutines that the subroutine is nothing more than another program of instructions for the microprocessor, except that provisions are made for the saving and restoring of information in registers that are common to both the main program and the subroutine. Therefore, if the routine shown in Figure 8-6 is to be used as a subroutine, it must be modified to perform the saving and restoring operations. A flowchart and subprogram designed to perform single input device polling is shown in Figure 8-9 and Table 8-2. It should be noted, however, that it is necessary to include a step in the main program that defines where in memory to place the information being inputted.

When the CALL instruction is executed, the address of the next instruction is pushed onto the stack, as was shown earlier. The first step of the subroutine is then executed. As was previously indicated, it is necessary to temporarily store the contents of those registers that see use in both the main program and the subroutine. Since all information coming into the microprocessor is routed through the accumulator, it is necessary to save its contents. The 8-bit condition code register (program status word register) is companion to the accumulator and is saved when the accumulator is saved. Therefore, when the PUSH PSW instruction is executed as the first step in the subroutine, the accumulator and condition code register contents are pushed onto the stack.

The second step in the subroutine saves the main program *H* and *L* register data. Status channel input is performed and the status word is shifted 1 bit right and inspected as in the fundamental parallel input program. In the polled input situation, though, we do not wish to continue inputting status information if the data available flag is not active, since we wish to continue with the main program. Therefore, if the data available flag is not active, the subroutine jumps to location 000322 in order to continue. The H and L main program data are restored at this step, followed by restoration of the accumulator and condition code register data for the main program. Control to the main program is supplied by the RETURN instruction. Thus, if data were not available, the subroutine merely looked *one* time and went back to main program execution until the subroutine was called again.

If, however, the data available flag was active, data were inputted to the accumulator by the IN instruction at location 000310. The location for storing the input data is retrieved from memory and placed in the H and L registers by the LHLD xxx,xxx instruction. Recall that the H and L registers may be used to identify a memory location in the MOV M, A instruction, which follows. Therefore, when the MOV M, A instruction is executed, the data that were in the accumulator are placed in the memory location identified by the H and L registers. The H and L registers are now incremented by 1 (INX H) and the incremented number stored back in memory at the location it originally held (SHLD xxx,xxx). When the LHLD xxx,xxx instruction is next executed, the address that is returned to the H and L registers will thus be 1 greater than originally stored so that the next data to be stored will be placed adjacent to the data just stored.

The H and L registers are now restored to their main program configuration (POP H), as are the accumulator and condition code register (POP PSW). Control to the main program is supplied by the RETURN instruction, and the subroutine remains inactive until called again.

Polling more than one device is relatively easy. The flowchart of Figure 8-9 may be modified as shown in Figure 8-10. Up to eight input devices may be serviced by using specific bit positions in the status word as data available flags. A binary 1 in any bit position indicates that device has information ready to input.

The same technique used for single-input polling is used for multiple input. After the input subroutine is CALLed and the necessary registers are saved, the status word is inputted. A check is made to determine if the data available flag is present for input unit 1. If it is, the flowchart continues as in single-input polling; if not, the flowchart shows that bit 2

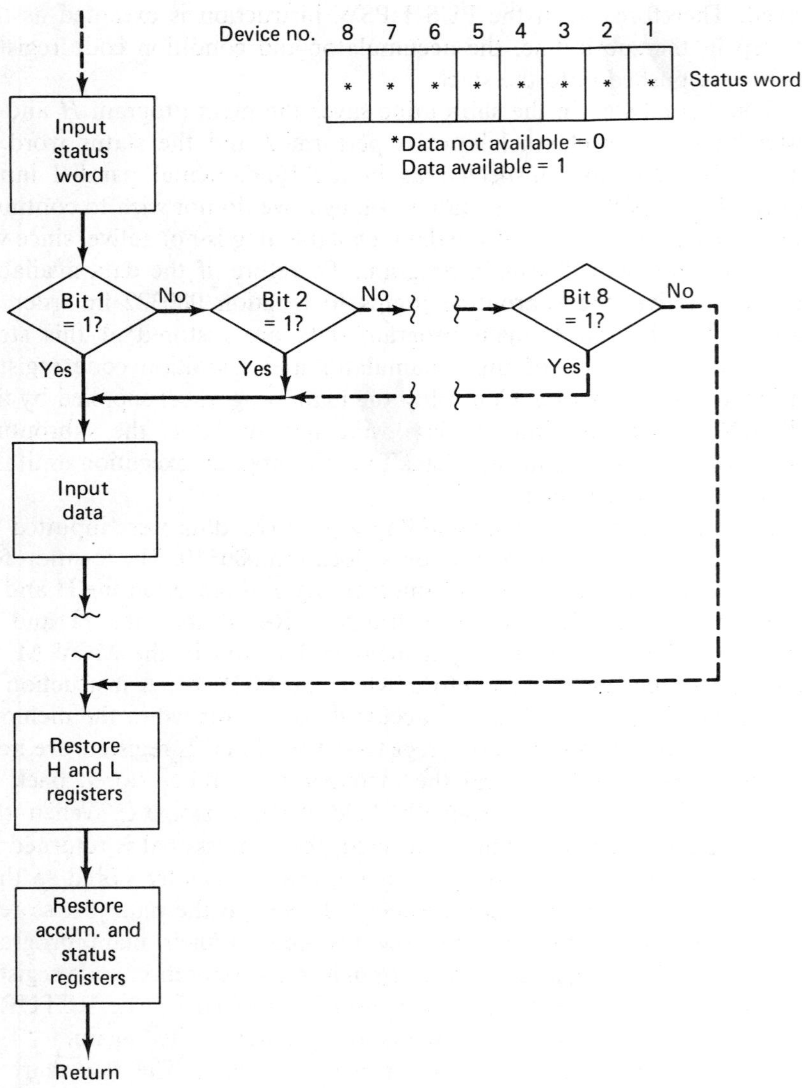

Figure 8-10 Multiple Polling Flowchart

position is checked. Each bit position is checked, in sequence, until an active input channel is detected. Should no active channels be located, the subprogram restores all necessary registers and returns control to the main program.

Data available flag testing in the program written to implement the flowchart of Figure 8-10 is performed in the same manner as in the single-input program. The status word is obtained, stored in the accumulator, and rotated (shifted) 1 bit right into the carry flip-flop to test for input unit 1 activity. Subsequent input unit testing is accomplished by additional right shifts and checking for data available flags. It should be noted that the input unit assigned as unit 1 has the highest priority, with order of importance decreasing as later inputs are checked.

Furthermore, if an input unit becomes active *after* it has been checked during an input subprogram execution, it will have to wait until the next subprogram CALL. A technique that combats this problem, plus a method to identify which of the input units originated the data available flag, is shown during the discussion of *interrupt* input.

Interrupt Input. Even though polled input does not require *all* the microprocessor's time, it does require that timing be generated to make periodic checks or that the programmer insert numerous CALL steps as the program is written. Most microprocessors have the capability to be "interrupted" from their regular operation without special hardware or software. Interrupts behave in different manners for different microprocessors, but the method discussed in this text is a composite, showing the best aspects of the most common interrupt schemes.

The simplest interrupt systems consist of a single additional input line to be used in conjunction with the usual input port. An *interrupt input* operates in the following manner. The main program proceeds with its usual functions until the input device provides an indication that it has information ready to transfer. Generation of the data available flag mentioned earlier is the usual indicator of availability of information for transfer, and it is applied to the *interrupt* input of the microprocessor. The interrupt input will be recognized by the microprocessor only if interrupts have been enabled, and it is assumed that a step has been included in the main program to perform this operation. Immediately upon recognition of the interrupt input, the microprocessor disables any further interrupts to prevent interference during servicing of the existing interrupt. The present instruction is completed and the present contents of the Program Counter are stored in the stack so that the program can resume upon completion of the interrupt operations.

It is now the responsibility of the interrupting device to provide one byte to the microprocessor. Most microprocessors interpret this byte as the *address* of the first step of the interrupt service routine, although occasionally the interrupting byte will supply the address of a *jump* to interrupt service routine, or in some cases even a pseudo-CALL type of instruction. In our case, the interrupting byte will be considered as the address of the first byte of the interrupt service routine.

Once the interrupt service routine is entered, operations are very much like the polled input subroutine. Necessary registers are first stored in the stack to save main program information. The buffer storage pointer is obtained to identify the location for storage of input information (it is assumed that the buffer storage pointer has been previously identified in the main program) and the input information is obtained and stored. Following increment of the buffer storage pointer, it is placed back in memory for later use with other input information. All registers are restored to the main program configuration, interrupts are reenabled, and control is returned to the main program. Thus, as far as the main program is concerned, it starts again as though it had never been interrupted. Input of information has been accomplished with a minimum of overhead time and with very little additional hardware.

The basic premise in any type of interrupt input operation is that the input device or some associated hardware supplies the information required to identify itself and point the microprocessor to the location in memory where the interrupt service software is located. In the single level interrupt scheme, this type of information can effectively be hardwired into the microprocessor. With multilevel input, however, many new problems appear. Not only must the device be identified, but some priority scheme must be arranged so that the microprocessor can service multiple interrupts in the sequence desired. A good case would be the situation where the microprocessor is serving not only numerous input devices, but also must perform a given sequence of operations to protect certain vital information in case of a power failure. Obviously, the power-failure-detection circuitry must be able to interrupt *any* other operation being performed, whether it be main program or any of the input devices servicing routines. Such priority methods of servicing interrupts usually requires not only additional hardware, but also dedication of certain locations in memory to store the interrupt service routines for each of the input devices or possible interrupt conditions.

The same general approach used in Figure 8-10 is used in multiple-input interrupt systems, although expansions must be made to accommodate the additional inputs. Of major importance is the realization that there

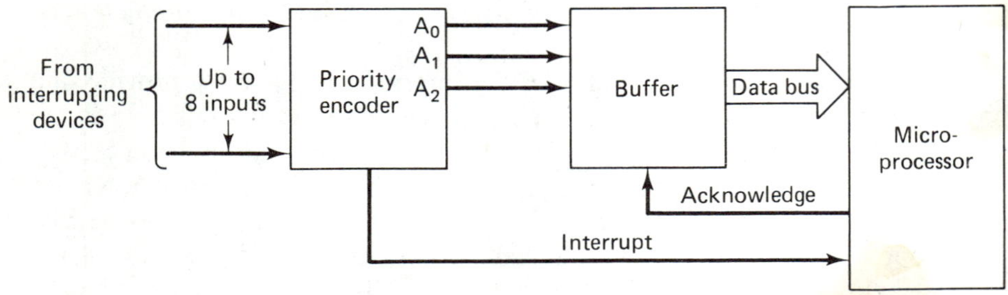

Figure 8-11 Priority Interrupt Hardware

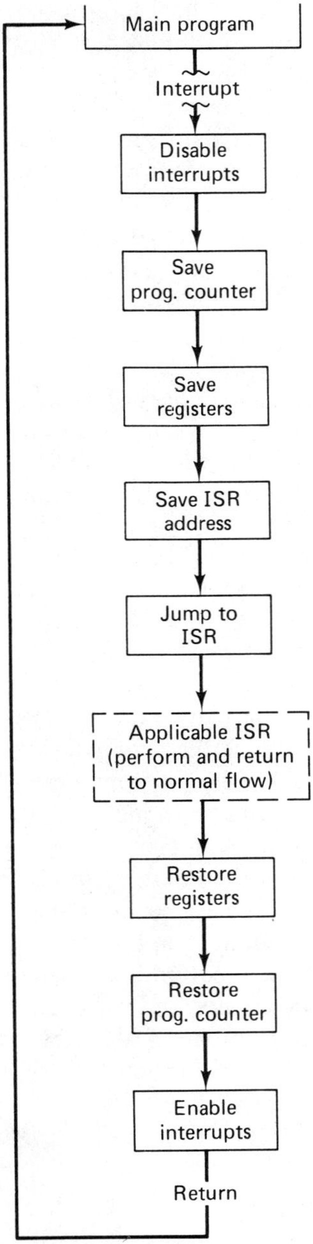

Figure 8-12 Interrupt Flowchart

may be only *one* interrupt input to the microprocessor. When *any* of the input devices generates an input, an interrupt must also be generated. Identification of the interrupting device is still via an 8-bit byte "jammed" onto the data bus. Actually, the interrupting device may not be identified, but the byte "jammed" onto the data bus must be the address of the location in memory that stores the interrupt service routine for that device. By virtue of the routine selected, we can identify the interrupting device.

Figure 8-11 is a simplified block diagram of a scheme to input information from eight separate devices in a selected priority order. Operation of the multilevel interrupt depends on the programmer's reserving of specific memory locations for each interrupt service routine, and assurance that those locations are not assigned to any other functions.

The priority encoder is a common LSI assembly and performs two basic functions. First, any input device that becomes active will not only activate its interrupt input line but will also cause an interrupt output to develop. Second, the A0, A1, and A2 outputs of the priority encoder will contain a 3-bit binary code corresponding to the highest priority interrupt. The encoded output of the priority encoder is combined with fixed values for five other bits in the buffer to form an 8-bit byte that will identify the location of the first byte of the interrupt service routine. The 8-bit output of the buffer is not applied to the data bus until interrupt is acknowledged by the microprocessor. Acknowledgment is furnished to the buffer at the time that the microprocessor normally expects to see an address on the data bus. Therefore, when any of the input devices has information to transfer, the main program of the microprocessor is interrupted, the address of the interrupt service routine is jammed into the microprocessor, the interrupt is serviced, and the main program is resumed.

Since many of the steps in the interrupt service routine are common from one device to the next, it may be most practical to provide subroutines for interrupt start and interrupt return. The only part of the overall interrupt service routine that must be unique is that part that specifies which device is to be serviced and where that device's information is to be stored. The previous program defining information input is adequate for the unique portions of the routine, while a common routine for interrupt start and return can be written. A simplified flowchart showing the operation described is shown in Figure 8-12.

8-2 PARALLEL OUTPUT

Concepts. If the operations of parallel input are understood, parallel output becomes easy. An 8-bit byte is placed on the data bus by the microprocessor for use by the output device rather than an 8-bit byte being placed on the data bus by an input device for microprocessor use.

The accumulator in the microprocessor is used in both input and output operations, and very similar software is employed. In actuality, the only real difference between input and output operations is the direction of information flow. It will be shown in later sections of this chapter that input and output interfaces are often combined. The same interface register is even used; different buffers are merely employed for input and output.

Hardware. Evaluation of a simple output interface (Figure 8-13) will demonstrate the simplicity of output operations. In the most primitive form the output interface requires inputs only from the microprocessor data bus and address bus. The *status register* output (via the status buffers) is either constantly or periodically checked for indication that the output device is requesting data in a manner similar to input processing (see Section 8-1). Detection of the output request causes data to be placed on the microprocessor data bus for storage in the *data register* via the *buffer amplifier*. An output ready signal is generated as data are stored in the data registers and the register output is sent to the output device via *driver amplifiers* (used to match characteristics of the output interface to the output device).

Software. A flowchart showing the relationship between the output interface and the microprocessor software is shown in Figure 8-14 and the accompanying program is shown in Table 8-3. The purpose of the routine is to output one line of memory-stored characters to a printer and prepare for the next line before returning to the main program. It is assumed that the line of characters is already stored and that the storage location is known. The occurrence of a carriage return character is used to signal the end of the line, at which time the carriage return will be output, followed by a line feed to prepare for the next line of printout.

The printing routine is entered from the main program by an appropriate CALL within the main program. As the flowchart shows, the first operations to perform are those that detect the fact that the output device is ready to accept characters for printing. The *input status* block and the *status bit present*? decision, it will be noted, are operating in a loop as long as the status bit is not present, and once the routine is entered, this loop continues until the status bit appears. At that time, the first character in the line to be printed is extracted from memory and placed in the accumulator. The accumulator is checked to determine whether or not the character is a carriage return. If not, the character is output to the output interface, the location in memory where the character was stored is incremented, and the input status loop is again entered. As soon as the output device has printed the character, status will again be present, and

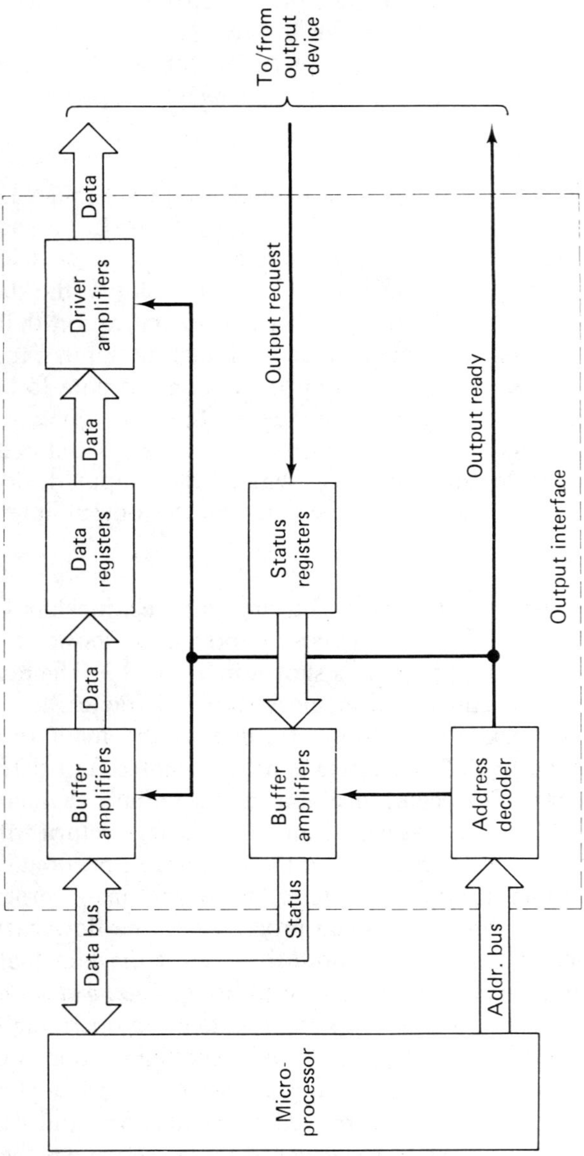

Figure 8-13 Output Interface

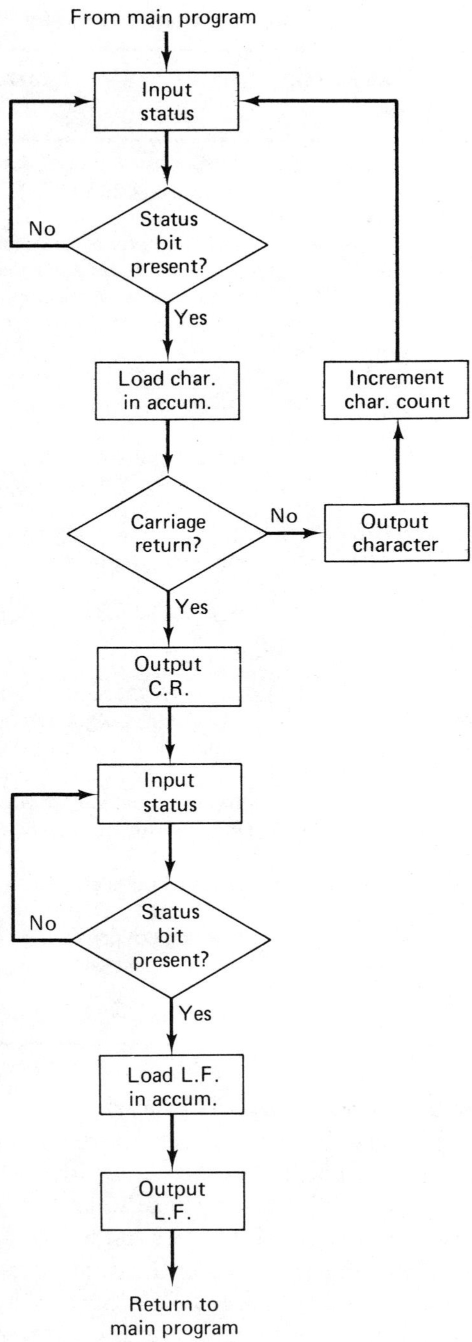

Figure 8-14 Output Program Flowchart

Table 8-3 Output Program

Location *Radix* 8	*Operation*	*Comments*
200	CALL	Perform Status Input Subroutine
201	231	at location 000 231
202	000	
203	MOV *A*, *M*	Get character to be output from memory
204	CPI	Is the character a carriage return?
205	015	015 is ASCII representation of carriage return
206	JZ	If carriage return, jump to complete output
207	217	routine
210	000	
211	OUT	Output character to device 002
212	002	
213	INX H	Advance character counter
214	JMP	Go back to the beginning
215	200	
216	000	
217	OUT	Output carriage return to device 002
220	002	
221	CALL	Perform Status Input Subroutine
222	231	
223	000	
224	MVI *A*	Move ASCII line feed character to accumulator
225	012	012 is ASCII representation of line feed
226	OUT	Output line feed to device 002
227	002	
230	RET	Return to calling program
231	IN	Input Status from device 002
232	002	
233	ANI	Mask all but status bit
234	001	
235	JZ	If status not present, go back and look for it
236	231	until it does appear
237	000	
240	RET	Return to calling program

the next character is recovered and printed.

When a carriage return is detected, the main loop is exited and the carriage return is performed. Another status input loop is entered to allow the printer time to perform the carriage return, and when the next status bit present indication is noted, the last status input loop is exited. A code equivalent to a line feed is loaded in the accumulator, and then output to the printer. The complete line has now been printed and the printer is ready for the next line. The routine returns control to the main program so that other required operations may be performed before reentering the print routine to print another line.

Of course, all the improved methods for input processing may be applied to output processing with adequate hardware. Polled output, interrupt output, and so on, are well within the realm of the modern microprocessor, and will be discussed shortly as a complete input/output interface operation is shown.

Detailed discussion of specific output devices is best left to instruction manuals provided by individual manufacturers. Many different schemes are used to convert the parallel ASCII characters into printed form, and most of them are electromechanical in nature. Such discussions are left to those materials dedicated to operation of output and input *devices*.

8-3 SERIAL INPUT/OUTPUT

Concepts and Devices. Although the microprocessor is basically a parallel device, it must accommodate input and output devices that furnish information appearing in serial (one bit following the other) form. Much of the communication that takes place between microprocessors separated by any appreciable distance uses serial transmission to reduce the number of lines necessary to carry information. One line can carry the complete word of information when it is serialized, whereas if an 8-bit computer word was transmitted in parallel form, eight lines would be required.

The Teletype* is a serial input/output device. The keyboard resembles that of a conventional typewriter, but the internal operation is much different. When a key on the Teletype is depressed, it selects mechanical levers that are used to produce marks (1s) and spaces (0s) corresponding to the common ASCII representation of the character. After the mechanical levers are set up, their positions are scanned by a motor-driven switch that closes when a mark is to be generated and opens when a space is desired. Motor speed is closely controlled so that each mark and/or space is exactly the correct length. Figure 8-15 shows the format of a Teletype word. Note that additional information is included in the word. A *start* bit to identify the beginning of the character and 1 or 2 *stop* bits to identify character completion are added by the mechanism of the Teletype keyboard. The start and stop bits are required to synchronize transmission, since there may be gaps between characters in serial transmission and it is difficult to include a means of clocking without self-contained synchronization.

The printer associated with the Teletype is also an electromechanical device. An electromagnet (the *selector magnet*) is energized by each mark (1) pulse received. At each bit time the selector magnet positions one of a number of notched levers, and upon reception of the complete character, a

*Trademark registered by Teletype Corporation.

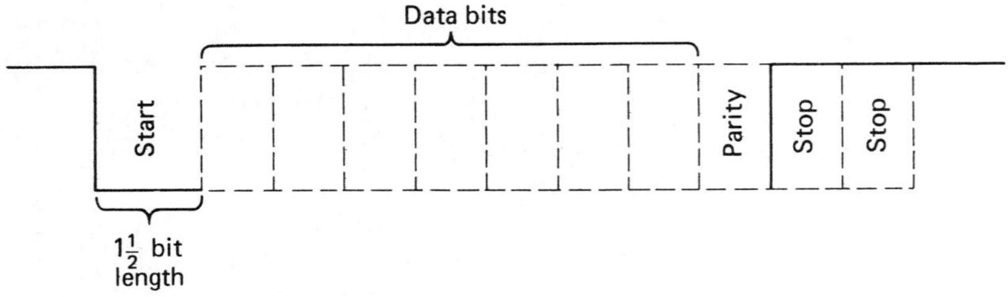

Figure 8-15 Teletype Word Format

hammer is released to impact each character's striker as in a conventional typewriter. However, only one character striker will be allowed to move as a result of the alignment of the notches in the notched levers. After the character is printed, all notched levers return to a neutral position to await reception of the next character.

Another increasingly popular serial input device is the digital tape cassette. Information is stored on the magnetic surface of the tape in serial form, quite often in the same format as the standard Teletype word. The tape cassette appears very much like the common audio tape cassette, with a few minor differences. The drive mechanism of the digital tape cassette is more precise so that data words will not become garbled, and the record/play heads are of better quality, as is the actual magnetic tape used. The recovery and recording of information on magnetic tape is discussed in Chapter 10, but once recorded or recovered, it is handled like any other serial form of data transmission.

Hardware. The primary task of hardware associated with serial data input/output is that of conversion from serial to parallel or parallel to serial form. The input/output interface should operate so that the microprocessor "thinks" it is working with a parallel device. Serial Teletype keyboard output, for example, must be converted to parallel form before it can be used within the microprocessor. Likewise, the parallel data from the microprocessor must be converted to serial form before the printer can use it. Since many input/output devices contain both keyboard and printer in the same enclosure, it is only natural that an IC assembly that performed both input and output data modification be developed.

Two MSI assemblies are found in common use. The *asynchronous communications interface adapter* (*ACIA*) and the *universal asynchronous receiver/transmitter* (*UART*) both perform the same functions, but in somewhat different manners. Both the ACIA and the UART accept a character from the keyboard in serial form, storing it in a register. Start and stop bits are stripped, and the data bits are made available to the microcomputer in parallel form. If a character is to be printed, the ACIA

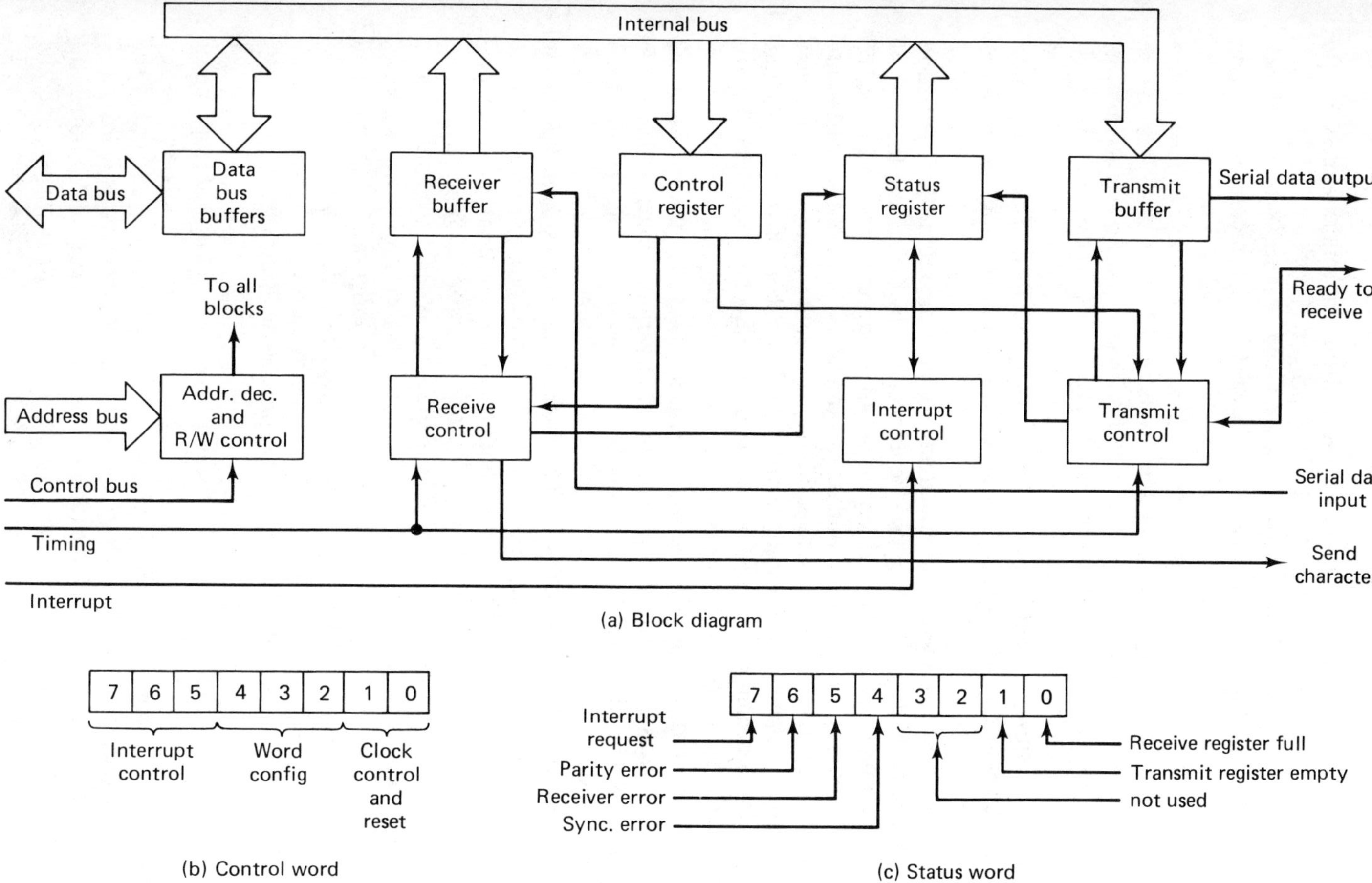

Figure 8-16 Asynchronous Communications Interface Adapter (ACIA)

and UART accept the parallel character from the microprocessor, add start and stop bits, and shift it out to the printer.

The ACIA will be discussed in this text. Figure 8-16 is a block diagram of a typical ACIA assembly, shown with the microprocessor and input/output device connections. Most of the blocks' functions are implicit in their titles. For example, the bidirectional *data bus buffer registers* isolate the microprocessor data bus from the internal data bus of the input/output interface, the serial-to-parallel *receive buffer register* accepts serially formed characters from the Teletype keyboard, places parallel-formatted characters on the internal data bus, and so on. *Receive* and *transmit controls* establish timing, check parity, remove or add start/stop bits, develop control signals, and sends status information to the *status register*.

Operation of the *control* and *status registers*, as well as the interaction of the remaining blocks, may best be understood by examining the flow of data through the ACIA. The key to ACIA operation is the handling of information in the control and status registers. The 8-bit, write-only control register establishes clock timing for both receive and transmit functions, selects word length, parity, and number of stop bits, and provides for control of receive and transmit interrupt enabling. Figure 8-16b shows the bit assignments of the control word.

The status register, a read-only register, indicates the status of the transmit register and the receive register, and various inputs to the ACIA. Figure 8-16c identifies the 8 status bits of the status register and their functions.

Read and write operations for an ACIA can be explained by using the block diagram of Figure 8-16 and the flowchart of Figure 8-17. Prior to actual use of the ACIA, certain initialization operations must be performed. The control register supplies this function by microprocessor-loaded control words. For example, the ACIA is reset by loading, at a minimum, bits 0 and 1 with binary 1s. Another control word is loaded following the reset word to establish word configuration, timing, and interrupt operations. The flowchart of Figure 8-17 describes a dedicated operation, where the microprocessor is programmed to only read or write information. Interrupt operation is also possible with the ACIA, but is left as an exercise to the reader.

Following initialization, the main program will set up either a read or a write routine. In the read routine, the status word is checked at the beginning of the routine. Bit 0 of the status word is checked, and if the receiver register is not full, status is checked again. The status word check loop is continued until an indication of "receiver register full" is noted. Bit 5 is now checked to verify that no receive error has occurred, and the routine continues. Detection of a receive error causes an error routine to be entered. The status register is also checked for existence of both synchronization and parity errors. If the receiver register is full and no

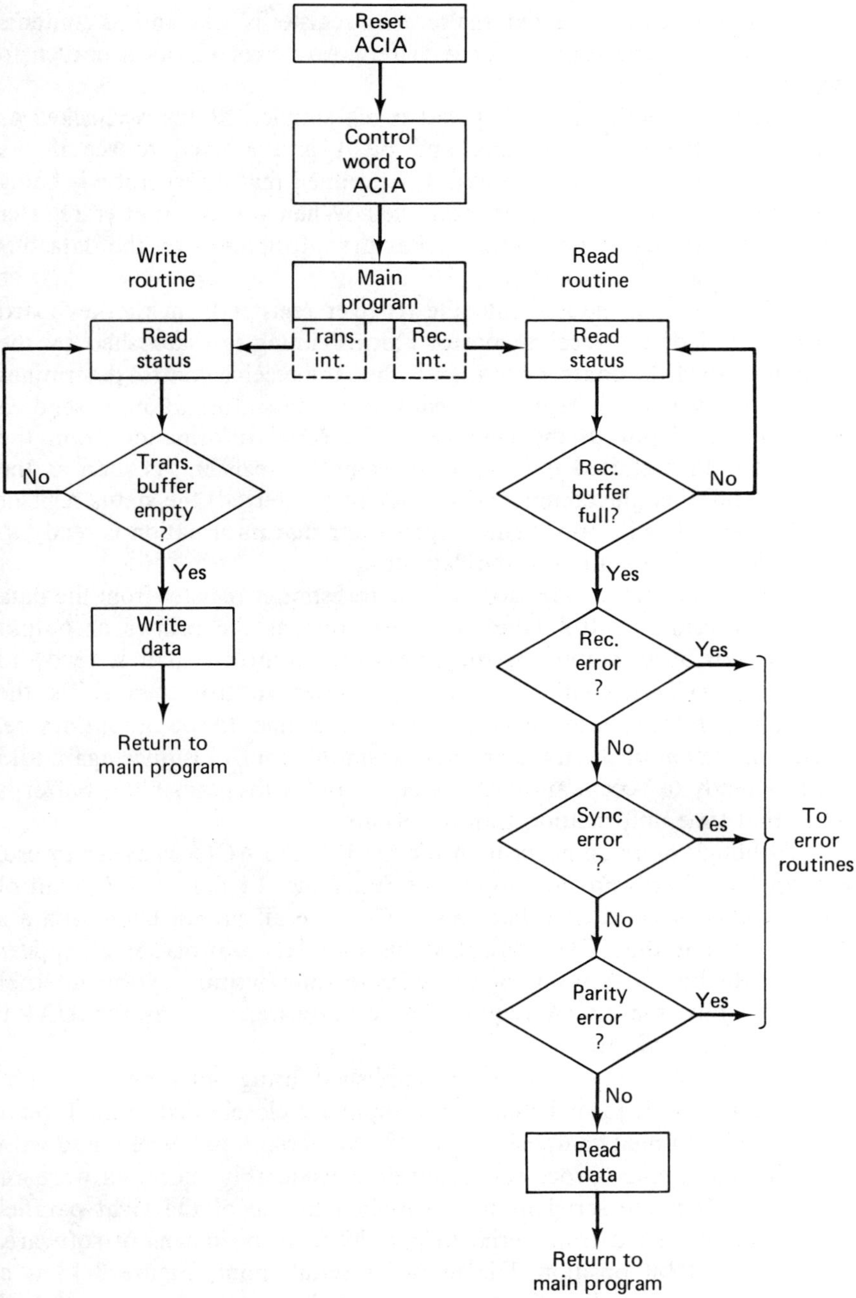

Figure 8-17 ACIA Operation Flowchart

errors of any kind are detected, the receiver register is read and its contents transferred to the microprocessor data bus. Any error causes a branch to an error routine.

The write routine is similar, but much simpler. Status is checked as required by the main program, and bit 1 is examined to see if the transmitter register is empty. If not, it is assumed that information is being transferred, and status is again examined. When the transmitter register becomes empty, the write routine writes the information on the data bus into the transmitter register.

Information is entered into the receiver register from the keyboard under control of the receive control block. Timing is established by the configuration of the control word, and when the receive control determines that the receiver buffer register is ready to accept information, it sends a *send character* input to the keyboard. The serial information from the keyboard is then shifted into the receiver buffer register. As soon as the register is full (i.e., the complete character is transferred) the status register bit 0 is changed to inform the microprocessor that information is ready to be transferred (in parallel) to the data bus.

When a character is loaded into the transmitter register from the data bus, the transmit control takes over. As soon as the printer or output device is not busy, it informs the transmitter control that it is ready to accept the ACIA's character. The transmitter control then shifts the information in transmitter register, one bit at a time, to the output device. Upon completion of the transfer, the transmitter control is once again told that it is ready to accept another character, and if the transmitter buffer is full at that time, information transfer occurs.

Although more complex than the UART, the ACIA is easier to use, requiring less hardware to support its functions. In the UART, control inputs, status outputs, and data bus buffers are all provided on separate lines. As seen, in the ACIA control, status, and data information all appear on the data bus as a result of software reading/writing of the internal registers. Thus, the ACIA requires more software, whereas the UART requires more hardware.

Serial input may also be accomplished using software (i.e., subroutines). The basic ideas behind serial input are closely tied to the format of the serially formed character (Figure 8-15). Serial input, when used with a parallel-input microprocessor, requires considerably more software or hardware. Often the serial input is supplied to *one* of the eight parallel inputs and converted from serial to parallel form by means of software. Assuming that bit position 7 is used for serial input, Figure 8-18 is a flowchart of a subroutine that may be used. Since many interesting and useful programming techniques are used for serial input, the flowchart is discussed in detail. A complete list of instructions for the subroutine is not shown, but selected portions are detailed for the reader's information.

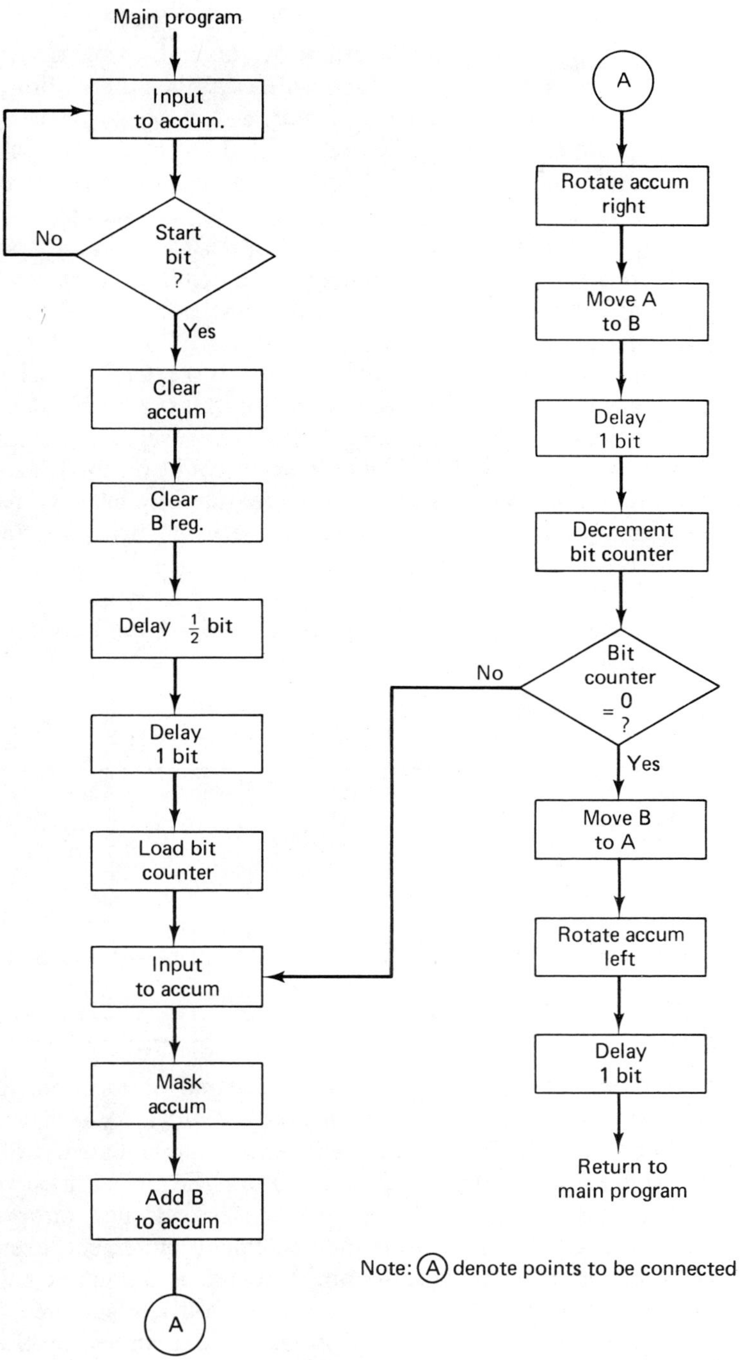

Figure 8-18 Serial Input Subroutine Flowchart

The serial input subroutine is called by the main program in much the same manner as parallel-input subroutines. When the subroutine is entered it is first necessary to determine if a start bit is present. Input information is routed to the bit 7 position of the accumulator when the INPUT instruction is executed. Bit 7 is then examined to see if it contains a 1 or a 0. If a 1 is found, the start bit has not yet occurred and the program loops back to the input again. When a 0 is found in bit 7 position, the status loop is exited and the accumulator cleared in preparation for the input of actual data bits. Register B is also cleared for use as a temporary store as the serial word is processed bit by bit.

Since the start bit is $1\frac{1}{2}$ information bits in length, it becomes necessary to delay any additional operations for that period of time. In the flowchart of Figure 8-18 this is accomplished by two separate subroutines, called HALFBIT and ONEBIT. A $\frac{1}{2}$-bit delay could be used three consecutive times in this application, but since full bit delay is required numerous times during the subroutine, two separate delay routines are used to improve program efficiency. The general concept of delay routines may be seen in the short program that follows.

It is assumed that a delay of approximately 4 milliseconds (4000 microseconds) is needed.

Address	*Mnemonics*	*Data*	*Cycles*
400	MVI R_0	006	11 (*not part of delay*)
401	330	330	
402	DCR R_0	005	7
403	INR R_0	004	7
404	DCR R_0	005	7
405	JNZ	302	16
406	402	402	
407	000	000	
			37 (*total cycles in delay*)

The short routine is loaded into memory starting at location 400_8. A constant (330_8) is initially placed in R_0 by the MVI R_0 instruction at locations 400_8 and 401_8. R_0 is then decremented, incremented, and again decremented (locations 402_8, 403_8, and 404_8). The JNZ instruction at location 405_8 checks to see if R_0 has reached zero. If not, the program returns to location 402_8 and repeats the decrement, increment, decrement sequence. When R_0 reaches zero, the program exits and continues with its original intent. Thus, a delay in program execution has occurred due to execution of this routine. The amount of delay is determined by the cycle time of the microprocessor and the number of cycles required for each instruction. Assuming a 0.5-microsecond cycle time and using the information shown with the program described above, the following calculations may be performed:

no. of loops $\times$ cycle time $\times$ no. of cycles = delay time (μsec)

$$244 \times 0.5 \times 37 = 3996\ \mu\text{sec (approx. 4000)}$$

Such delay techniques are commonly encountered in microprocessor applications.

Another important programming technique is seen in the delay routine just discussed. It should be noted that the overall program that inputs the serial word is a subroutine that is called by a main program. Within the subroutine other subroutines (such as the delay subroutine) are called. Such a technique is called *nesting* and is accomplished using the same stack that stores main program registers and program counter information while the input subroutine is being executed. Each time the CALL instruction is executed, the stack stores information required to return to the instruction that was being executed when the CALL occurred. Thus, the address of the step just prior to the beginning of the delay routine would be pushed on the stack, followed by any other information occurring in the delay routine that required saving. Information already on the stack is pushed farther onto the stack. Therefore, as subroutines within subroutines are executed, more and more information appears in the stack. The first information off the stack is, of course, the information required for returning to execution of the instruction that was being executed prior to the CALL. The stack therefore grows and shrinks according to the number of subroutines in operation simultaneously.

When both the $\frac{1}{2}$-bit and the full-bit delay have been accomplished, the program is ready to start looking for the value of the first information bit following the start bit. A bit counter is set to the value representing the total number of information bits in the word, and an INPUT instruction is executed. The logic value (1 or 0) of the first information bit is placed in the accumulator by the INPUT instruction. So that it can be assured that only the information in bit position 7 is used, the technique of *masking* is performed on the accumulator. By ANDing the contents of the accumulator with the binary word 10000000, any information in bits 6 through 0 in the accumulator is effectively erased, since ANDing any binary digit with 0 results in 0. If bit 7 is a 0, it will remain 0, since ANDing 0 with 1 results in 0. Likewise, if bit 7 is 1, the accumulator continues to have a 1 stored in bit 7, since ANDing 1 with 1 results in 1.

The contents of register B, which was just cleared, are added to the accumulator, and the first time through this portion of the input subroutine, the accumulator contains the first bit of the serial information in the input word. The accumulator is now shifted right one bit position (ROTATED) to prepare for the next bit input, and the accumulator is moved to register B for temporary storage. A 1-bit delay is inserted as the subroutine readies for the next information bit. The bit counter register is decremented 1 count, and the counter tested to see if the bit counter is at 0. If not, the subroutine loops back to the information-input step and repeats all the input operations. If the bit counter does equal 0, the contents of the B register are moved to the accumulator for final adjustment. The accumulator is shifted one bit left to adjust for the excessive shifts required by the input loop and a 1-bit delay is inserted prior to return to the main program. From a timing standpoint, the second stop bit should now be in process, and the main program can go on about its normal business until some indication of a new start bit occurs.

Actual microcomputer input/output operations may use any or all of the techniques discussed in this chapter. A typical input/output operation is shown in detail in Chapter 11, where a commercially available microcomputer is discussed.

8-4 SUMMARY

Chapter 8 has been an extremely long chapter, but it is this way for a good purpose. One of the least understood segments of the microcomputer is the input/output operation, and much detail has been included in this chapter in an attempt to clarify some of the gray areas. When one realizes that the only function of the input/output operations is to match the peripheral devices to the microprocessor, some of the mystery disappears. The many variations of input and output to the microprocessor have been shown, and the hardware purposely closely integrated with the software. As should be apparent by now, in the microcomputer the hardware and software cannot be easily separated. Since, as pointed out in Chapter 7, the microcomputer is wired to perform a specific function by computer words, the words required to input or output information must also be understood.

Mechanical details have purposely been deemphasized in Chapter 8. A separate text would be required to properly explain the many schemes found in modern input/output devices. One must recognize that the mechanics of the input/output devices results in an electrical representation of a character, computer word, and so on. Our interests in this text are in the electrical operations only.

QUESTIONS

1. What is the primary purpose of the input function; the input device?
2. What are the functions of the interface?
3. Explain how a keyboard output could be tailored to match microprocessor input.
4. Draw a functional logic diagram showing how a keyboard could be interfaced to a microprocessor.
5. What is the purpose of a data available flag?
6. Explain how a flowchart can be converted to a computer program.
7. What is polled input?
8. Define the terms subroutine; subprogram; stack; stack pointer.
9. What is the purpose of a memory map?
10. Why is a RETURN instruction required for each CALL instruction used in a computer program?
11. Why are subroutines used extensively in large computer programs?
12. What are the purposes of the PUSH and POP instructions in most microprocessor instruction sets?
13. Explain the sequence of events that are encountered when a subprogram is called by a main program.
14. How is multiple-input polling implemented?
15. What is interrupt input? What are its advantages and disadvantages?
16. Explain how an interrupt is processed by a typical microcomputer.
17. What is the function of a priority encoder?
18. Draw a functional logic diagram of a typical parallel-output interface device and discuss its operation.
19. Discuss the software necessary to support the parallel-output interface device of Question 18.
20. What is the difference in requirements between parallel-output devices and serial-output devices?
21. Define the terms ACIA; UART.
22. Draw a functional logic diagram of an ACIA and explain its operation.
23. Show the software required to support an ACIA.
24. Draw a functional logic diagram of a UART and explain its operation.
25. Show the software required to support a UART.

CHAPTER 9

Internal Memory

Another feature that distinguishes a microprocessor from a microcomputer is its internal memory capability. The general concepts behind memories were initially discussed in Chapters 4 and 5. In this chapter these concepts will be expanded to include the various methods of implementing the *internal* or *high-speed* memory, that part of the microcomputer's memory that usually exists as an integral part of the microcomputer's basic set of assemblies. External, high-volume memory is discussed in Chapter 10.

9-1 INTRODUCTION TO INFORMATION STORAGE

Classification by Location. Defining "memory" in terms of actual hardware is difficult, since "memory" implies storage of information, and storage functions exist throughout, and external to, the microcomputer. Actually, memory (storage) exists in all the functional sections of the microcomputer. *Memory* is used in the functional diagrams to designate the *main memory* of the microcomputer, where information that is *immediately* needed to operate the computer is stored. The capacity of this portion of storage depends on the microcomputer under study and can vary from as little as 1K (1024) 8-bit words to as many as 64K (65,536)

16-bit words. This type of *internal storage* is characterized by access times* on the order of less than 500 nanoseconds to 2 microseconds, with the 500-nanosecond time being steadily reduced as technology advances. Semiconductor storage cells are used in internal storage.

Even faster storage elements exist in other functional sections of the microcomputer. A typical example is the semiconductor-implemented Instruction Register in the CONTROL section of the microprocessor, where the op code from an instruction word is temporarily stored. Information is available at the output of the Instruction Register within a few nanoseconds of the time it is initially stored. This type of fast, temporary, extra-small capacity storage (one word or less) is characterized by access times on the order of 10 to 100 nanoseconds. Examples of register storage are seen throughout the INPUT/OUTPUT, ARITHMETIC/LOGIC, and CONTROL sections of the microprocessor.

As microcomputers grow larger and more powerful, the need for storage capacity grows. Since the rate at which a microcomputer can use information is generally limited by the rate at which instructions can be processed (and only *one* instruction can be processed at a time in most microcomputers), the need for massive amounts of short-access-time storage is not great. As long as sufficient storage is available to keep the microcomputer from having to *wait* to get information from memory, storage methods with longer access time may be employed. Slower memory devices are characterized by capacities on the order of many millions of bits and access times from microseconds to seconds. The number of words stored per cubic foot of space is much greater, and cost per bit drops dramatically as slow access times are allowed. Magnetic disks and tape cassettes (Chapter 10) are the most common *auxiliary* storage devices. It should be noted, however, that technological advances are providing cost and size reductions in the short-access-memory devices, and the cost dividing line between medium-capacity and small-capacity memories is less well defined than it was in the past.

Needs exist in larger-scale computers for the storage of extremely large amounts of information. A bank, for example, stores all its savings, checking, and investment account information along with day-to-day operating requirements. Much of this information may be needed at monthly intervals, and there is no need to have it available within microseconds, or even milliseconds. Access within seconds, even minutes, is quite adequate. Large-capacity, slow-access-time storage devices such as magnetic tape result in considerable economies in price and space.

Thus, there exists within, and adjunct to, the modern microcomputer

*Access time is the time interval between information request and information availability.

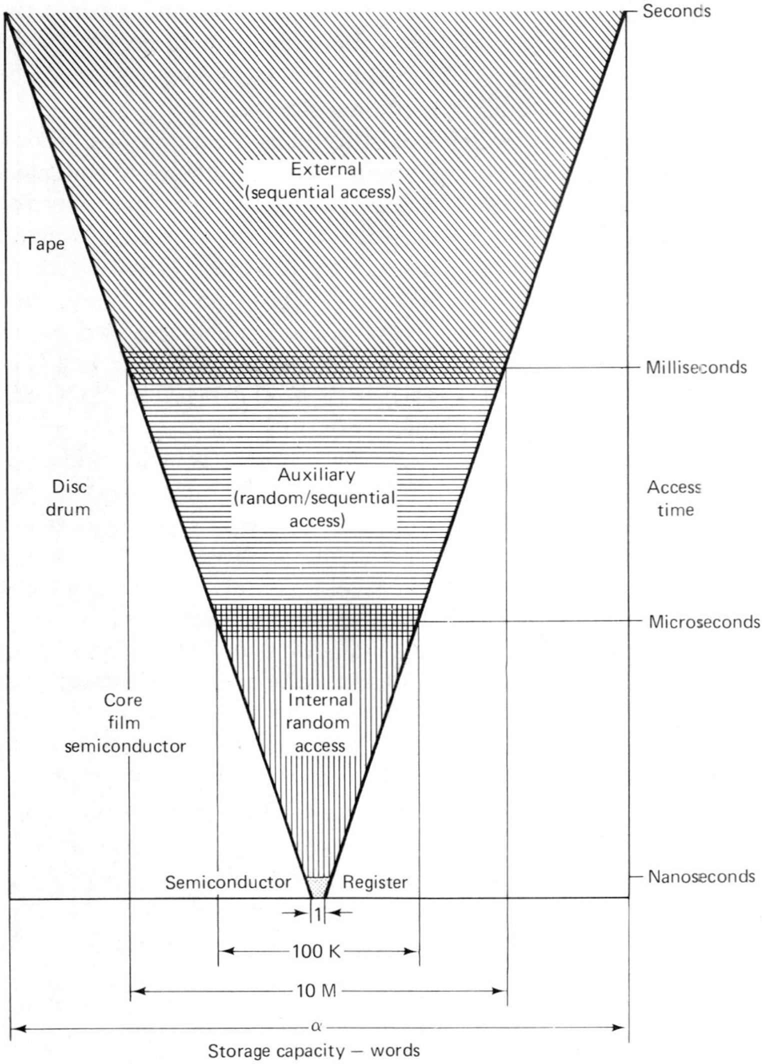

Figure 9-1 Computer Storage Device Hierarchy

information-storage capabilities ranging from capacities of one word or less to many millions of words. Access times range from tens of nanoseconds to minutes, generally determined by the capacity of the storage device. Figure 9-1 depicts the hierarchy of storage devices in microcomputers.

Classification by Access Methods. Storage devices are often classified in terms of the manner in which information is made available to or from the memory. Two methods of accessing storage devices are commonly employed: random access and sequential access.

Random-access methods allow *any* of the storage locations to be accessed at will without cycling through other locations. *Fixed* access time is an identifying characteristic of a random-access storage device. A "jukebox" uses random-access methods. The desired record is selected by identifying its storage location and the retrieval mechanism goes from its resting place directly to the selected record. Each record in between does not have to be played to reach the desired record. The conceptual memory discussed in Chapter 4 is an example of a random-access memory. Internal memories generally are random-access devices.

Sequential-access methods require cycling through all previous storage locations before arriving at the desired information. This type of storage is characterized by a *variable* access time, since the time it takes to obtain a given computer word is determined by its location in the storage device. An inexpensive cassette tape recorder/player is an excellent example of a sequential-access machine. If the music desired to be played is the first recording on the tape, the access time is very short; yet, if it is the last recording, the access time could be quite long. Most external storage devices are sequential in nature.

Some memory devices employ both random- and sequential-access methods. Such devices have access times which fall between sequential-access and random-access limits and are usually classified in the *auxiliary* memory area. Figure 9-1 includes access classification information.

Classification by Operational Mode. Internal memory in particular is often classified in terms of its mode of operation. For example, some memory applications merely require that information be made available *to* the microcomputer. Once the information is initially placed in memory there may be no further requirements to change that information. *Read-only memories* (*ROMs*) that operate in this mode are discussed in detail in Section 9-3.

Other requirements exist within the internal memory to temporarily store information and to place new information in the memory. *Read-write memories* (*RWMs*) perform this operation. It should be noted that although the Read-Write terminology is most technically correct, it is not a common term. *Random-access memory* (*RAM*), discussed in Section 9-2, is the generally accepted nomenclature for RWM. Actually, ROM is also random access in nature, as are other types of memory, but RAM is employed only to describe the type of memory that can both supply information and accept information. Both RAM and ROM are found in the internal memory, and it appears that as technology advances they will encroach on the auxiliary memory field.

Concepts of Information Storage—General Operation. Internal or "working memory" can be viewed as shown in the conceptual diagram of Figure 9-2. Almost all modern working memories have as their principal component a *storage matrix*. The actual storage medium will be discussed shortly. A matrix arrangement (row and column organization) is used to reduce the number of input lines required to define a location (address) in memory. For example, 16 input lines may be required to identify 16 addresses if locations are arranged in tabular form as in Figure 9-2a. Information is obtained from the memory by *linear selection* when the tabular form of organization is used. ROMs often use a variation of linear selection. However, when arranged in the matrix form of Figure 9-2b, only two row and two column inputs are needed (there are 16 combinations of four inputs). The matrix form is called *coincident selection* and finds application in both static and dynamic RAMs and some ROMs.

The only additional hardware requirements for matrix-organized storage are *row* and *column decoders* to translate the input lines into separate row and column lines, as shown in Figure 9-2b. Note that storage locations are not only identified by a decimal number but may also be located by intersection of row and column inputs. The correlation between decimal and binary identifications is easily seen.

Actual data to be placed in locations in the storage matrix (WRITTEN) or data to be retrieved (READ) are routed to/from the matrix via the INPUT/OUTPUT (I/O) control. The READ/WRITE (R/W) input controls the direction of data flow, enabling circuits that recover information from the addressed location for the READ operation. During the WRITE operation READ circuits are disabled and data are accepted by the memory. The CHIP ENABLE (CE) input serves effectively as an additional address input line and is discussed in more detail in Section 9-2.

Storage Cells. Memories constructed with semiconductor devices generally fit into three categories: *static*, *dynamic*, and *read-only*. Each is discussed below in adequate detail so that the reader may appreciate the techniques associated with its application.

Static. Static memories use storage cells that resemble the flip-flop circuit, storing information in one of two stable states. Such cells are *nondestructive* during information readout. *The stored information remains in the cell even though it is transferred to other parts of the microcomputer.* In addition, as long as power remains "on," the memory retains its information. Unfortunately, the static semiconductor cell is *volatile* in nature, *losing its information when power is removed* unless special techniques are employed. (Some storage media are nonvolatile.)

Address				Location
A_3	A_2	A_1	A_0	
0	0	0	0	00
0	0	0	1	01
0	0	1	0	02
0	0	1	1	03
0	1	0	0	04
0	1	0	1	05
0	1	1	0	06
0	1	1	1	07
1	0	0	0	08
1	0	0	1	09
1	0	1	0	10
1	0	1	1	11
1	1	0	0	12
1	1	0	1	13
1	1	1	0	14
1	1	1	1	15

(a) Tabular form (Linear selection)

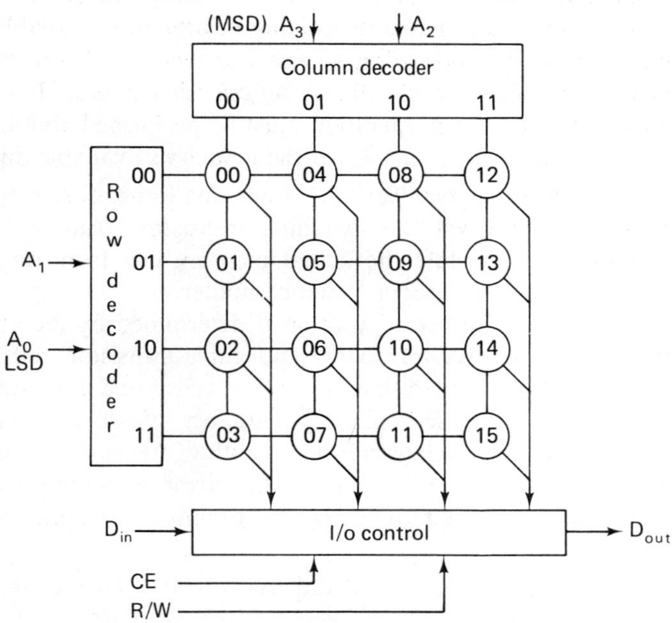

(b) Matrix form (coincident selection)

Figure 9-2 Concept of Internal Memory

Although numerous technologies are used to manufacture memories composed of static cells, the important point to remember is that each of these cells responds very much like the simple gated RS flip-flop discussed in Chapter 2. The different technologies merely result in memories with different access times, power requirements, and so on. Such comparative information may be found in current trade publications, and is purposely not discussed in this text due to the widely varying and expanding technologies.

A memory constructed with static cells would then contain one gated RS flip-flop at each storage location. The row and column inputs of the memory assembly are connected to the input gates to enable either READ or WRITE operations. Setting the flip-flop generally stores a binary 1, while resetting stores a binary 0. Information stored in the cell is retrieved by examining the flip-flop's output and gating that output to the data output lines.

Dynamic. Dynamic memory cells store information using the absence or presence of an electric charge on a capacitor*. Figure 9-3 shows the concepts of a dynamic memory cell and the types of circuits required to support its operation. The cell that is to store information is selected by the address circuits of the memory, enabling the storage capacitor to either store new information or to make its existing information available to the microcomputer. Since the charge on a capacitor tends to deteriorate with time, it must be *refreshed* periodically. A *refresh input* is provided for this purpose. Typically, the refresh operation must be performed about every 2 milliseconds and requires less than 1% of the memory's available time.

A bidirectional *sense amplifier* allows information to flow into or out of the storage capacitor via the switching transistor. *Input* and *output buffers* are provided to isolate the actual memory cell from outside influences within the microprocessor/microcomputer.

The actual storage cell configuration is determined by the manufacturer. Single-transistor, dual-transistor, and three-transistor cells are in common use. It should be noted that most static cells use a minimum of six transistors, and it can be seen that the number of storage cells in a dynamic memory can be much greater as a result of the smaller number of semiconductor devices required. However, the refresh requirement tends to generate additional circuit requirements, and timing in dynamic memories

*A capacitor is a device that stores electric energy (charge) based on the physical parameters of the capacitor, the amount of energy available, and the amount of time the source of energy is applied to the capacitor. In the dynamic storage cell, the physical parameters are determined by the construction of the cell. The source of energy is the power supplied from the computer to the memory. Timing originating in the microprocessor determines the amount of time the energy is applied to the capacitor.

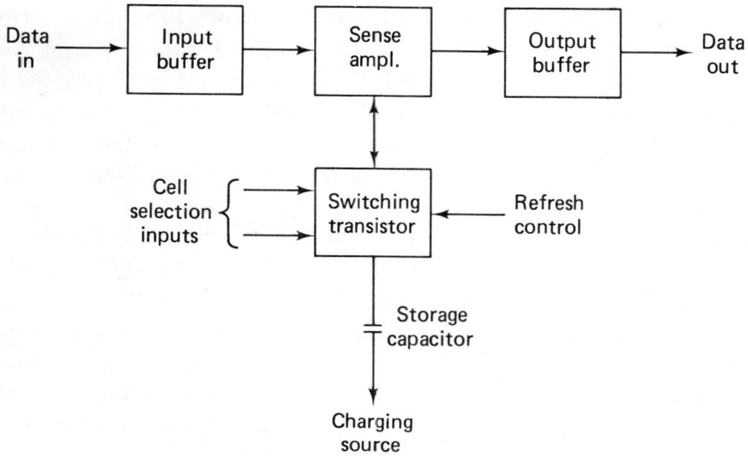

Figure 9-3 Dynamic Memory Cell

is much more stringent. Dynamic cells are nondestructive during information readout as long as refresh operations occur as required. As with static cells, dynamic cells are volatile.

Read-Only-Memory Cells. A number of different cell configurations exist in ROMs. Those that do not require the capability of erasing and rewriting are relatively simple. An unprogrammed ROM is manufactured with storage locations containing either all 1s or all 0s. As each location is programmed, a Nichrome or polysilicon fusible link is blown if the bit at that location is to be changed. Figure 9-4 is a simplified logic diagram of a typical fusible-link type of ROM storage cell.

When an address is selected, the *select* input is at a HIGH level. CHIP ENABLE is made HIGH when the chip in use is required. If the fusible link has not been blown, the NAND gate will have a LOW and 2

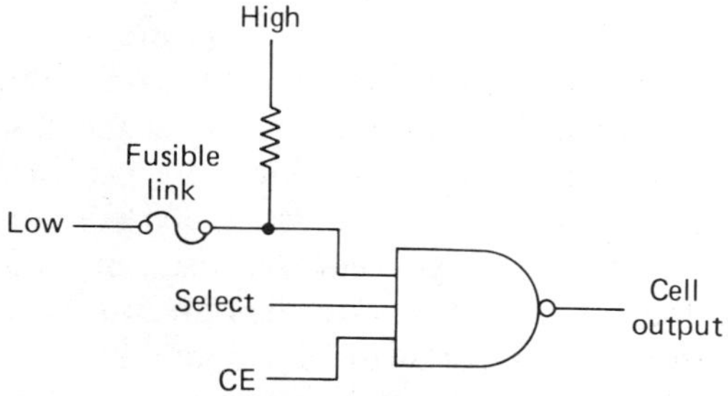

Figure 9-4 ROM Cell

HIGH inputs, resulting in a HIGH output. When the link is blown, all NAND gate inputs are HIGH, and the output becomes LOW. Thus, if LOW is equated to binary 0 and HIGH to binary 1, blowing the link is equivalent to placing a 0 in memory at that location. The ROM may be obtained with the bit positions already programmed, or the programming may be accomplished by the user with a device designed to open the fusible link.

If the cell must be erasable and rewritable, other techniques are used. Although different in the manner in which they are implemented, most erasable ROMs store data as an electrical charge. Complex principles of semiconductor physics are needed to explain the special charge storage phenomenon and they are beyond the scope of this text. Data are placed in the ROM cell by an electrical current and retrieved by noting the effect of the stored charge on the elecrical characteristics of the semiconductor cell. EAROMs (electrically alterable ROMs) can be erased by using an electrical current to remove the stored charges, while EPROMs (erasable programmable ROMs) require application of ultraviolet light for erasure.

Organization of ROM cells into read-only memory is discussed in Section 9-3.

9-2 RANDOM-ACCESS MEMORY

Static. Now that the basic concepts of working memory have been established, it is possible to examine actual, or near-actual, memory applications. The general block diagram of Figure 9-2 is adequate for initial explanations. Figure 9-5a, however, more accurately (and in more detail) reflects the thinking of memory device designers. Although not an actual memory device, it does incorporate the concepts of most existing static memory chips. Dynamic memory chips are discussed later. Initially, the size of the cell matrix is retained at 16 cells to speed explanation, but larger arrays will be discussed subsequently.

The reader should once again verify that the X and Y decoding scheme does indeed result in selection of one and only one cell at a time. That is, any given combination of A_3, A_2, A_1, and A_0 will cause coincidence at only one location in the memory cell matrix. Both the X and Y decoders accept two address inputs (A_3 and A_2 at the Y decoder; A_1 and A_0 at the X decoder). The buffer amplifiers are provided to isolate the memory from the remainder of the microcomputer. Two inputs to a 1-of-4 decoder such as the X and Y decoders will result in an active output on only one of the four output lines. Decoders are merely combinations of gates connected so as to operate as discussed above.

Figure 9-5a deviates somewhat from the simplified block diagram of Figure 9-2 in that the input/output path is combined with the Y decoder

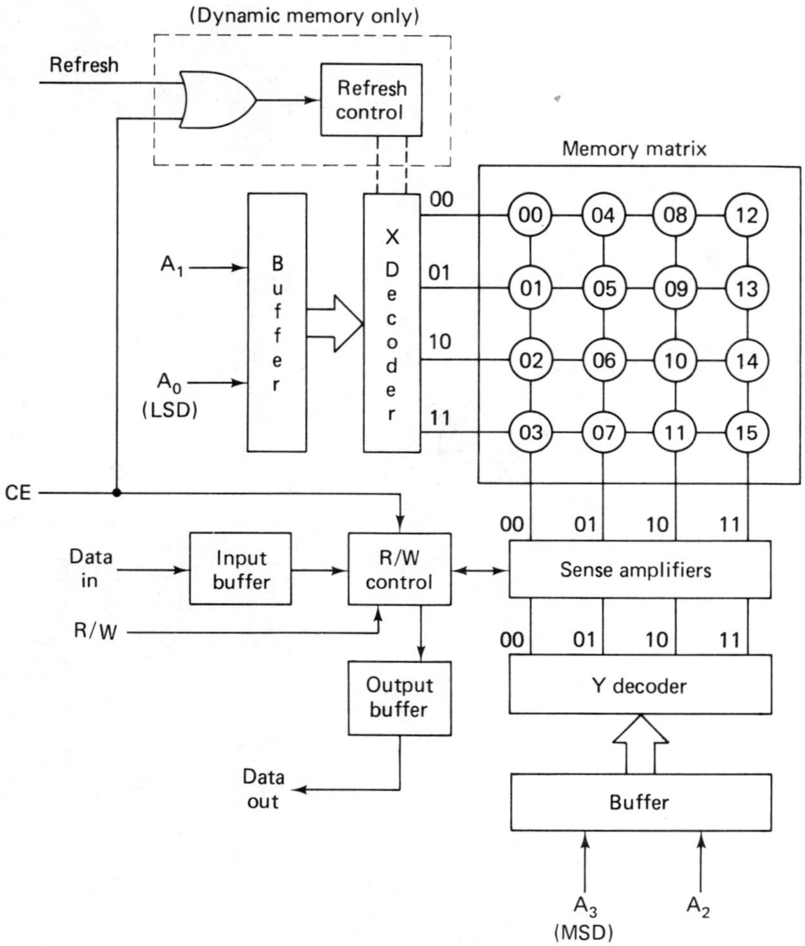

(a) Generalized layout

Figure 9-5 Random-Access Memory: (a) Generalized Layout; (b) Input/Output Control

lines. The manner in which the input/output path can be combined with decoder lines is shown in simplified logic form in Figure 9-5b. Assume that a single bit of information is to be placed in memory at a location identified by the address 1001. The most-significant digits (10) enable column 3, while the least-significant digits (01) decode row 2. Memory cell 9 is located at the intersection of column 3 and row 2. It merely remains to determine how the output of the *Y* decoder can be combined with the input path to store information at the selected location. Input information

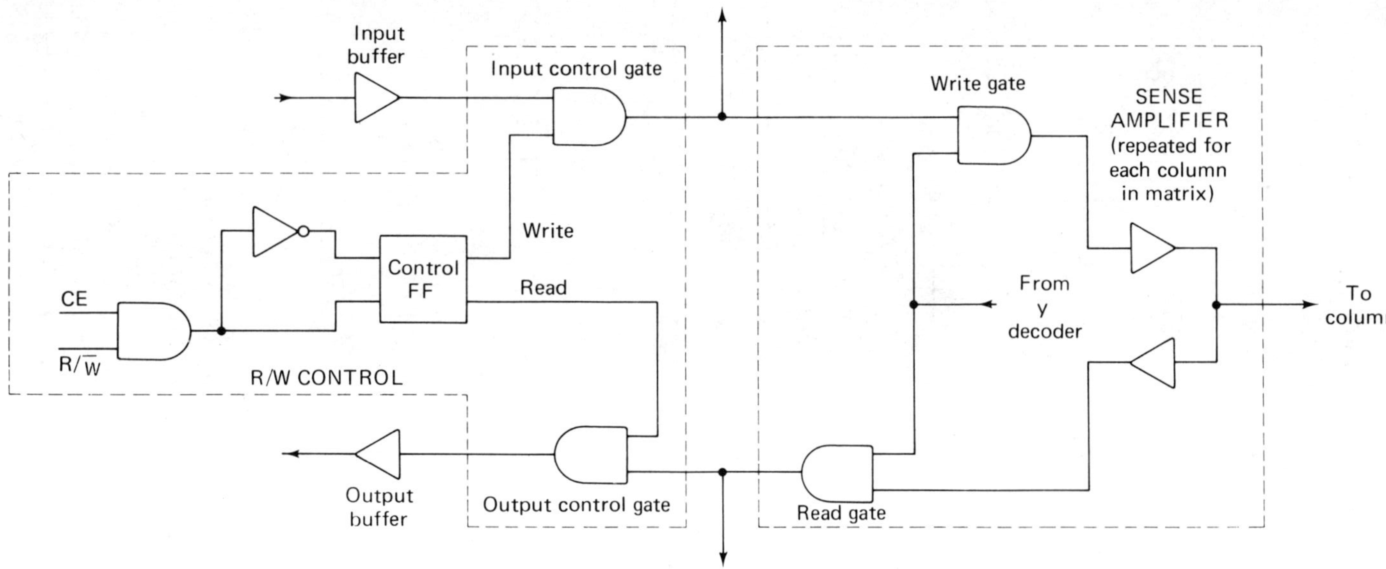

(b) Input/output control

Figure 9-5 (Continued)

is supplied to the input buffer amplifier and then to an AND gate in the R/W control. The AND gate is enabled by the control flip-flop, which is either set or reset by the control input. R/$\overline{\text{W}}$ is applied to the control flip-flop via another AND gate, which allows the R/$\overline{\text{W}}$ signal to be effective only when the specific memory chip is enabled (CE). When the WRITE output of the control flip-flop is HIGH, the input control AND gate is enabled, and the input information is applied to the write AND gate in the sense amplifier. Enabling of the write AND gate is accomplished when the appropriate *Y* decoder output is HIGH, and the input information is fed through the sense amplifier buffer to the selected column. Although applied to all cells in the selected column, only the cell that is enabled by a row decoder input will accept the information provided. The READ operation is identical to the WRITE operation except for direction of information movement. It should be noted that one sense amplifier assembly is needed for *each* column of cells. All sense amplifiers receive their information from the common R/W control.

The operation of a single-bit memory has now been described, and with it the concepts of storage functions required within a microcomputer. The only difficulty remaining is that very few single-bit microcomputers are encountered. Expansion of these concepts to include more common microcomputer situations is not too difficult. For the sake of simplicity, however, a small-scale, 16-word memory will be retained, although the word length is now expanded to 8 bits. Expansion of memory size to more than 1 bit per word does not really complicate matters. Instead of representing a package of information with only 1 bit, that information is represented with, in our example, 8 bits. Therefore, at each location, or address, there must be provisions for storage of not 1 but 8 bits. Instead of enabling one storage cell when the computer word is desired, eight cells must be enabled.

The single-bit, 16-word memory previously discussed is sometimes called a *memory plane*. Properly interconnecting a number of memory planes can result in multiple-bit memory. The number of planes used determines the number of bits per word, while the number of cells per plane determines the capacity (number of words) of the memory. By using 16 cells per plane, interconnecting eight planes will result in a 16-word, 8-bit memory device. This concept is shown in Figure 9-6.

When the computer word at a specific location is desired, it is necessary to enable the appropriate *X* and *Y* inputs for *all* eight planes. This is accomplished by connecting X_1 of plane 1 to X_1 of plane 2 to X_1 of plane 3, and so on, to include all eight planes. Thus, all the remaining *X* lines are connected together, as are all of the *Y* lines. Therefore, when an *X* line is selected, it enables that *X* line on all eight planes. The same actions exist for the selection of a *Y* line. Since 8 bits instead of 1 bit are now

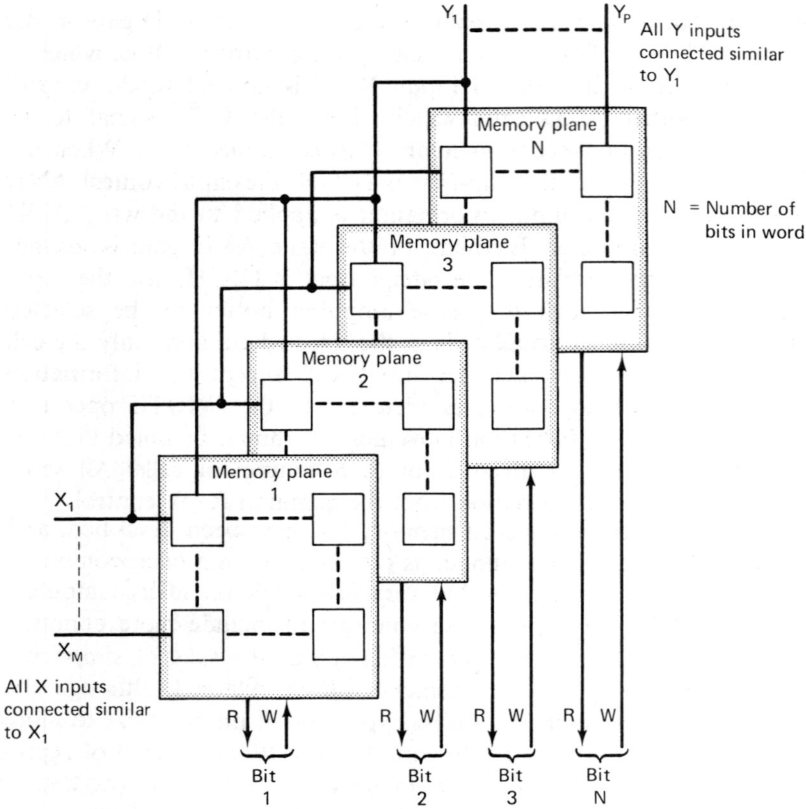

Figure 9-6 Concept of Memory Planes

made available to the READ/WRITE operation, 8 bits are either written or read. In the practical 8-bit memory, then, each time an address is selected, a complete 8-bit computer word is either written or read. Thus, information is handled in groups of bits rather than from the much slower single-bit-at-a-time approach.

Complete lists of instructions, made up of many computer words, may be stored in the type of memory just described. It is necessary, of course, to make available thousands of words of storage, but it should be apparent by now that this is easily accomplished by merely increasing the number of cells on each plane and expanding address decoding capability.

Semiconductor memories are available in a number of configurations and capacities. Rather than generate confusion, those memories that are organized in the manner already discussed will be initially covered before going on to other organizations. Common static RAM configurations include 256×1, 1024×1, and 4096×1, where the first number denotes the number of storage locations and the second number denotes the number of bits storable at a location. Therefore, the simple memory previously discussed could be considered 16×1. The smaller memories (256×1 and 1024×1) are rapidly giving way to 4096×1 configuration, and 8192×1 and 16,384×1 chips are available in small quantities. For the sake of simplicity, a 1K memory (1024×1) is discussed. It should be apparent that, since the 1K memory is merely an enlargement of the 16-bit memory, the jump to 4K or even 8 or 16K is easily accomplished.

The 1K chip is organized in exactly the same manner as the 16-bit chip shown in Figure 9-5. Since there are 1024 storage locations, the 1024 memory cells are arranged in a 32×32 matrix. The 1-of-4 decoders must, of necessity, then, become 1-of-32 decoders in order to select each of the 32 rows and columns. READ/WRITE control circuitry remains essentially the same, while 32 instead of 4 sense amplifier assemblies are required to accommodate the 32 columns. Additional gating and buffering is necessary to combine the outputs of the 32 columns into one output, but those details are best left to the memory-chip designer.

If one is to use a practical memory, it will be necessary to determine how to combine devices such as the 1K static memory chip into, for example, a 1K×-8-bit memory rather than the 1K×1-bit memory. It is common practice in microcomputer diagrams to use block diagrams rather than the detailed logic diagrams of semiconductor assemblies. Therefore, as the 1K×1-bit memory is expanded to 1K×8 bits, the technique of block diagramming will be used. It has been shown that a 16×8-bit memory is possible by combining individual 16-bit memory assemblies. The same technique is used for 1K×8-bit memories. A block diagram of such a memory is shown in Figure 9-7.

The 10 address lines A_0 through A_9 are separated into row addresses (A_0 through A_4) and column addresses (A_5 through A_9). A_0 inputs from all 1K static memory chips are connected together, as are the A_1, A_2, and so on, inputs. Thus, when an address is placed on the address bus of the microcomputer, the addressed location is accessed on all eight memory chips at the same time. In order for action to occur, however, all chips must be enabled (CE inputs), and all chips must be in either the READ or the WRITE mode ($R/\overline{W}$). Note that the CE inputs are all connected together, as are the $R/\overline{W}$ inputs. However, the data input (DI) and data output (DO) lines are *not* paralleled. Since each memory chip represents 1 bit in the 8-bit computer word, each data input or data output must be separately available to make up the computer word.

Most static memory assemblies found in microcomputers have more than 1K×8-bit capacity. Common configurations are 4K×8 and 8K×8,

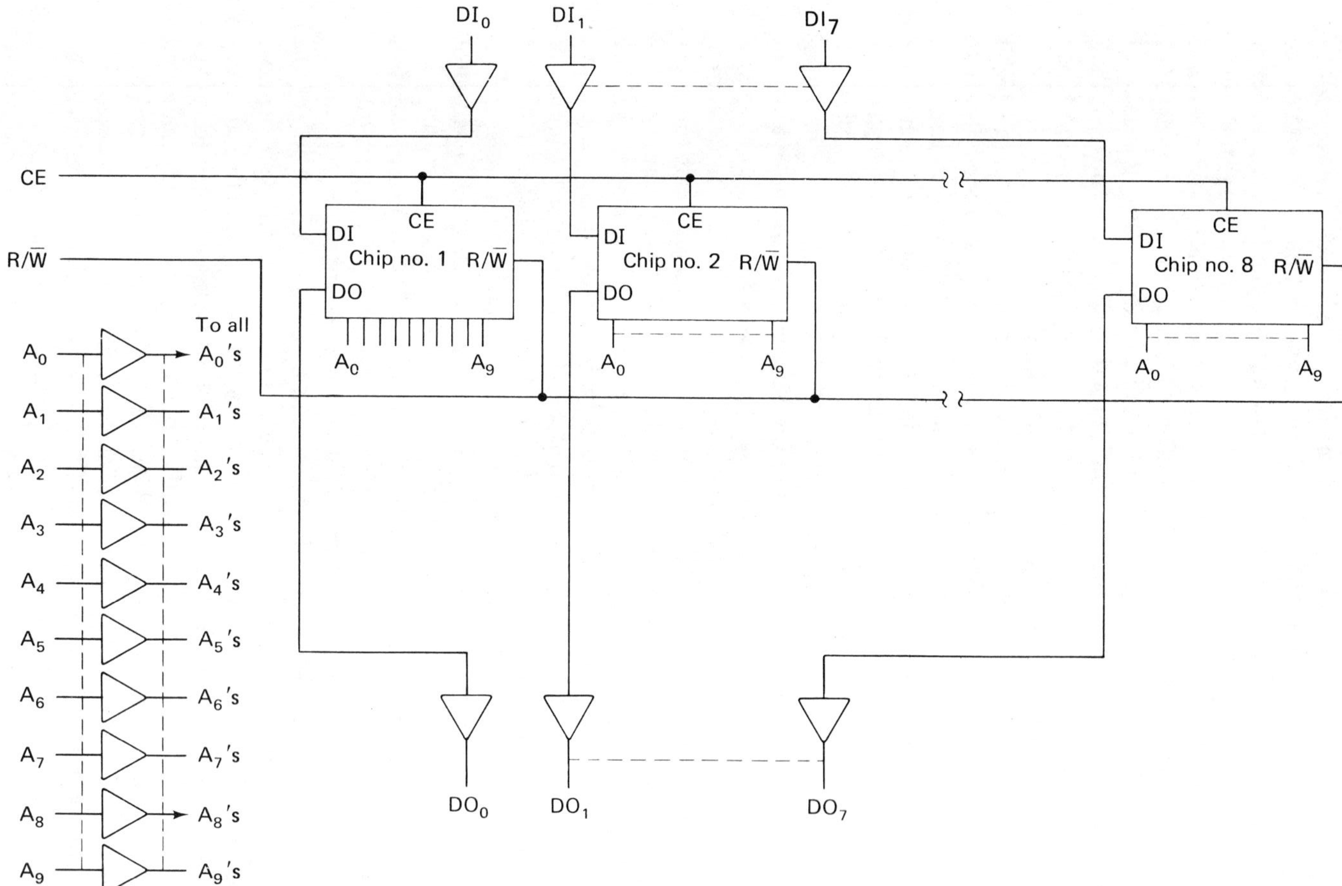

Figure 8.7 1K Memory Organization Using 1K × 1 Bit Memory Chi

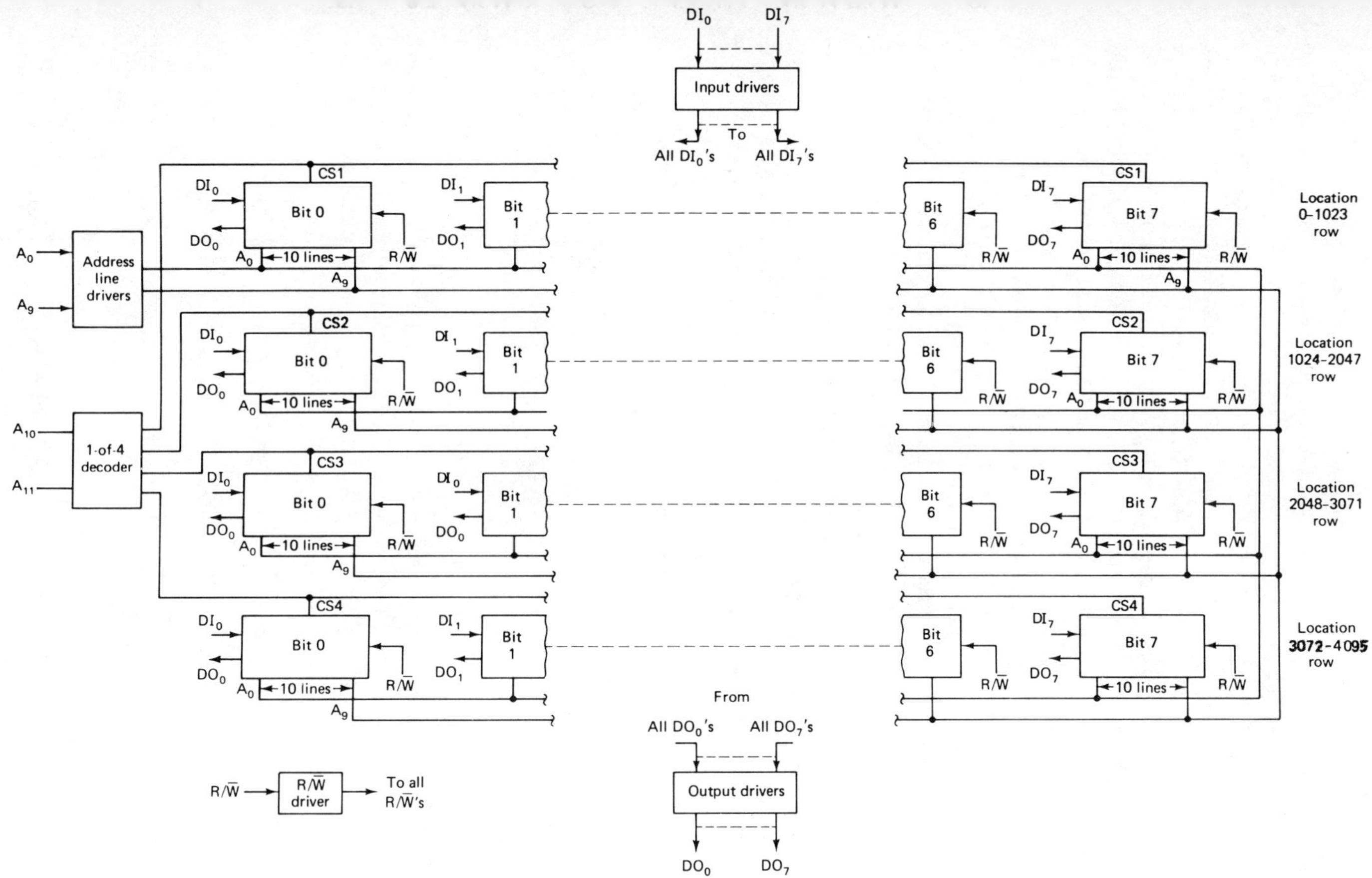

Figure 9-8 4K Memory Organization Using 1K×1-Bit Memory Chips

with 16K×8 assemblies becoming commonplace. Expansion from 1K×8 to 4K×8 using 1K×1 chips is shown in Figure 9-8. It is only necessary to expand this concept to obtain 8K×8 or 16K×8 capacity. Since a row of 8 1K×1 memory chips gives the 1K×8 capacity, it is only necessary to *add rows of chips* to increase overall capability. For example, if two rows are used, 2K×8 results; and so on. The four rows of 1K×1 chips in Figure 9-8 thus give 4K×8-bit storage. The $R/\overline{W}$ input is still common to all 32 chips because all chips must be placed in the same mode. Data inputs and outputs are connected so that all bit-0 chips (first column) are connected to the same buffer amplifier, all bit-1 chips (second column) are connected to the same buffer amplifier, and so on. Thus, regardless of which *row* of chips is selected, bit-0 input or output will occur from column 1, bit-1 input or output from column 2, and so on.

Selection of the appropriate location in memory is still achieved by the address input, with the least-significant 10 bits of the address selecting 1-of-1024 locations just as in the 1K×8 memory. A 1-of-4 decoder is used to select the appropriate *row* of 1K × 1 chips using the two most-significant digits in the address and the CE (chip enable) inputs of the memory chips. Many microcomputers using modern microprocessors are capable of addressing up to 64K of memory, requiring 16 address lines. Table 9-1 shows how any 1-of-65,536 storage locations may be accessed using 16 address inputs, assuming a 64K memory made up of 4K memory assemblies.

Use of Table 9-1 is shown in the following examples.

EXAMPLE 9-1 What storage location is selected by the address 0000000100100100?

SOLUTION

Using the principles of binary-to-decimal conversion discussed in Chapter 2, determine the decimal value of each of the 1s by reference to Table 9-1. Adding each of the decimal equivalents results in a decimal address of 292:

$$256+32+4=292$$

The location is thus Group 0, Assembly 0, Row 0.

EXAMPLE 9-2 What storage location is selected by the address 0011100100100100?

SOLUTION

$$8192+4096+2048+256+32+4=14{,}628$$

The location is Group 0, Assembly 3, Row 2.

Table 9-1 Storage Location Addressing

Address bus lines	A15	A14	A13	A12	A11	A10	A9	A8	A7	A6	A5	A4	A3	A2	A1	A0
Decimal equivalence	32768	16384	8192	4096	2048	1024	512	256	128	64	32	16	8	4	2	1

A9–A0: Select one of 1023 locations

A11	A10		
0	0 = 0 – 1023	(Row 0)	1K ROW SELECT
0	1 = 1024 – 2047	(Row 1)	
1	0 = 2048 – 3071	(Row 2)	
1	1 = 3072 – 4095	(Row 3)	

A13	A12	
0	0 = 0 – 4095	4K ASSEMBLY SELECT
0	1 = 4096 – 8191	
1	0 = 8192 – 12287	
1	1 = 12288 – 16383	

A15	A14	
0	0 = 0 – 16383	16K GROUP SELECT
0	1 = 16384 – 32767	
1	0 = 32768 – 49151	
1	1 = 49152 – 65535	

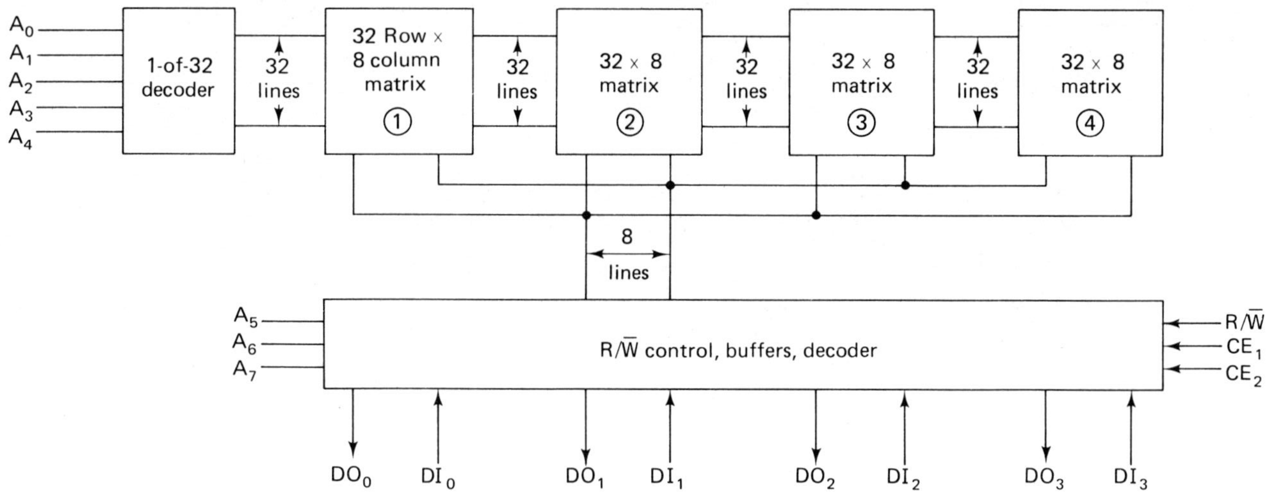

(a) Chip organization

Figure 9-9 Memory Organization Using 256×4-Bit Chips

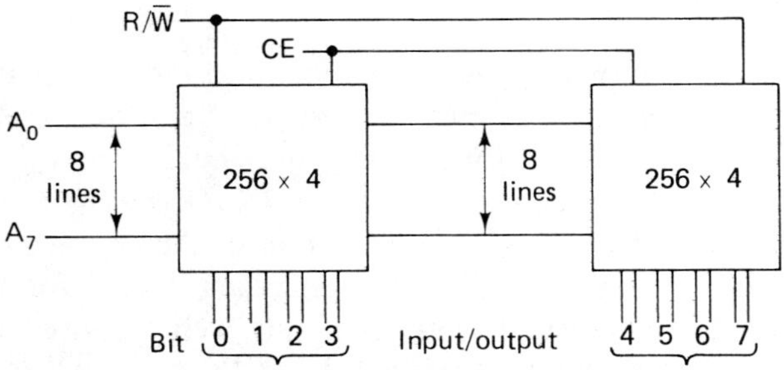

(b) 256 × 8 memory

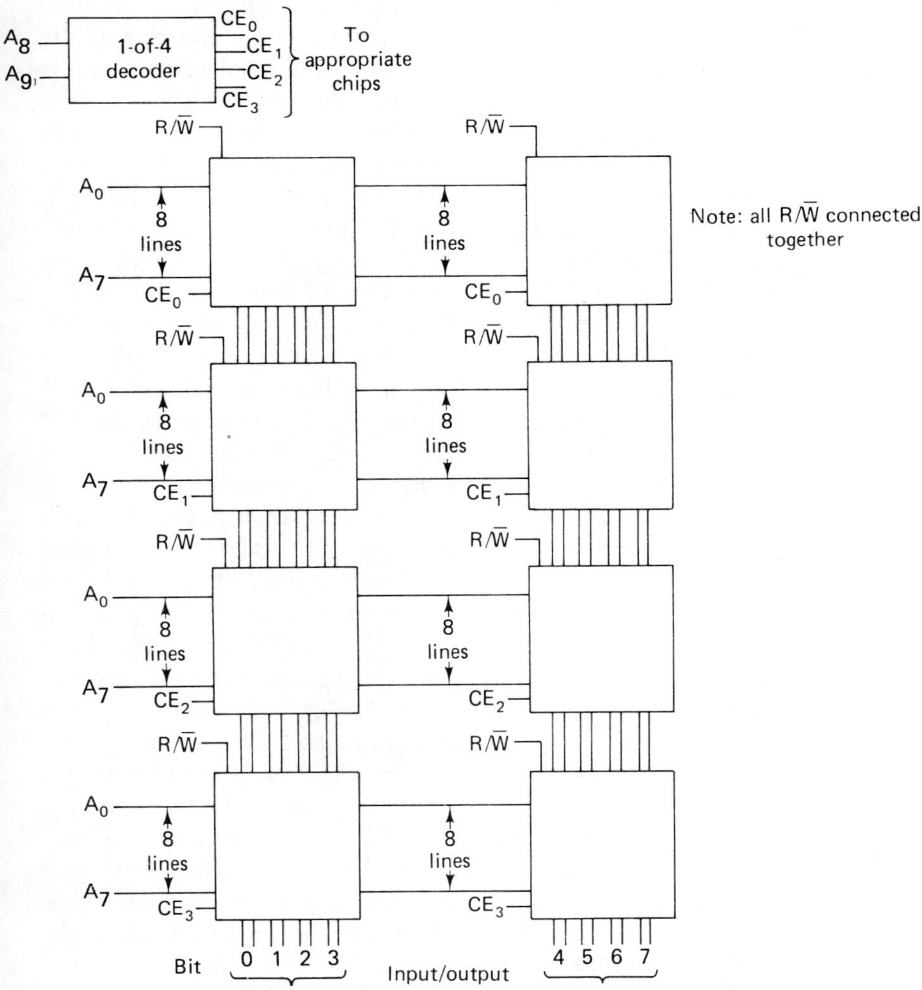

(c) 1024 × 8 memory

Figure 9-9 (Continued)

Thus, it may be seen that by merely decoding the binary address that is placed on the address bus of the microcomputer, the exact group, assembly, row, and cell that is selected in memory may be determined.

The denser 4K×1 static memory chips allow 4K×8 configuration with only 8 instead of 32 memory chips. Organization of the chips into a full-scale memory is identical to the 1K×8 organization except for the additional address inputs. Whereas 10 address inputs were required for 1K-chip addressing, 12 address inputs are needed for 4K chips. The two added address inputs may be implemented by two more pins on the IC chip, or by combining functions of other pins on a *multiplexed* or time-sharing basis. Obviously, adding two additional pins on the chip is the easiest implementation to understand. The two additional address inputs are integrated with the rest of the address decoding on the memory chip, thus allowing the 8-chip 4K×1 memory to operate in the same manner as the 8-chip 1K×1 memory. If increased capacity is to be achieved without extra pins, it is only necessary to time-share *one* of the chip pins. The chip enable can become one of the two required address pins, and enabling (if required) can be accomplished by external circuitry. One other pin is not required. Since the memory chip cannot both read and write at the same time (the $R/\overline{W}$ input controls that function), *one* pin can serve both functions. Internal circuitry on the chip combines the cell input/output, so a data input *and* a data output are not required. Thus, two extra pins have been obtained, and 4K×1 capacity can be obtained with the same number of external connections as the 1K×1 chip. Other multiplexing schemes are discussed when dynamic memory components are explained.

Another common static memory chip organization is one in which the package is 4 bits wide rather than 1 bit wide. 1K capacity is obtained with a 256×4 configuration, while 4K capacity is achieved with a 1K×4 matrix. The 256×4 configuration is used to explain chip operation. Cell organization is in the form of four 32×8 matrices (Figure 9-9a). Each matrix thus stores 256 bits. Matrix 1 stores the bit associated with bit 0 of the word, matrix 2 stores bit 1, and so on. The exact storage location is determined by the 5 least-significant bits of the address (A_0 through A_4) selecting the appropriate row on each matrix, while the 3 most-significant bits (A_5 through A_7) select the appropriate column on each matrix. READ/WRITE and input control, decoding, and buffering are very similar to those functions described in the 1K×1 organization. Note that two CHIP ENABLE inputs are shown. These are used as additional address inputs when larger memory organizations are employed, just as the CE inputs on the 1K×1 chip.

A 256×8 memory may easily be implemented using two 256×4 chips as shown in Figure 9-9b, while 1K×8 organization is shown in Figure 9-9c. 256×4 memory chips are most often used in efficient design

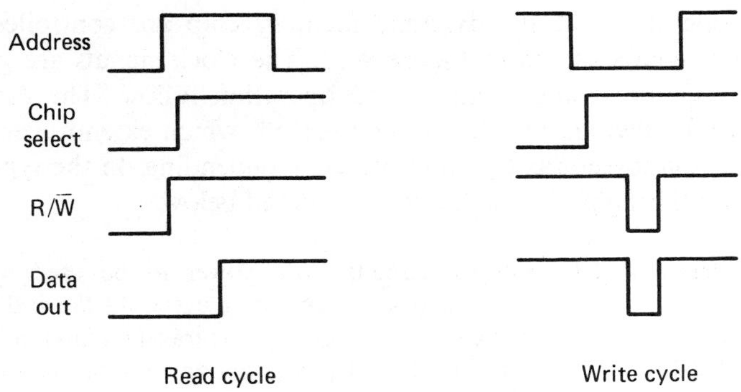

Figure 9-10 Static Memory Timing

of small memory systems and/or in systems where selection of 256-word increments of memory is important.

Little has been said about timing requirements with static memory systems, but they are extremely important. As with any type of logic circuit, finite times are required for devices to change state when inputs are applied. A READ/WRITE-cycle timing diagram is shown in Figure 9-10. In general, the address and chip enable inputs must be present and stable for a given period of time before data may be read or written. Typically, the address inputs are applied first, followed by chip enable, then data input or output. Depending on the technology employed to manufacture the memory chip, overall cycle times range from less than 100 nanoseconds up to greater than 1 microsecond. The reader should consult individual memory-chip specifications for exact timing requirements.

Dynamic. Dynamic memories appear in somewhat the same configuration as static memories, with just a few exceptions. As indicated in Section 9-1, the mode of storage is not permanent (even with power constantly applied), and each cell must be periodically *refreshed* in order to retain its information. In addition, since the cell requires much less space, many more cells may be placed in the same chip area. 16K×1 chips are not uncommon, while 4K×1 chips are the run-of-the-mill device used. For the sake of simplicity, however, the older 1K×1 dynamic memory chip is used for explanation. Although actually more complex, the block diagram of the static memory chip (Figure 9-5) is adequate for the dynamic memory chip functional explanation with the addition of the *refresh control* shown in dashed lines. The major functional difference between static and dynamic memory operation lies in the area of timing requirements. It is the function of the *refresh control* block to perform these requirements.

All operations of the dynamic memory chip are controlled by the three clock signals shown in Figure 9-11. The clock signals are generated when CE occurs. General timing and operation follow. The three clock signals may be divided into five timing periods which extend over a period of from 200 nanoseconds to 1 microsecond, depending on the type of chip in use. Each timing period is discussed in detail below.

Period 1 (P1). Internal circuits store power to be used during the remainder of the operating cycle. At the end of P1 the rising edge of clock 1 (CL1) transfers input address and chip enable (CE) information into the internal buffer stores.

P2. Row and column decoders select the desired bit.

P3. The information in the bit is made available to the ouptut terminal of the chip. Output information reaches a stable value near the end of P3.

P4. Write enable circuitry is activated.

P5. If a "write" is desired during this period, data are written into the selected bit. Information stored in all other bits sharing the *same* addressed column is refreshed.

There are three basic operating cycles for the memory (see Figure 9-11). The READ/REFRESH and READ/WRITE cycles are the same length and use identical clock patterns. Only the READ/WRITE input level determines the functions that are performed. The SHORT READ cycle omits CL 3 and therefore does not perform WRITE or REFRESH operations. Any number of SHORT READ cycles may be used as long as the column refresh requirement is met.

It is generally the responsibility of external circuitry or software to determine how often and which column of cells require refreshing. The memory matrix, decoders, READ/WRITE control, and so on of the dynamic memory cell may all be arranged similarly to Figure 9-5.

One of the major advantages of dynamic operation is the reduced power requirements. Power is required only during the time a memory operation is being performed in a dynamic memory, whereas power is continuously used in a static memory.

Larger capacity memories are achieved in much the same manner as with static memory chips. Increased word length (1 bit per word to 8 bits per word, for example) is accomplished by adding more columns of chips. Greater storage capacity (i.e., more words) is obtained by adding more rows. As with static memory chips, however, changing technology resulted in greater cell density, and dynamic memory chips easily became 4K×1. Larger storage matrices were required, along with more address decoders, just as with the static memory cells. When the 16K×1 chips entered the

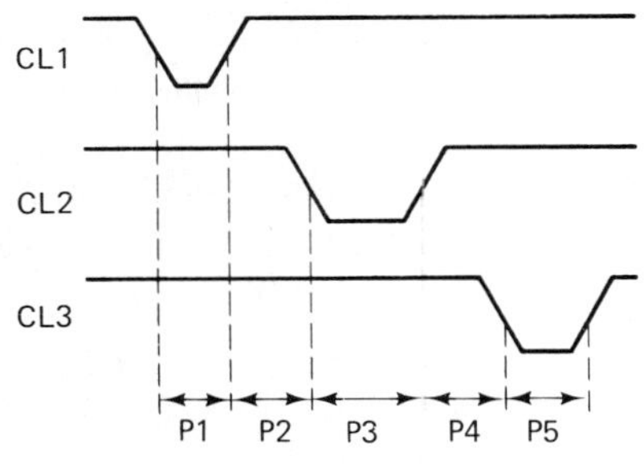

(a) Chip timing

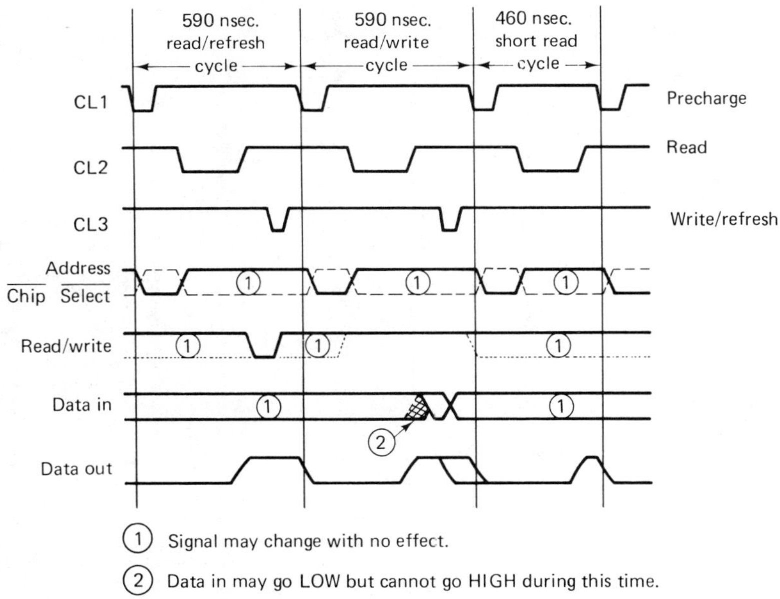

(b) Chip operating cycles

Figure 9-11 Dynamic Memory Timing

scene, however, the problem of extra address line inputs had to be faced. A certain amount of standardization within the industry dictated the use of 16-pin integrated-circuit packages, and 14 address lines are required to obtain 16K of address. The problem was solved by *multiplexing* the address lines.

Figure 9-12 is a simplified block diagram of a 16K dynamic memory chip. Note that only seven address lines (A_0 through A_6) are shown. Software of the microcomputer places the 7 address bits representing the

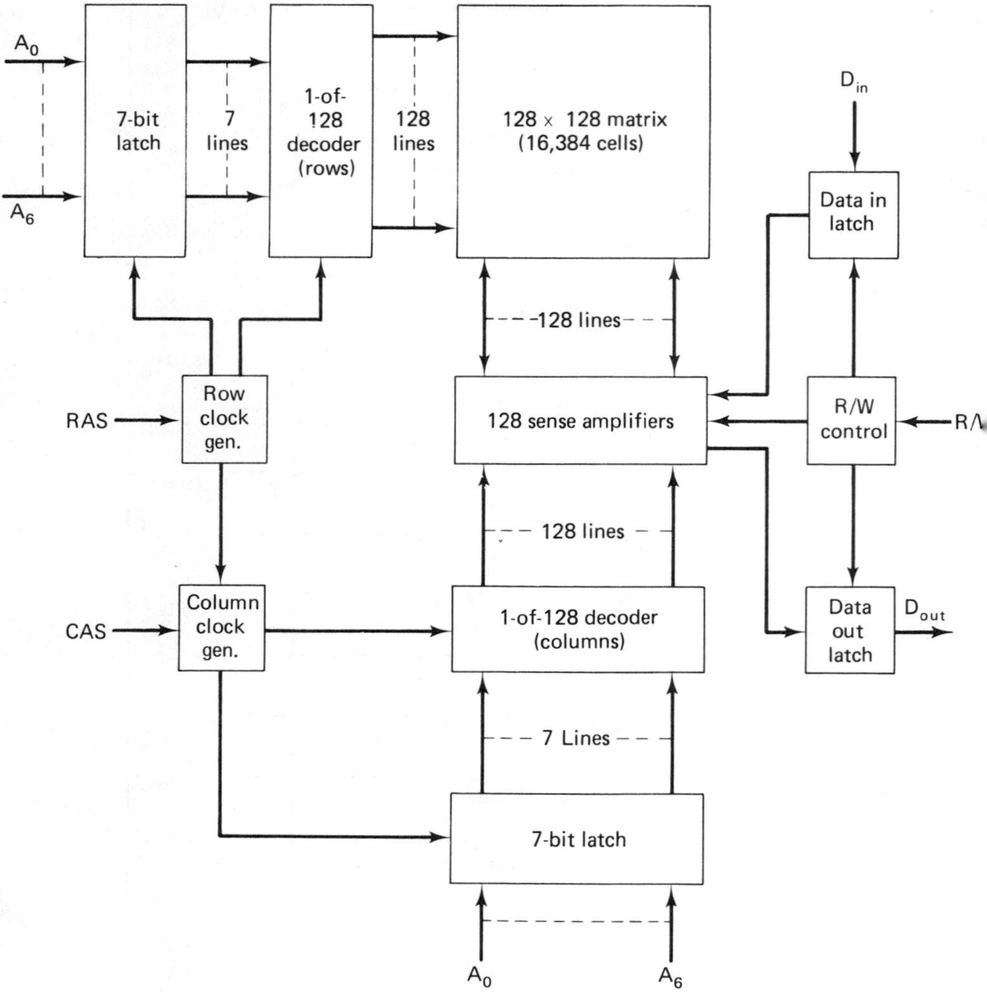

Figure 9-12 16K Dynamic Chip

row to be selected on the address bus and generates a *row address select* signal. The row clock generator is activated, supplying the necessary timing signals to latch the address information into the 7-bit row latches and enabling the row latches and the row decoder. One of the 128 rows in the 16K matrix is selected. The 7 address bits representing the column to be selected are now placed on the address bus and the *column address select* signal is generated. Activation of the column clock generator latches the column address and the appropriate column is selected. The READ/WRITE control determines direction of information flow (into or out of the chip), just as was accomplished with the static memory chip.

Timing requirements for the 16K chip are similar to the 1K dynamic memory chip and are met by the row clock generator and column clock generator. Refresh operations are also very similar, with a special refresh cycle available when *only row address select* and *row address* are provided by the software. It should be noted that Figure 9-12 is merely a conceptual 16K dynamic memory chip. Various manufacturers implement their memories with different technologies and design concepts. However, the important factors associated with large-capacity dynamic memory chips have been discussed, and the reader should be prepared to investigate the many different design approaches and apply the information as required.

The problem of volatility with semiconductor memories can be met by providing battery backup. Normal power is constantly monitored, and when it falls below a critical level, the standby mode is entered. With some static memory design technologies, it is only necessary to maintain a given minimum power, while with others it is necessary to perform complex switching to disable those circuits not used during data retention on standby power. Dynamic memories are faced with the additional requirement to refresh periodically, and power-failure detection and switching techniques are far more stringent.

9-3 READ-ONLY MEMORY (ROM)

ROM, as originally intended, was to mean *read-only memory*, a system for storing information in permanent, nonvolatile form. Since the original concepts of ROMs were developed, however, many different types of read-only memories have appeared. ROM has evolved today to mean a specific type of circuit, one that is programmed by the manufacturer to the user's specifications. Permanent storage of information offers many advantages. Most ROMs are faster than the read/write memories discussed in the previous section. Furthermore, the information stored in ROM is protected from loss as the result of unintentional write operations or power loss. However, there are also disadvantages. Stored information changes mean creation of an entirely different ROM, requiring perhaps a completely new production run. Needless to say, such an operation is quite costly.

A memory device much like the ROM, the *programmable read-only memory* (*PROM*), possesses all the advantages of the permanent storage of ROM without the disadvantages. The user programs his/her own PROM, thereby circumventing the manufacturer's programming charges. PROMs are relatively inexpensive when compared to manufacturer's costs. However, it should be realized that mistakes are costly, even with PROMs. A mistake will require another PROM and associated programming effort. PROMs also tend to be less reliable than ROMs because of the mechanics of programming.

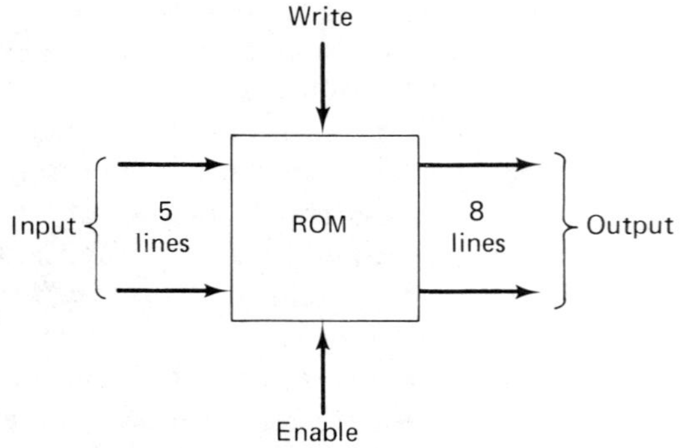

(a) Symbol

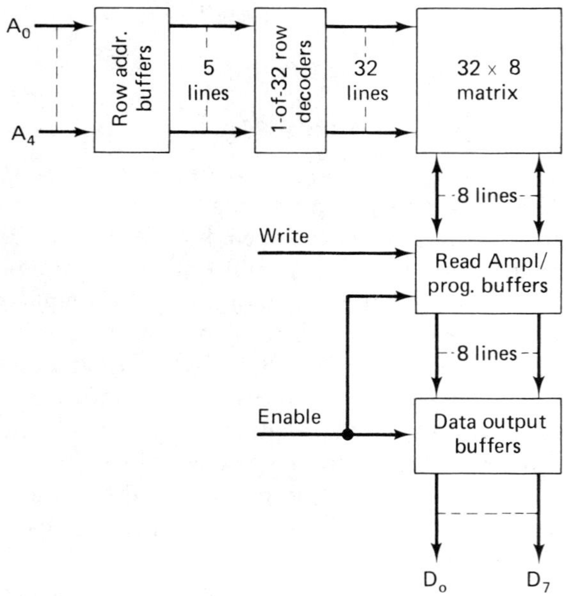

(b) Block diagram

Figure 9-13 ROM Chip: (a) Symbol; (b) Block Diagram; (c) 512×8-Bit ROM

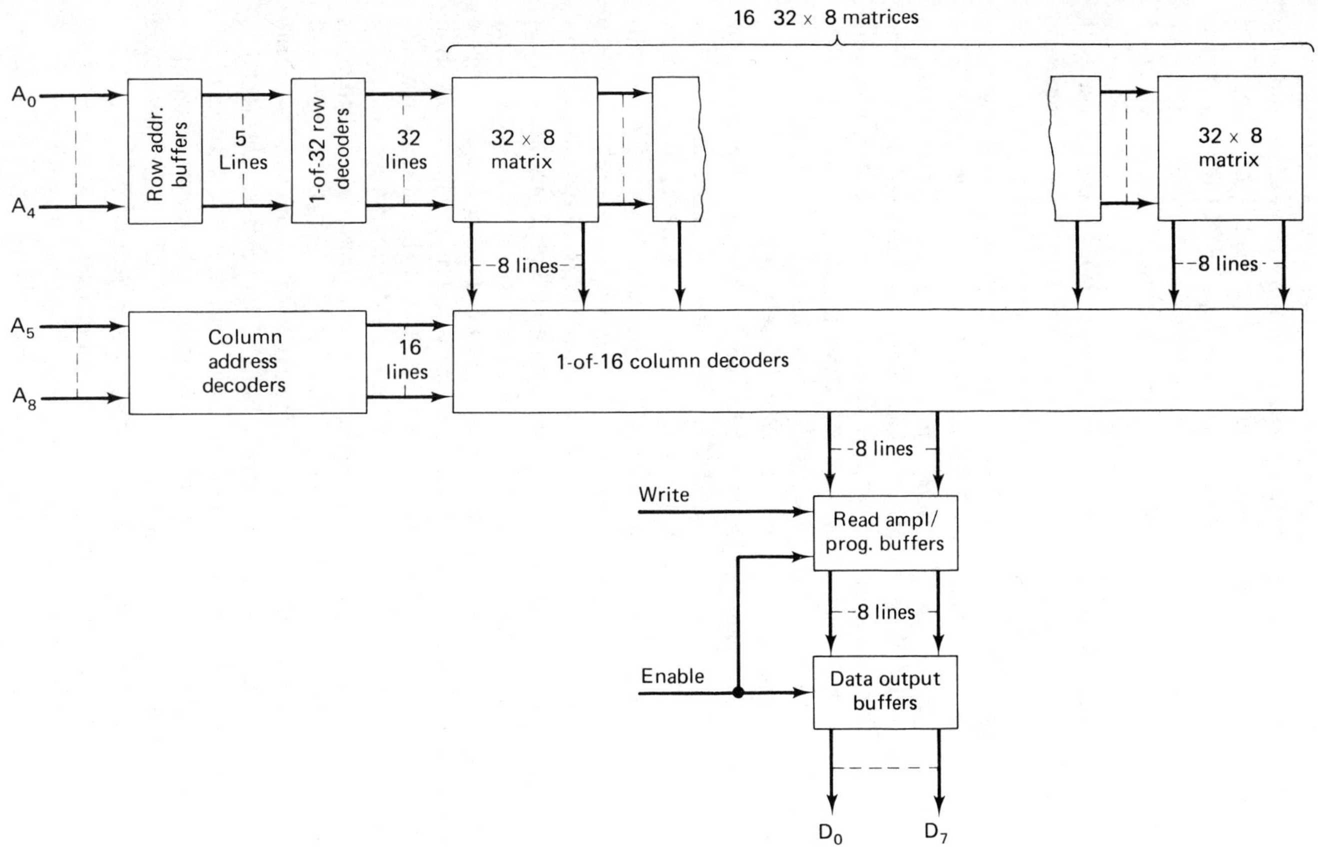

(c) 512 × 8 bit ROM

Figure 9-13 (Continued)

A group of memory components capable of erasure and rewriting also exist. The term *EPROM* refers to an *erasable programmable read-only memory* that can be erased by exposure to ultraviolet light. Thus, if a program requires change, it can be erased and the EPROM reprogrammed by conventional means. *Electrically alterable read-only memories* (*EAROM*) are similar to EPROMs except that information is erased electrically and, in many cases, selectively.

For the purpose of this section, the acronym ROM is used generically to describe *all* types of read-only memories. Techniques and applications differ little among all types of ROMs, thereby simplifying explanation of ROM operation. ROMs can be considered as fixed logic blocks or functions with several inputs and several outputs. Consider the simplified ROM shown in Figure 9-13a. The five inputs A_0 through A_4 correspond to address inputs and allow 32 different locations to be accessed. Each of the 32 locations is capable of permanently storing an 8-bit word, accounting for the eight output lines D_0 through D_7. Thus, the storage matrix consists of 256 cells similar to those discussed earlier, arranged in 32 rows by 8 columns. The block diagram of such a ROM is seen in Figure 9-13b.

One of the major differences between ROMs and RAMs is the arrangement of the cell matrix. RAMs are often organized in an $X \times 1$ configuration, while ROMs are commonly found in $X \times 4$ or $X \times 8$ matrix form. In smaller ROMs, such as the 32×8 chip shown in Figure 9-13b, it is convenient to employ the *linear-selection* technique, whereby a single decoder is used to select a specific row which contains all the storage cells. As ROM capacity increases, a combination of linear and coincident selection is used. Consider a 4K ROM organized in a 32×128 cell matrix to furnish 512 8-bit words. Such an array may be visualized as 16 separate 32×8 matrices (Figure 9-13c). Each row is selected by the linear-selection method just shown. Furthermore, each of the 32×8 matrices may be considered as columns to be addressed by the column address bits of the address word. The 1-of-16 decoder selects the appropriate matrix and gates all 8 bits contained in the enabled row of the matrix to the read amplifiers/programming buffers. If information is being read from the ROM, it is routed through the output buffers. If, however, the ROM is being programmed, the information on the output lines is treated as input, and the WRITE command supplies the additional power required to permanently program the memory location. The ENABLE input acts similarly to the CHIP SELECT input discussed in Section 9-2.

ROMs are available in capacities ranging from as small as 32×8 (256 bits) to as large as $2K \times 8$ (16,384) bits. Some are arranged as $X \times 4$ matrices, although the tendency seems to be toward the $X \times 8$ matrix. Application of ROMs is virtually endless. They are used to store complete programs in microcomputers, generate codes that correspond to letters and symbols for printout or graphic use, translate one type of code to another, and so on. Applications in microcomputers are limited only by the imagination of the user, and as prices drop, ROMs are replacing more and more of the heretofore discrete logic component functions.

9-4 MAGNETIC CORE

Although semiconductor ROMs and RAMs are the most common of the internal memory devices, other technologies are encountered. Foremost of these is magnetic core storage, which continues to be used despite the almost overwhelming onslaught of semiconductor devices. Magnetic core storage was the first of the high-volume, high-speed storage devices to be used in the computer field, and continued improvement has kept them in contention.

The magnetic core used for storage of binary information is a tiny toroid (doughnut-shaped piece) of ferromagnetic maderial. It can be magnetized very easily and will retain its magnetism unless purposely changed. Cores are pressed or stamped from an iron oxide compound material containing a binder to hold the shape. The core toroids are ceramic materials and require firing in a kiln at high temperatures to achieve the necessary chemical, physical, and magnetic properties. Many different sizes of cores are used, although modern toroids are 18 or less mils in outside diameter.

Any magnetic material may be studied by means of a graph showing the flux density* (B) that results from the magnetizing force (H) applied to the material. The graph is commonly called a *B–H curve*, and Figure 9-14a shows the reaction of typical ferrite-memory-core material to changes in magnetizing force. A magnetizing force can be generated by causing electrical current to flow through a wire, since a magnetic field surrounds a wire in which current is flowing. The strength of the magnetic field is a function of the amount of current flow, while the direction of the magnetic field is a function of the direction of flow of current. This is depicted in Figure 9-14b.

If a wire carrying electrical current is threaded through the center of a magnetic core toroid, the magnetic field surrounding the wire will transfer to the core in the manner shown in the B–H diagram of Figure 9-14c. Assume that in some way a magnetic core has been manufactured entirely devoid of any magnetic flux and that a wire carrying zero current is threaded through the core. If current is caused to increase in the direction equivalent to $+H$, the magnetic flux of the core will increase, as shown, from X in an increasing $+B$ direction. Magnetic flux will only increase so far and then a point of saturation is reached where increase in

*Flux density is a measure of the concentration of a magnetic field; hence it is concerned with magnetic field strength.

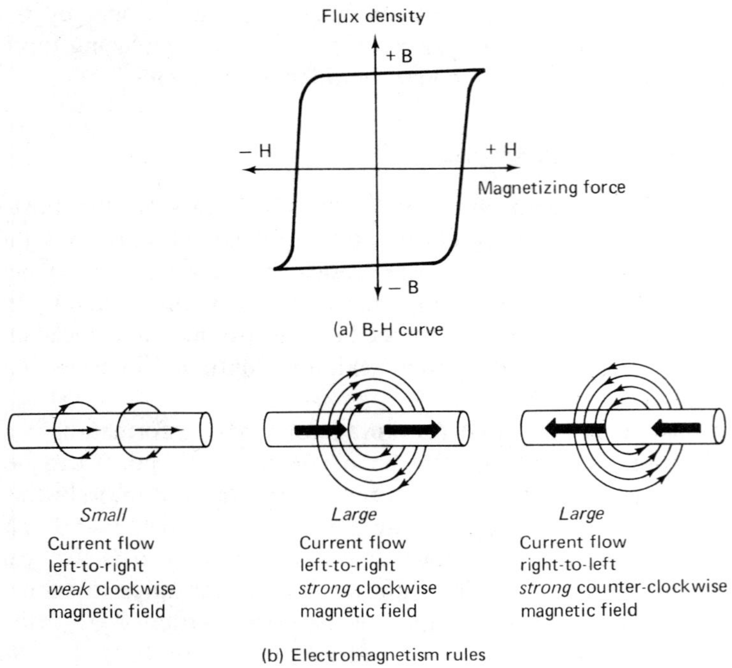

Figure 9-14 Magnetic Core Memory Concepts: (a) B–H Curve; (b) Electromagnetism Rules; (c) Magnetic Cores and B–H Curve

current in the wire will cause very little or practically zero increase in magnetic flux. This point is labeled *a* on the *B–H* curve.

Now a strange thing happens. As the magnetizing force is returned to zero (*b* and *c*), the magnetic flux does not return to zero. In fact, the change in magnetic flux from the saturating magnetizing force back to zero magnetizing force is barely perceptible. In other words, the core has "remembered" the direction of current in the wire, despite the fact that the current has returned to zero. If the current in the wire is now increased in the opposite direction (*d* to *e*), no change in magnetic flux in the core occurs until sufficient magnetizing force (*e*) is applied to cause the core to switch to the opposite direction of magnetic flux (*f*). Returning the current in the wire to zero (*g* to *h*) now leaves the core magnetized in the opposite direction to that previously held, and the core once again remembers, but this time something different than before.

Thus, the core is capable of storing the fact that the current in a wire through the center of the core has been in either one direction or the other by retaining a direction of magnetic flux that is dependent on the direction of the center wire current. And, since the current direction can be only one direction or the other, the core stores *binary* information. For discussion

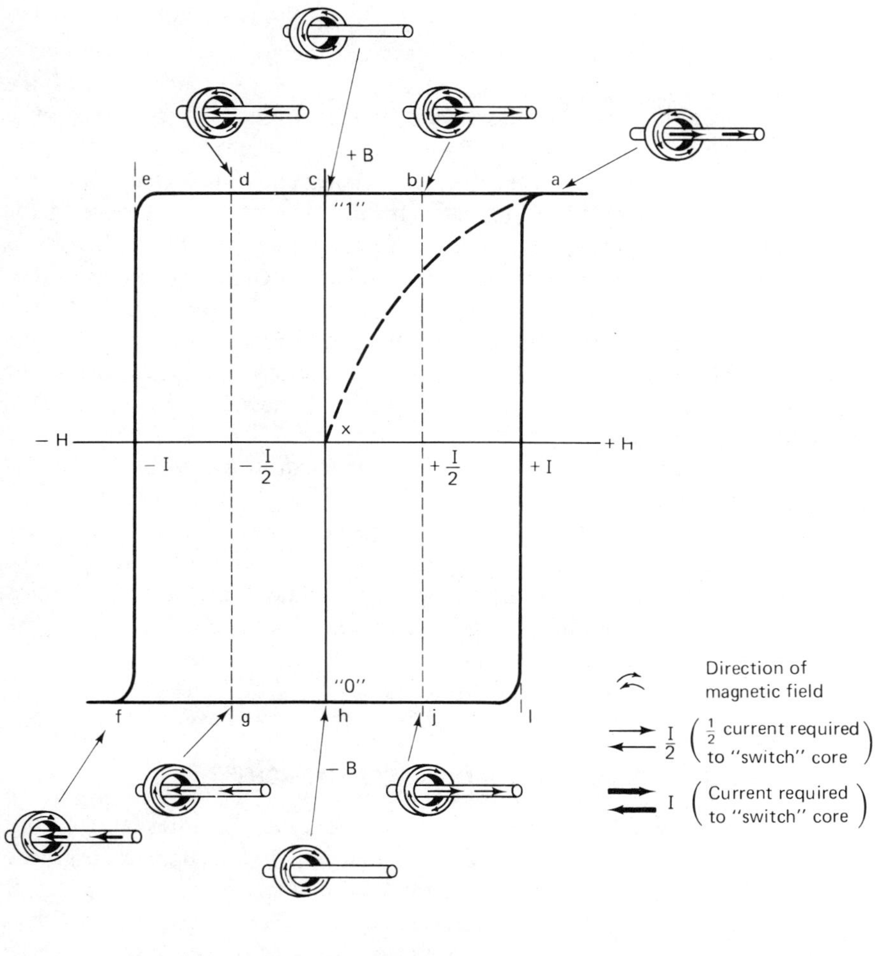

(c) Magnetic cores and the B-H core

Figure 9-14 (Continued)

purposes, when the core is at the $+B$ portion of its curve, storage of a binary 1 is assumed. At the $-B$ portion, a binary 0 is assumed to be stored. The important aspects of the B–H curve and storage of binary information are thus summarized in Figure 9-14c.

Now that binary information has been stored in a magnetic core, how can it be retrieved? Another fundamental property of electromagnetic circuits is applied. When a magnetic field collapses or changes direction, any wires in the vicinity will have a current induced in them. Actually, a flow of current cannot exist without an attendant magnetic field, and vice versa (a changing magnetic field cannot exist without causing a flow of current in adjacent wires). Another wire, in addition to the wire placing information into the core, is threaded through the core. The second wire is used to sense any change that might occur in the magnetic field of the core. A commonly employed method to determine the contents of a magnetic core is to write a 0 into the core using the input wire. If a zero is *already* in the core, very little change in magnetic flux occurs, and the sense wire detects very little induced current. However, if a 1 were in the core, the core would switch direction of magnetic flux when the 0 was written. The sudden change in flux direction would cause a large current to be induced into the sense wire, and it is then known that a 1 had been written in the core. Unfortunately, this method of reading information in a magnetic core destroys the information. In modern core memories, the information that was in the core is temporarily stored in a register and written back into the core via the input wire.

Magnetic cores are organized in matrix form in the same manner as the semiconductor cells. One wire is placed through each core in a row, while a second wire is threaded through the columns. Thus each core has two input/output wires, which represent row and column intersections, and can be addressed in the same manner as semiconductor cells. A third wire is added to sense changes in magnetic flux so that information may be recovered from each cell. Decoders, amplifiers, and buffers are used in a manner similar to those already discussed.

9-5 A TYPICAL MICROCOMPUTER MEMORY

Hardware. Now that the two basic types of internal memory found in the microcomputer have been discussed, it is time to organize these memory chips into a usable storage subsystem. A microcomputer, if it is to operate efficiently, must have both ROM and RAM. ROM is required, of course, to store information that is to be used a large percentage of the time. Bootstrap programs to assure that the microcomputer starts in a known configuration when power is applied, assemblers and compilers to allow the user the freedom of nonmachine language programming, and routines to operate input/output devices all fall into the category of candidates for ROM storage applications. External data (information brought in from the outside world), results of calculations and data movement, information to be printed, and so on are stored in RAM.

Many microcomputers are organized around plug-in modules. The user determines how much memory capacity is required, for example, and plugs in as many memory boards or modules required to gain that capacity. A typical $4K \times 8$ memory module was shown in Figure 9-8, and it was noted that expansion to $8K \times 8$ or even $16K \times 8$ was only a matter of adding more rows of $1K \times 8$ chips or using 2K chips in place of 1K chips for 8K×8 capacity and 4K chips in place of 1K chips for 16K×8 capability. Additional address decoding is necessary, of course, as memory capacity is increased. In the interest of maintaining simplicity, it is assumed that both ROM and RAM modules are identically organized as 4K×8 modules. Figure 9-15 shows how a typical microcomputer memory system might be implemented.

Referring back to Chapter 6, it will be recalled that the microcomputer is organized around a microprocessor which possess an address bus, a data bus, and a control bus. As a result of program steps, the microprocessor places an address on the address bus, and either waits for information at that address to be supplied to the data bus for a READ operation, or places information on the data bus to be stored at the

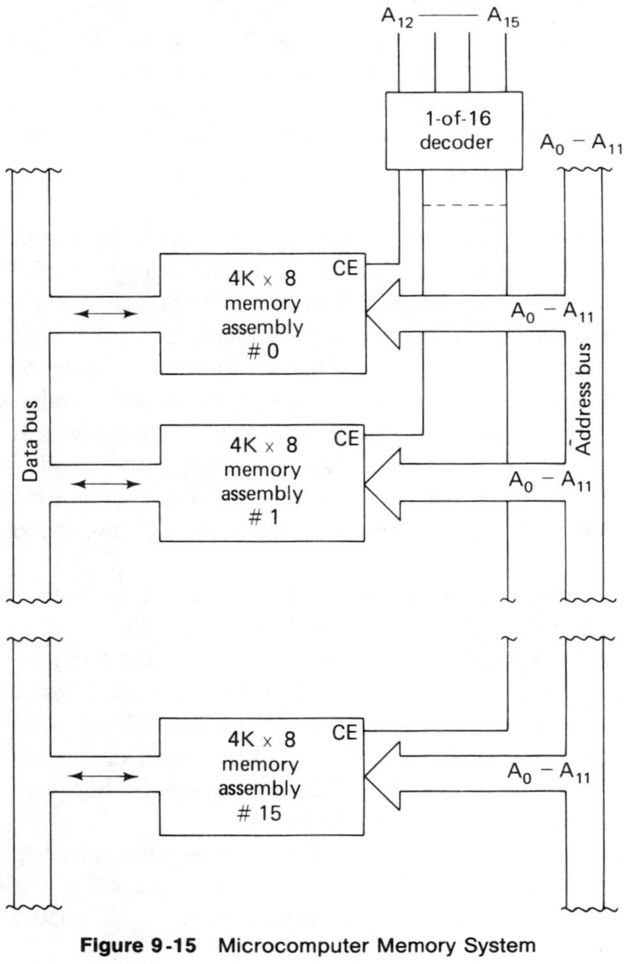

Figure 9-15 Microcomputer Memory System

addressed location during a WRITE operation. READ/WRITE and ENABLE signals are typically found on the control bus. The addressing scheme is one that is compatible with Table 9-1. Assume a total memory capability of 64K 8-bit words.

Address bits A_0 through A_{11} are applied to *all* memory assemblies for selection of 1-of-4096 storage locations on each memory assembly. A separate 1-of-16 decoder is used to activate the CHIP ENABLE (CE) on the appropriate assembly for selection of the desired assembly. Address bits A_{12} through A_{15} are used for the CE function decoding. Since the assumption has been made that both RAM and ROM assemblies are identically organized, either type of memory may be used in any of the 16 assembly locations. The single 1-of-16 decoder does somewhat structure the memory as a whole. A common answer to this limitation is to place a 1-of-16 decoder on each memory assembly and provide on-assembly switches or jumpers so that the address of each board is selectable. All address lines (A_0 through A_{15}) are then routed to each memory assembly and switches or jumpers adjusted to determine the actual address. Using the 1-of-16 decoder on each assembly thus allows assignment of ROM or RAM to any 4K block desired.

Software. The microprocessor instructions associated with internal or high-speed memory use actually extend throughout all the instruction groups discussed in Chapter 7 because of the scheme of considering the complete memory as register 6. It is highly advisable that the reader review Chapters 6 and 7 at this time so that the integration of microprocessor instructions with hardware will be clearly understood. Any instruction that requires access to addressable memory will cause the microprocessor to initiate a timing sequence that either reads information from memory or writes information into the addressed location. General timing requirements were discussed in Section 9-2 and shown in Figure 9-10.

Consider the sequence of events that transpires when the STORE ACCUMULATOR (STA) instruction is executed. The STA instruction is accompanied by a two-byte address that specifies the memory location in which the contents of the accumulator are to be stored. Four subcycles are required to complete the processor cycle (Figure 6-2) associated with the STA instruction. Operations conducted during each of the timing periods for each subcycle that integrate the microprocessor hardware, software, and memory functions are discussed below.

The first subcycle is always used to obtain the instruction from its location in memory so that the op code may be placed in the instruction register of the microprocessor. Each timing period is detailed as follows:

P_1. Program counter contents are placed on the address bus so that the instruction may be obtained. A memory-read cycle is initiated.
P_2. The program counter is incremented while the read operation is being performed.
P_3. Contents of memory at the location defined by the program counter at P_1 are placed in the instruction register.
P_4. Instruction is decoded.
P_5. Not used.

The second subcycle is three timing periods in length, and is a memory-read cycle used to obtain the first byte of address that indicates the location in memory for storage of the contents of the accumulator.

P_1. Program counter contents are placed on the address bus so that the first byte of address may be obtained. The memory-read cycle is initiated.
P_2. The program counter is incremented while the read operation is being performed.
P_3. Contents of memory at the location defined by the program counter at P_1 are placed in a temporary register.

The third subcycle is identical to the second subcycle, except that the second address byte is obtained and placed in a temporary register. The fourth subcycle is a memory-write operation.

P_1. Program counter contents are placed on the address bus so that the first byte of address may be obtained. The memory-read cycle is initiated.
P_2. The program counter is incremented while the read operation is being performed.
P_3. Contents of memory at the location defined by the program counter at P_1 are placed in a temporary register.

Thus, it can be seen that the major part of the operation of the memory of a microcomputer is directed by the microprocessor itself. It only remains for the memory assemblies to be capable of responding quickly enough to react to the demands of the microprocessor. In some cases it may be necessary for the microprocessor to wait for the memory to complete its operations before initiating the next memory cycle. Most microprocessors have a wait state initiated by an indication from the memory that it is busy. However, since this text considers memory assemblies matched to their microprocessors, this function has not been detailed.

9-6 SUMMARY

Internal memory is one of the most important parts of the modern microcomputer. Its capacity determines the *scope* of operations, that is, the level of computer language and amount of data that can be used. Therefore, if the microcomputer is to be applicable to a wide range of applications, as much memory as possible should be included. The increasing density of memory chips promises to further reduce the physical space and power requirements, and smaller, more powerful microcomputers are in the offing. Just as the 4K memory chip gave way to the 16K chip, so will they be replaced by 64K chips.

Memory speed is also being improved as new technologies appear. The former 1-microsecond access time has been replaced by an easily obtained 500-nanosecond figure, and 100- to 200-nanosecond times are not uncommon. Improved access time, of course, facilitates more complex operations in less time.

As RAM/ROM technology advances, more applications appear and microprocessors with integral memory are appearing in many noncomputer fields. The person familiar with the interplay of microprocessor and memory is sure to find many challenges.

QUESTIONS

1. What factors distinguish internal storage from external storage?
2. Define the term access time.
3. What are auxiliary storage devices?
4. Explain the differences between random-access and sequential-access methods of obtaining information from memory.
5. Explain the differences between linear selection and coincident selection of a memory location. Discuss advantages and disadvantages of each.
6. Explain the differences among static, dynamic, and read-only memory cells.
7. Discuss the $X-Y$ decoding scheme used in many RAMs.
8. Can the input/output path be combined with either the X or the Y decoder operation? If yes, explain how.
9. Define the term memory plane.
10. How can a 16-word single-bit memory be expanded to, for example, a 256-word, 8-bit memory?
11. Discuss expansion of a 256-word, 8-bit memory to $4K \times 8$ configuration.
12. Using Table 9-1, determine the group, assembly, and row in which the following addresses are located:
    a. 1010101010101010
    b. 1100110011001100
    c. 0000000011111111
    d. 0011100111001111
13. How can high-density memory chips be implemented on 16- and 18-pin packages?
14. Define the term cell refresh.

15. Why must dynamic cells be periodically refreshed?
16. Explain the differences in operation of dynamic and static memories.
17. Define the terms ROM, PROM, EPROM, and EAROM.
18. Explain the differences in organization between ROMs and RAMs.
19. What is a B–H curve?
20. Explain how binary information is stored in magnetic toroids.
21. Discuss how microprocessor instructions are related to memory operation.

CHAPTER 10

External Memory

As the requirements for large amounts of information storage grow, it soon becomes impractical and not economical to continue using high-speed, internal memory systems. Minicomputers and full-scale digital computers offer many choices for external memory, but in order to maintain the cost and size advantages of microcomputers, only a few of the multitudes of mass storage devices are employed. Most microcomputer users employ some form of magnetic recording media in which the surface of the medium moves past read and recording stations. In some cases both the surface and the read/recording stations move. In still other cases, no moving parts are needed, and some of the more modern mass storage systems use advanced semiconductor techniques. In this chapter, only those mass memory devices particularly effective for use with microcomputers will be examined.

10-1 PRINCIPLES OF MASS STORAGE DEVICES

A Universal Memory. Large-scale storage devices may be approached as a separate system, even though they may be directly linked with a microcomputer. The link is most often software (programming) rather than hardware, since the organization of the storage devices tends to differ radically from that of the microcomputer. Most of the devices that store large masses of information have many principles in common and can be viewed as shown in Figure 10-1. The *storage medium*, most often a magnetizable surface, retains the information presented to it by the *write heads*. Conversion of the electrical representation of microcomputer information to a form that will activate the storage media is the function of the write heads. The *read heads* convert the stored information back to electrical form. A *drive* provides the motive force to physically move the storage media. The *controller* is a multifunction device, formatting and converting data from one form to another, checking for errors, maintaining constant drive speed, and so on.

Although some large-scale memory devices use other than magnetic techniques, by far the majority depend on some aspect of magnetism. Semiconductor memories have already been discussed, but, as noted, large-scale storage capability has been too expensive in the past using this method. At the expense of access time, less expensive magnetic techniques may be used. Practically all of the large-scale magnetic storage devices use a common recording/reproducing technique. It should be noted, however, that cost reductions in semiconductor memory manufacturing are reducing the cost gap between magnetic and semiconductor large-scale memories.

A magnetic material that has not previously been magnetized may be considered as made up of millions of tiny magnets randomly scattered throughout the material (Figure 10-2a). The random arrangement results in the magnetic fields cancelling each other. If, however, an external magnetic

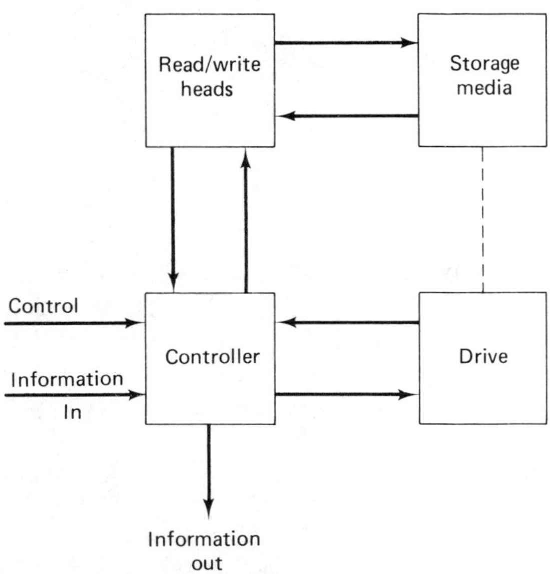

Figure 10-1 General Large-Scale Memory Device

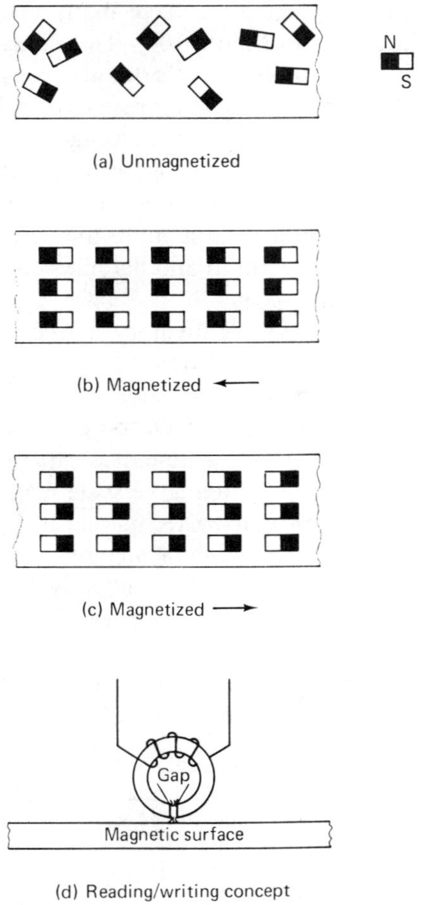

Figure 10-2 Recording Data Magnetically

field is applied to the material, the tiny magnets align themselves with the field, and the material becomes "magnetized." The direction of magnetization (Figure 10-2b and c) depends on the direction of the external field, which is physically designed to be one direction or the other; that is, it is binary in nature. Information may then be recorded by controlling the direction of the external magnetic field.

Information may easily be recorded on the magnetic media. A magnetizing force can be generated by causing electrical current to flow through a wire, since a magnetic field surrounds a wire in which current is flowing. The strength of the magnetic field is a function of the amount of current flow, while the direction of the magnetic field is a function of the direction of current flow. Furthermore, if a number of turns of wire are wound around a circular magnet, the magnetic field due to current flow in the wire will transfer to the magnet. The strength of the induced magnetic field is determined by the amount of current flow in the coil of wire, and the direction of the magnetic field is a function of the direction of current flow. If the *magnetic* path is opened by cutting a gap in the core, the field is not maintained when the electrical current causing the field is removed. However, the presence of magnetic material near the gap completes the magnetic path and the magnetic material becomes magnetized at the spot immediately adjacent to the core gap. Removal of the electrical current stops the magnetic field in the core, but the magnetic material near the core is now magnetized and information is stored (see Figure 10-2d). The physical configuration of the core (more commonly called the *head*) varies greatly, but all have the general properties mentioned above.

Data Recording Methods. In all magnetic storage devices information is stored by changing some characteristic of the storage medium and is retrieved by sensing that change. Many techniques exist, and only the most common methods are discussed. NRZ (nonreturn-to-zero), PM (phase modulation), and FM (frequency modulation) are the most popular techniques used in magnetic data recording devices.

As discussed earlier, passing current pulses through wires wound on the cores of recording heads causes the recording medium to become magnetized in a direction determined by the direction of the current. Figure 10-3a shows a typical bit sequence and the resultant WRITE head current in a technique called RZ (return-to-zero). RZ techniques are explained only so that groundwork may be provided for the more common methods of data recording and reproduction. Note that current directions are opposite ("+" for a binary 1 and "−" for a binary 0) and that no current flows between pulses. Thus, the surface is magnetized in a small area for each bit, and a nonmagnetized area exists between bits. Since the current causes an area to be magnetized, the current waveforms also represent direction of magnetization for each recorded bit.

A principle of electromagnetism states that the current that is induced in the winding of the READ head is proportional to the *rate of change* of the magnetic field. Furthermore, the polarity depends on the direction of the magnetic field and whether it is increasing or decreasing. As a magnetized area on the surface of the recording medium moves under the READ heads, a pulse of current is induced in the winding (See Figure 10-3b). The polarity of the pulse is determined by the magnetic field direction (1), and once the magnetic field is no longer changing, the pulses cease. No output is present from the READ head until the magnetized area moves away from the head. Since the field is now *decreasing*, a pulse of the opposite polarity appears at the READ head (2). Between points 2 and 3 there is no change in magnetic field, and the READ head output is

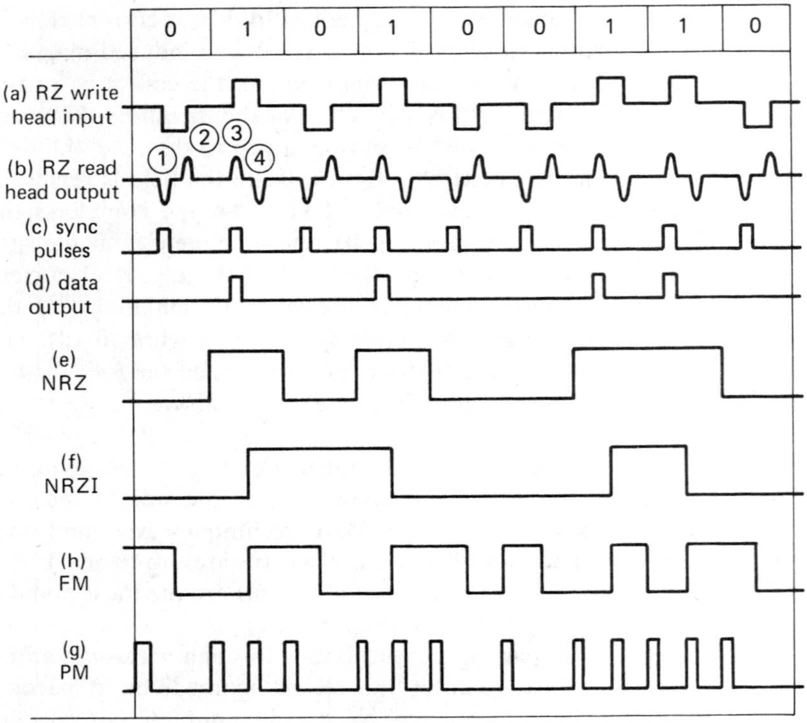

Figure 10-3 Data Recording Methods

zero. The recorded magnetic pattern between 3 and 4 is opposite to that at 1 and 2, and the READ head output is an opposite pulse. Note that *both* a positive-going and a negative-going pulse appear for *each* stored bit. It is necessary to supply some form of synchronizing pulse, perhaps from a timing track, that assures sampling of the READ output at the time the bit location *starts* moving under the READ head. Figures 10-3c and d complete the waveform drawings for RZ recording/reproduction techniques. All other techniques use the same principles of recording/reproduction, and only the input waveform to the WRITE head will be shown.

RZ methods have given way to NRZ, PE, and FM techniques, all of which have many variations. The basic characteristic of all NRZ techniques is that the recording medium is *always* saturated in one direction or the other. Each variation is based on the interpretation of the *change* in magnetic field (flux transitions). In normal NRZ recording a flux transition indicates a change in bit significance—that is, 1 to 0 or 0 to 1. Groups of 1s or groups of 0s do not cause a flux transition within the group. Figure 10-3e shows that the NRZ code records binary 1s with a positive pulse during the entire bit period. NRZ recording is quite efficient in terms of number of flux transitions, since only one transition is required per bit. However, since no flux changes from bit to bit indicate no change (i.e., if a 0 follows a 0, the pulse does not change), it is difficult to distinguish between the *intentional* absence of a signal or an accidental data dropout. Furthermore, the NRZ code is not self-clocking, since flux transitions do not occur at regular intervals.

NRZI (inverted) uses a flux transition to indicate a 1; lack of a flux transition indicates a 0. The waveforms of Figure 10-3f portray NRZI operation. Binary 1s are recorded by a flux reversal at the center of a bit period, while 0s appear as no flux reversal. NRZI is also efficient in terms of flux transitions since it requires only one transition per bit. Self-clocking is also a problem with NRZI, owing to the irregularity of flux transitions.

Both NRZ and NRZI are relatively easy to implement from a circuitry viewpoint but must be supplemented with circuitry and/or space on the storage medium for clocking information. PM, phase modulation (Figure 10-3g), contains its own clocking information, but suffers from the requirement to use two flux reversals per cell in some cases. In phase modulation, a binary 1 is recorded as a positive flux change at the center of a bit while a 0 appears as a negative change. Furthermore, the converse of this rule is applied if and only if the next bit is the same as the preceding bit. During readback, a positive pulse at the beginning of the bit indicates that the preceding period contained a 0. A negative pulse indicates the preceding bit contained a 1, and if no pulse appears, the present bit contains a bit opposite to the preceding bit. It is thus seen that perhaps the advantages of self-clocking are outweighed by the complexity of circuitry required to implement phase modulation.

Frequency modulation (FM) techniques record a 1-bit as a flux reversal at the center of a bit and a 0-bit by a constant polarity during the entire bit. A flux reversal is always initiated at the beginning and end of each bit. Therefore, during readback two pulses during a bit signify a 1-bit, while a single pulse signifies a 0-bit. (The 1-bit frequency is *twice* the 0-bit frequency; hence, FM is sometimes called *double-frequency coding*.) Since a pulse always appears at each bit beginning, clocking is greatly simplified. Two flux reversals per bit may be required, however, and maximum density suffers with FM recording methods. Figure 10-3h shows FM compared with other recording techniques.

New coding and recording methods are constantly appearing, but generally those that allow maximum density (minimum number of flux reversals per bit) result in increased circuit complexity. It should be noted, however, that LSI manufacturing techniques are resulting in complete coding and decoding circuits in a single integrated circuit, and that complex circuitry may soon be an unimportant factor.

Error Detection and Correction. Any movement of information from one location to another is a potential source of error, and if highly reliable data movement is a requirement, some method of error detection and correction must be provided. Three methods of error detection are in common use: parity, checksum word, and cyclic redundancy check (CRC).

One of the common and relatively inexpensive methods of ensuring data-movement accuracy is the addition of a standardizing, or *parity*, bit at the end of each byte. It is arbitrarily determined that all data bytes will contain either an even or an odd number of 1s. If the standard, or parity, is an *even number of 1*s, it is said that *even parity* is being used. If an *odd number of 1*s is required to meet the standard, *odd parity* is the name applied to the standardizing system. Prior to movement of a data byte, the number of 1s in the byte is determined. For an odd-parity system, if the number of 1s is even, a 1 is placed in the parity position (usually a ninth bit position). When received, the number of 1s is again determined. If an odd number of 1s is present, it is assumed that the data word has been moved correctly. If an even number of 1s is detected, a warning is provided that the transferred data is *not* correct.

It should be noted that this type of standardizing or error-detection system is only effective in detecting an odd number of 0s which have changed to 1s—that is, 1, 3, 5, and so on. Furthermore, parity systems tend to be inefficient, since an extra bit position is required. The extra bit results in over 11% of the data-storage capability being used for error detection.

The *checksum* error-detection technique sums all the data in the block being checked, byte by byte, using modulo-2 addition. (*Modulo* or *modulus* is a term that describes the maximum number that can be used in the system. For example, modulus 10 has available only the digits 0 through 9, while modulus 2 has only 1 and 0. Thus modulo-2 addition sums each digit but ignores the carry.) Space is provided at the end of the block of data being checked to store the resultant checksum. During a read operation the data are checksummed and the result is compared with the checksum stored as part of the data block. If the two checksums do not agree, an error exists. Large blocks of data may generate relatively long checksums and require excessive extra storage space.

Although *cyclic redundancy checking (CRC)* is a bit more complex, it provides a highly reliable method of error detection. Any block of data may be considered as a binary polynomial of degree N, where $N+1$ is the number of data bits in the block. (A *polynomial* is merely a general definition for algebraic/numeric expressions containing two or more terms. For example, 10101 is a fourth-degree polynomial that may also be expressed as X^4+X^2+1.) When the block of data is divided by a known polynomial, a quotient *and* remainder will result. The remainder will be one degree *less* than the divisor, and will be a unique number for any given data block. Therefore, if a 17-bit divisor is used, the remainder will be a 16-bit number called the *cyclic redundancy check character* (CRCC). The 16-bit CRCC may then be stored in a two-byte location for later use. Example 10-1 shows a typical CRC polynomial division using the standard CR-16 polynomial $X^{16}+X^{15}+X^{2}+1$ as the divisor. A 17-bit polynomial may be used to detect single-bit and most multibit errors in blocks of data up to 4096 bits long. CRC is a relatively standard error-detection system for external storage devices. It may be implemented in either software or hardware, and since many manufacturers supply CRC capability on LSI chips, hardware implementation is common.

EXAMPLE 10-1

```
                   Quotient                   Remainder
                    1100 · · · · 100001  R  0001011000000110
                                            \______________/
11000000000000101 |1110 · · · · · 100011         (CRC)
(CR-16 polynomial)
```

10-2 DISK MEMORIES

Disk memories store information in concentric tracks on rotating disks coated with a magnetizable surface. The disks may be coated on one or both sides, rotating in either the horizontal or vertical plane, and be permanently mounted or removable. Information is recorded/reproduced by a single head per track or by movable heads that select the appropriate tracks. Minicomputers and large-scale digital computers use disks that provide bit densities greater than 4000 bits per inch, have access times that average as low as 2.5 milliseconds, and rotate at speeds up to 6000 rpm. Each surface of the disks is capable of storing well over 10 million bits. Microcomputer requirements, however, are not as stringent, and much smaller, less expensive disks and drives are employed.

The Storage Medium and Drive. Magnetic recording techniques are used in disk memories. Whereas the larger disks are rigid in nature to accommodate the high-volume and high-precision requirements, most small-scale disk memories employ the smaller "floppy" disk. The relatively standard floppy disk consists of a 0.003-inch-thick oxide-coated Mylar disk permanently enclosed in an 8-inch-square protective jacket. Openings in the jacket allow access for the recording/reproducing heads, index holes, and drive spindle mounting. Refer to Figure 10-4 for details.

Disk rotation is accomplished by physically mounting the disk to a spindle. The spindle is driven either by a belt and pulley arrangement or directly from a drive motor. Recording/reproducing heads are positioned

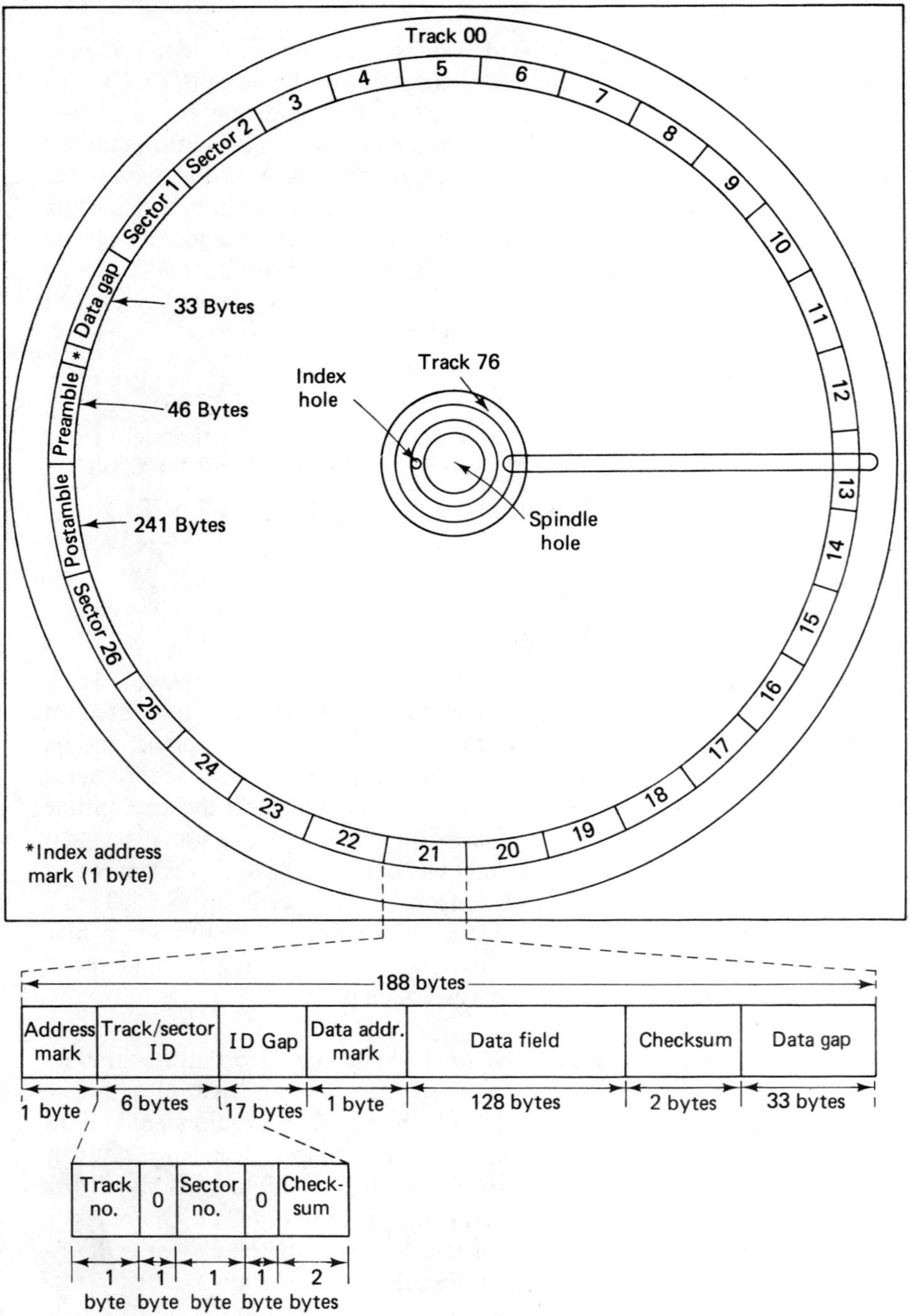

Figure 10-4 Floppy Disk Format

over the appropriate storage location by means of high-accuracy mechanical drives. One method employs heads mounted on a carriage which is moved by the rotation of a lead screw. The lead screw is driven by a stepping motor which rotates a fixed number of degrees every time it receives a pulse. Lead screw pitch is chosen so that each motor step translates the head carriage motion into the track-to-track radial distance of 0.0213 inch (standard track spacing). In normal operation the heads are retracted to track 00 and the necessary number of pulses supplied to the stepping motor to move the heads to the desired track.

It should be noted that the heads do not usually contact the recording surface of the disk at all times, so that minimum wear will occur both on the disk and the heads. A pressure pad may be actuated to place the disk in contact with the heads during read and write activity, and to retract the heads during nonactive times.

Linear actuators may also be used to position the heads over the desired track, offering a possible advantage of more precise positioning and the experience gained from the larger disk drives. A linear actuator moves a precise distance proportional to the electrical current applied. Since the current may be adjusted in very small increments, tracks may be recorded with synchronizing information to cause *exact* head positioning. Electromechanical feedback is also used with linear actuator-driven head assemblies for precision positioning. Other innovative techniques exist, such as stepping-motor-driven cams and metal bands wound on stepping-motor-driven drums. Linear actuators and stepping-motor-driven lead screws tend to dominate the field, however.

The recording surface of the floppy disk is divided into a number of concentric *tracks*. Each track is divided into *sectors* where information is stored in a predetermined format. Information is thus identified by track and sector, as shown in Figure 10-4. The standard floppy disk has 77 tracks, with either 26 or 32 sectors on each track when information is stored in *formatted* form. Although upward of 400,000 8-bit bytes may be stored on one side of a floppy disk, data-storage-reliability requirements may reduce that capability to less than 250,000 bytes. Variations of the IBM standard format shown in Figure 10-4 are widespread, but the IBM standard is shown in an attempt to explain one widely accepted method of information storage on floppy disks.

In the IBM standard, each track is identical, containing a *preamble* and *data gap*, 26 *sectors*, and a *postamble*. The track begins with the first bit of the preamble and is identified by detection of the leading edge of the *index hole*, which is located in the center of the disk near the spindle hole. Four of the 77 tracks are reserved for housekeeping functions and spare tracks in the IBM standard, leaving 73 tracks available for information storage. The preamble and data gap prepare the mechanics and electronics of the disk drive for the read or write operations. Each of the sectors is detailed in Figure 10-4. Of the 188 bytes in each sector, only 128 bytes are used to store information. The other 60 bytes provide additional housekeeping functions. For example, the first byte in the sector is used as a mark to prepare the system to read the address information in the next six bytes.

The track/sector identification section uses one byte to identify the track, a 0 byte for spacing, one byte for sector identification, a 0 byte for spacing, and two bytes for a CRC checksum indication. A 17-byte data gap follows the track/sector identification portion of the sector information. The data gap provides a delay between the read operations performed in the track/sector section and the write operations to follow in the data field. A data address mark follows the data gap and supplies synchronization for the following data field. A total of 128 bytes of data may be stored in the data field, followed by a two-byte CRC checksum for error detection. The 33-byte data gap following the data field once again delays reading or writing to allow time for head function switching, head movement, and so on.

Normal rotational speed of the disk is about 360 rpm, which equates to 167 milliseconds for one complete rotation of the disk. Head movement requires about 10 milliseconds per track. Thus, it could take up to 1 second to find and read a specific record in the worst-case condition. Average time is much closer to $\frac{1}{2}$ second. To spend the minimum time to locate specific information, some sort of record is maintained in software so that the exact track and sector numbers of the desired data are known.

The *soft-sector* format just discussed is not the only method allowed. Sectors may also be identified by additional index holes punched directly on the disk (*hard sectoring*) to locate the beginning of each sector. This technique removes the requirement for much of the sector addressing and extends the sector capacity to 32 per track without sacrificing data reliability. If one is willing to risk a small amount of reliability, data may be stored in even longer data fields (fewer, but longer, sectors). Reliability is somewhat related to how often checking is performed, and longer sectors with less checking may reduce that reliability. The reader should refer to manufacturers' instruction manuals for the specific disk and drive in use to obtain actual sectoring and data-field information.

Disk recording is not limited to a single side, nor is the physical size necessarily restricted. Both sides of the actual disk may be coated with magnetic material, and with the addition of a second read/write head, additional storage space and reduced access time is available. Smaller disks and drives are also used for special-purpose applications where stringent standards cannot be observed.

Controller. The disk drive controller contains the electronic circuits that interface the serially recorded information on the disk to the parallel-organized computer. Figure 10-5 shows a typical disk drive controller. Data placed in the controller in parallel form are converted to the required serial form (after buffering) by the parallel-to-serial converter. Likewise, data from the disk (in serial form) are converted to parallel form by the serial-to-parallel converter and supplied to the microcomputer via the data buffers. To assure data accuracy, a cyclic redundancy check (CRC) is performed by the R/W control logic. CRC is discussed in Section 10-1.

The R/W control logic and drive control logic blocks function together to control the mode of operation (READ or WRITE) and to properly position the R/W heads so that data may be manipulated. Using such information as "drive ready," "track position," and "index detected," from the status buffer, the drive control logic informs the microcomputer

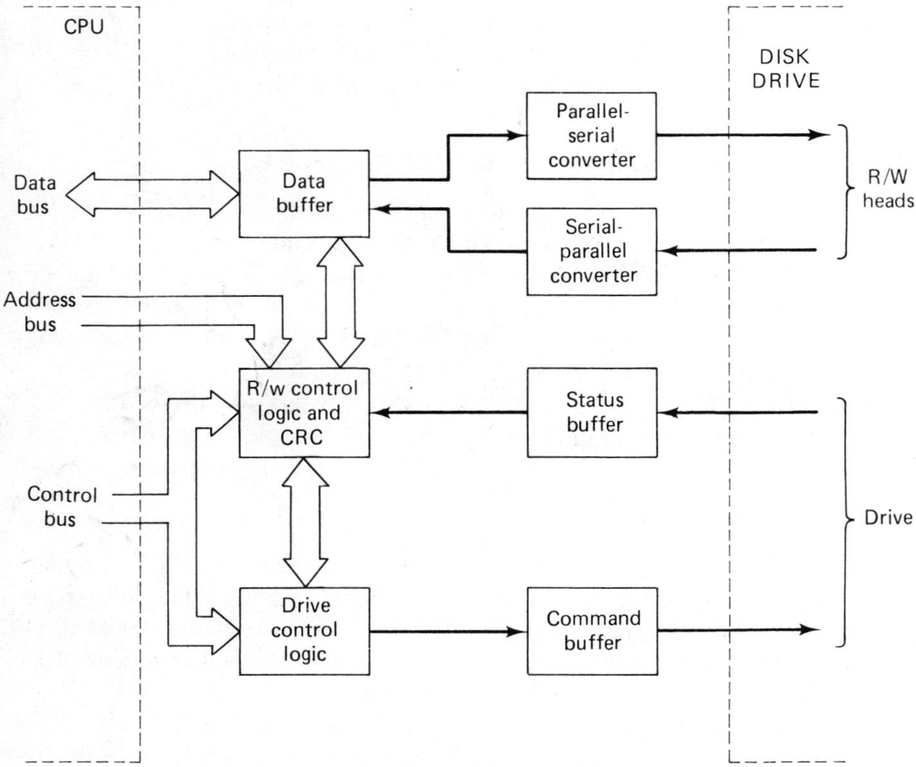

Figure 10-5 Typical Disk-Drive Controller

of the drive status. The command buffer uses control information from the drive control logic to select appropriate drives, position and enable R/W heads, and control the amount of current used to write information on the disk. Information is interchanged between the R/W control logic and the drive control logic so that accurate positioning and operation of the heads is assured. The address bus supplies information to the controller to select the appropriate drive if more than one disk drive is to be used.

Operation of the controller is generally as follows. The microcomputer supplies a byte to the controller, which indicates a general description of the type of operation to be performed. In many cases, additional bytes indicating more detailed information may be required, and they follow the initial command byte. When the last command byte has been received by the controller, it proceeds with implementation of the command and the microcomputer may go on to other tasks. When the command has been completed, the controller initiates a status byte to the microcomputer to indicate the outcome of the operation so that the next command may be issued. As may be imagined, software requirements can be quite complex when floppy disks are used with microcomputers. Such software is beyond the scope of this book, and the reader should consult the manufacturer of the specific floppy disk being used.

10-3 MAGNETIC TAPE CASSETTE MEMORIES

Information may be stored in serial fashion on moving Mylar tape coated on one side with a magnetic surface. The recording/reproducing heads are usually fixed in position, and the tape is moved across the face of the heads. Minicomputers and large-scale computers use reel-to-reel tape machines with data densities on the order of more than 4000 bits per inch and tape speeds from 1.875 inches per second to 200 inches per second. Each 2400-foot $\frac{1}{2}$-inch-wide tape stores millions of bits of information. Microcomputer requirements, however, are less demanding, and smaller, less expensive tape recording/reproducing devices such as magnetic tape cassettes are used.

The Storage Medium and Drive. The magnetic recording/reproducing techniques discussed earlier in this chapter are used in tape cassette information storage applications. Magnetic tape cassettes usually contain about 300 feet of 0.15-inch-wide tape permanently mounted on storage and take-up reels, and resemble the common audio tape cassette.

Most cassette recording/reproducing devices use fixed head(s) and the tape is moved across the head(s). Tape movement varies from relatively primitive techniques in inexpensive cassette devices to complex drive and speed control in the more expensive units.

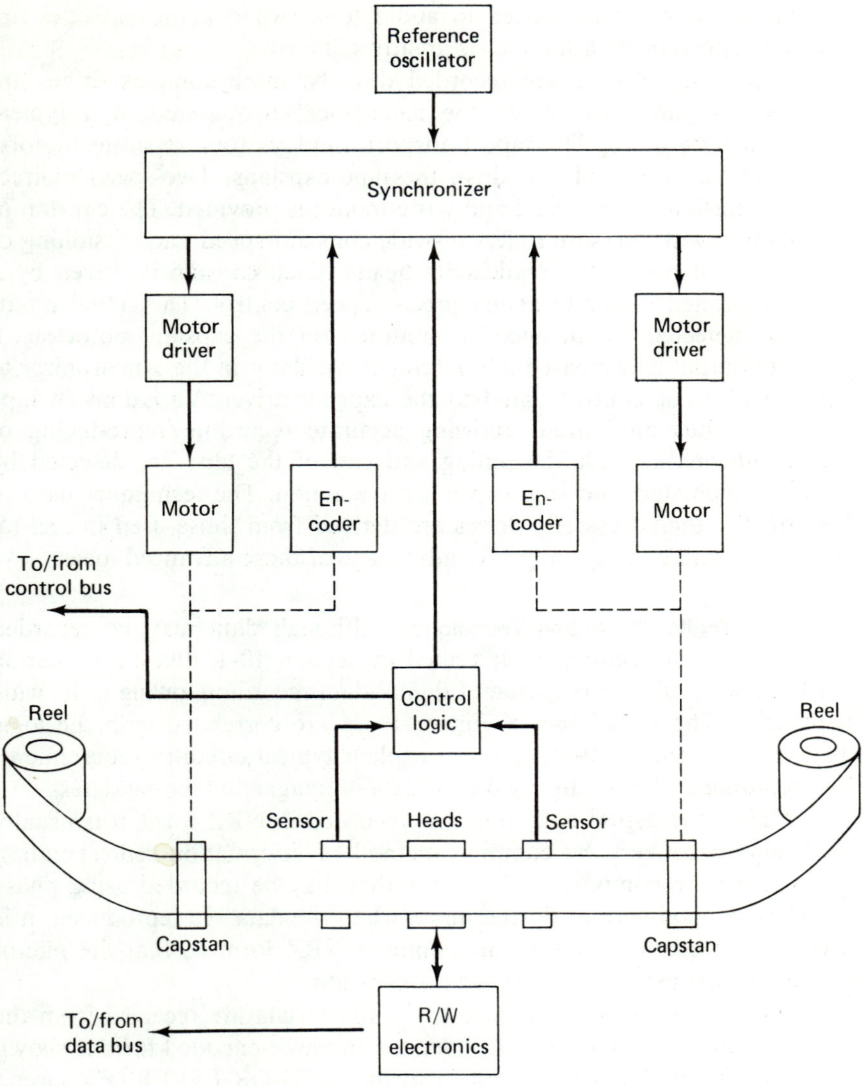

Figure 10-6 Digital Cassette Drive

Inexpensive units are like audio cassettes, and audio cassettes are in fact used in many low-cost information storage devices. The tape merely passes by guides that hold the tape against the heads under tension. Belt and pulley or direct spindle drive from a motor is used to rotate the take-up reel. These units are most often *not* speed-controlled, and speed variation-tolerant recording/reproducing methods must be used. Digital information is often converted to audio tones when audio cassettes are used and speed of data transfer is greatly reduced.

When digital data are recorded directly, more complex drives are necessary. Figure 10-6 shows the mechanics encountered in a typical digital cassette drive. The tape transport employs four separate motors: two drive the reels and two drive the tape capstans. Two-speed bidirectional operation in both read and write modes is provided. The capstan in conjunction with pressure rollers provide constant speed and tensioning of the tape as it passes the read/write heads. Each capstan is driven by a servo-controlled motor to assure precise speed control. The actual motor speed is detected by an encoder mounted on the capstan motor shaft. Encoder output is compared to a reference oscillator in the synchronizer to develop a motor control signal to the capstan driver. Variations in tape speed are thus minimized, allowing accurate recording/reproducing of digital information. The beginning and end of the tape are detected by sensors which start and stop capstan drive action. The techniques used in high-quality digital cassette drives are derived from those used in reel-to-reel drives, and consequently take advantage of those advanced ideas.

Digital Recording Techniques. Although data may be recorded using any of the techniques described in Section 10-1, phase modulation has been selected for discussion of digital data recording, owing to its wide popularity. The waveforms of Figure 10-7a are correlated with the logic diagrams of Figures 10-7b and c to explain typical circuitry requirements for recording and reproducing digital data on magnetic tape cassettes.

Unless the digital data are to be recorded in NRZ form, it is usually necessary to convert the common method of information representation within the microcomputer into a form that may be recorded using phase encoding or modulation. Furthermore, when the data are reproduced, it is necessary to convert back to the common NRZ form so that the microcomputer may use the reproduced information.

As the waveforms of Figure 10-7a show, data are received from the microcomputer in NRZ form. Conversion to phase-encoded form is shown in Figure 10-7b. Timing is supplied to the AND-OR-INVERTER circuit (encoder) along with the NRZ data from the microcomputer. The output of the encoder is a phase-encoded representation of the NRZ input. Power amplification is supplied to furnish adequate current to drive the write head into saturation in both the north and south directions. It should be noted that a high-to-low transition of the phase-encoded data causes a transition of the magnetic field on the tape from north to south, while a low-to-high transition results in a south-to-north magnetic field change.

As the tape moves by the read head during reproduction, only the magnetic field transitions are detected. Therefore, the output of the read head will be a pulse, negative for a north-to-south transition and positive for a south-to-north transition. The output of the read head is filtered to remove extraneous noise and amplified sufficiently to easily identify the positive and negative peaks. *Threshold detectors* supply a high-going pulse of constant amplitude for both positive and negative peaks which exceed a value of signal from the read heads that has a very high probability of resulting from the recorded data rather than noise. A *peak detector* identifies the maximum point and direction of change to supply a gating signal to develop a pulse for each + transition of data and a separate pulse for each − transition. The transition pulses as gated by the *one-shot* set or reset the *data flip-flop* and convert the phase-modulated data to NRZ data. A one-shot triggered by transition pulses reconstitutes the clock data from the recorded data. It should be noted that the $\overline{Q}$ output of the one-shot and the inputs to the data flip-flop are disabled when the one-shot is in the clear state. The time constant that controls the "on" time of the one-shot is selected to mask any transitions occurring at unwanted times, thereby preventing one-shot and data flip-flop changes when no change in data occurs. The method used to recover the phase-modulated data from the recording media results in a $\frac{1}{2}$-bit delay, as may be seen in Figure 10-7a.

When a block of data is recorded in digital form on magnetic tape cassettes, it is organized much the same as on disks. That is, each segment of the tape contains an address and identification section, the actual data section, a means of verifying the accuracy of the data, and a gap to separate data segments. Numerous detailed formats are used, and little standardization is found. The reader is urged to consult instruction manuals for specific equipment under study for actual data formats.

Audio Recording Techniques. Numerous methods of audio-frequency recording of digital data also exist. Such methods are employed when low-cost data storage is required, and as noted earlier, the price paid for low cost is slow access to the stored data. Characters are represented in the standard serial form (1 zero-start bit, 8 bits for character representation, and 2 one-stop bits), allowing use of existing microcomputer software and hardware. A commonly encountered scheme uses eight cycles of 2400 Hz to represent a logic 1 and four cycles of 1200 Hz to represent a logic 0. It should be noted that some systems use more or fewer cycles of 2400 Hz and 1200 Hz, while others employ entirely different, unrelated frequencies. Audio-frequency-represented digital data generally allow the use of existing audio circuits in inexpensive cassette recorders and require only minimal external circuitry. Figure 10-8a is the functional diagram of the circuitry that may be used to convert digital data from the microcomputer

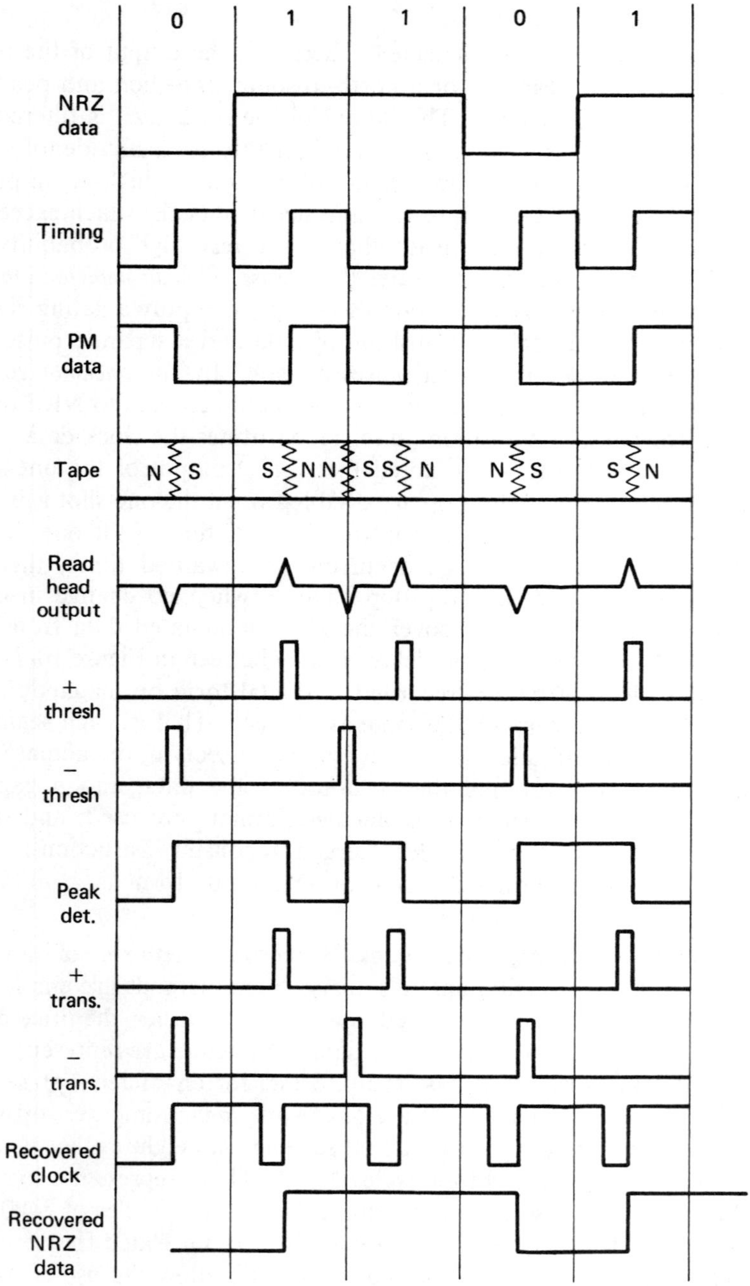

(a) waveforms

Figure 10-7 Digital Data Recording: (a) Waveforms; (b) NRZ-to-PM Conversion; (c) PM-to-NRZ Conversion

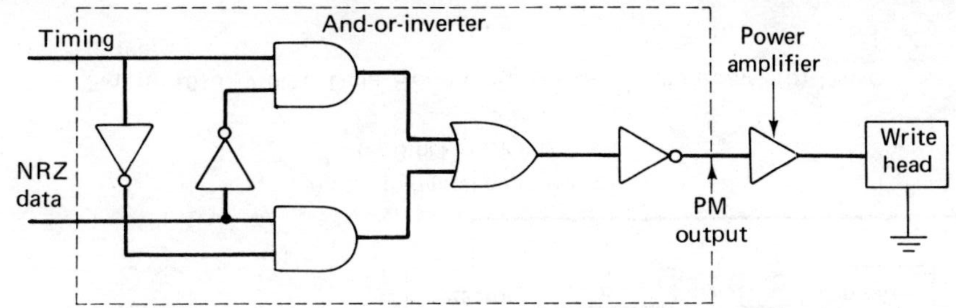

(b) NRZ to PM conversion

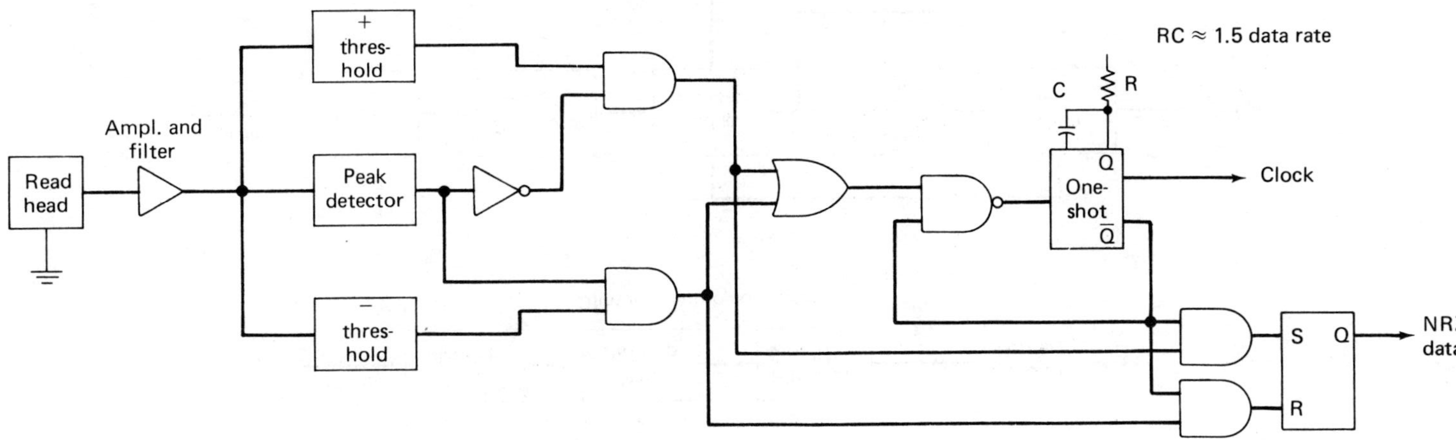

(c) PM to NRZ conversion

Figure 10-7 (Continued)

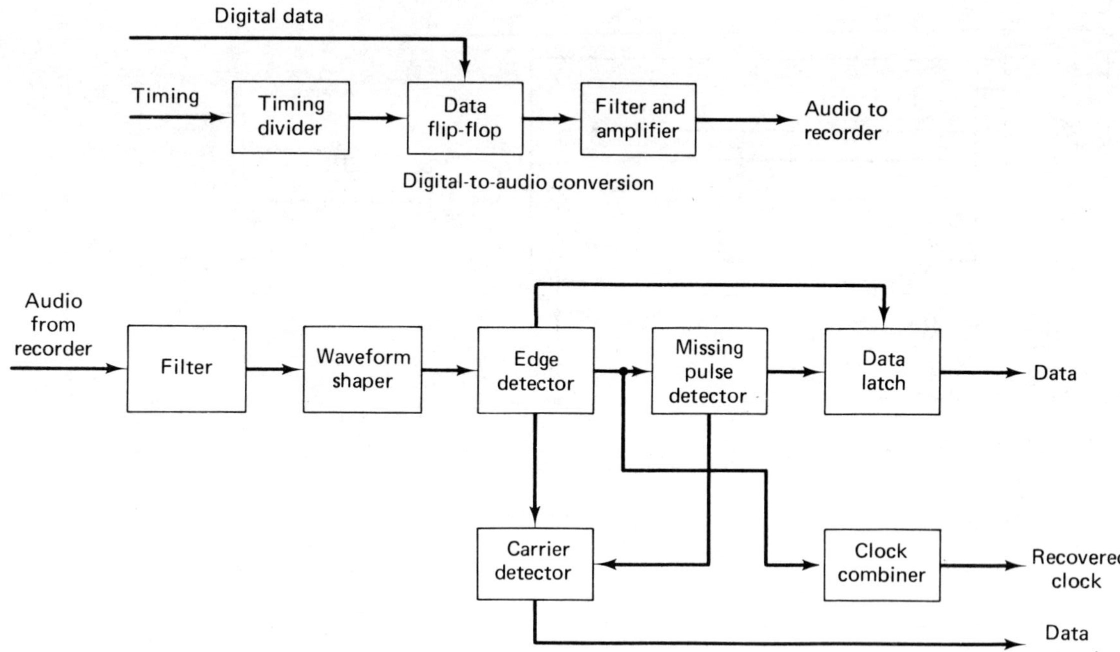

Audio-to-digital conversion

(a) Block diagrams

Figure 10-8 Audio Data Recording: (a) Block Diagrams; (b) Waveforms

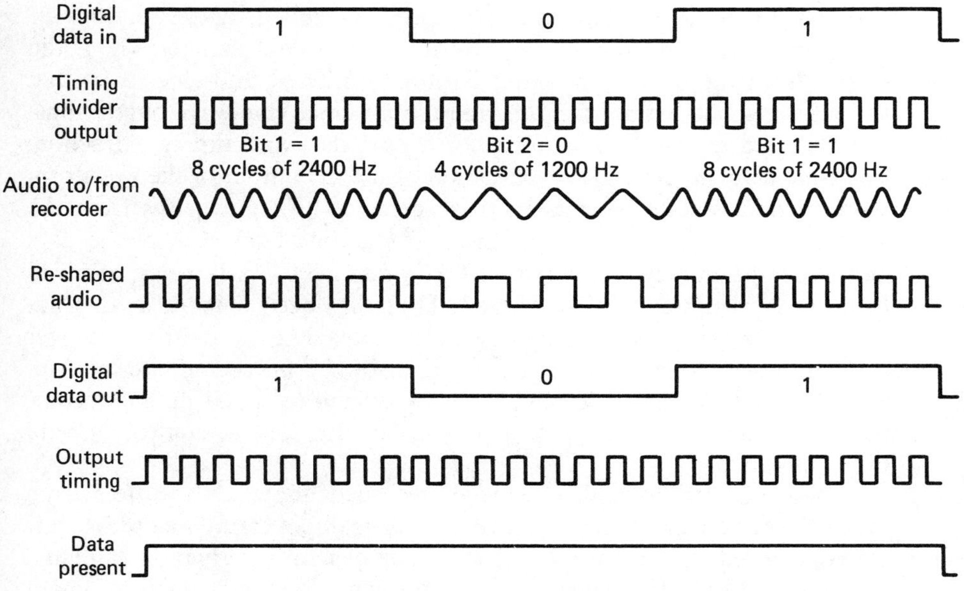

(b) Waveforms

Figure 10-8 (Continued)

to audio frequency for recording, and audio frequency to digital data for data reading into the microcomputer. Waveforms are shown in Figure 10-8b.

The timing divider in Figure 10-8a converts the timing provided by the microcomputer into a 2400-Hz square wave as required by the data flip-flop. When the digital data input is HIGH, the data flip-flop allows the 2400-Hz square wave to go directly through; when the digital data input is LOW, the 2400 Hz is divided by 2, to 1200 Hz. Thus, a logic 1 (HIGH) is converted to 2400 Hz, while a logic 0 (LOW) is converted to 1200 Hz. The number of cycles of either 2400 Hz or 1200 Hz that are output from the data flip-flop is determined by the length of time the data input remains at a specific level, which is a function of the microcomputer. Clock input frequency and data rate must be synchronized and controlled by the microcomputer. The filter and amplifier following the data flip-flop convert the square-wave output of the data flip-flop into sine-wave form for easy processing by the audio section of the cassette recorder.

Data transferred from the cassette into the microcomputer must possess a number of specific characteristics. For example, the data must carry not only their actual data representation but must also supply enough information to recover a timing signal to be used by the microcomputer. Most audio cassettes are not speed-controlled, and variations of transport speed could result in many errors during data transfer. The audio output from the cassette recorder is initially filtered and shaped so that logic circuits may extract the data and timing, since sine-wave output must be converted to square-wave form for proper data and timing extraction. The edge detector generates a pulse every time the output of the waveform shaper changes state. A missing pulse detector uses the pulses from the edge detector to control the triggering of the data latch so that the data output to the microcomputer is HIGH when 2400 Hz is being received from the recorder and LOW when 1200 Hz has been demodulated. Data latch outputs are further combined with the edge detector pulses to recover and synthesize the clock signals to be used by the microcomputer as data are processed. A carrier detector is also provided to signal the microcomputer that information is present on the cassette, and to control cassette drive operation.

Organization of audio-frequency-recorded digital data suffers from much the same lack of standardization as does direct digital recording. It is often recorded in block form with the same general information as digital recordings. The "personalized" nature of many of today's applications of microcomputers results in almost as many formats as there are users. Once again, the reader is urged to consult appropriate manuals for current data.

Controller. The controller used with magnetic tape cassette memories is relatively straightforward. Data modulation/demodulation functions discussed in the preceding paragraphs are probably physically contained with the rest of the functions of the controller, and are represented as such in the diagram of Figure 10-9. It should be noted, however, that serial input/output is assumed for the microcomputer. When such capability is not available, additional circuitry must be provided to convert the parallel input/output of the microcomputer to serial form for use with cassette memories.

Functions of the magnetic tape cassette controller closely resemble those of the flexible disk controller. The specific cassette to be used is selected by an address from the microcomputer's address bus. Read/write control logic decodes the address and, in conjunction with microcomputer directions on the control bus, selects the appropriate drive and either the read or the write function. Control signals from the microcomputer also direct the drive mechanism, turning it on and off at the proper time. The actual operation performed by the drive is returned via the status buffers to aid in control of the read/write function and the drive operation.

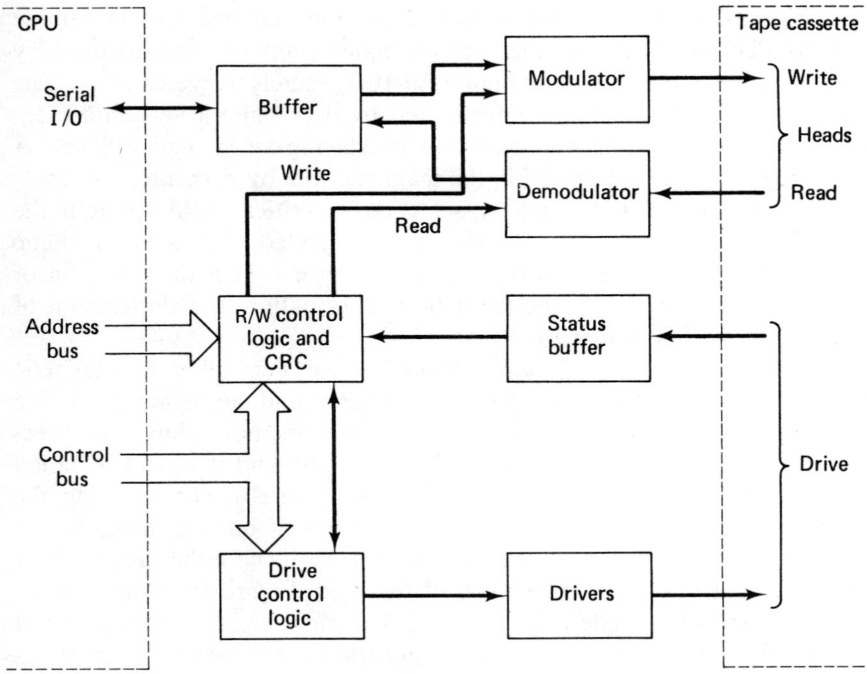

Figure 10-9 Typical Cassette Controller

10-4 MAGNETIC BUBBLE MEMORY

Concepts. Magnetic bubble memory is a fast-growing storage medium offering mass storage capability that is faster, less expensive, and more reliable than most moving-surface magnetic devices. Certain magnetic materials allow the formation and movement of tiny areas of magnetism that are different than surrounding areas in terms of magnetic field. The tiny areas of magnetism are called *domains*, or *magnetic bubbles*. Bubbles exist in certain magnetic materials that are exposed to a fixed magnetic field in the proper direction. By increasing and/or decreasing the fixed magnetic field, bubbles may be formed or annihilated at will. If the existence of a bubble in a known location on the magnetic material is used to represent a binary 1, the nonexistence of a bubble can be used to represent a binary 0. Furthermore, since a bubble may exist anywhere on the magnetic material, it may be moved by application of a moving magnetic field and may be combined with other bubbles to form a *serially oriented* memory device.

Figure 10-10 shows the basic construction and concepts of bubble memory devices. Numerous methods of implementing the bubble memory exist, and the information in Figure 10-10 is merely representative. The basic bubble memory chip is constructed by depositing a very thin magnetic film on a nonmagnetic base and overlaying an insulative cover. A fixed magnetic field is induced in the magnetic film by permanent magnets in proximity to the film, and by selection of critical field strength the capability to form bubbles is provided. As indicated above, the magnetic bubbles are discrete areas in the magnetic film in which the direction of magnetic field is opposite the fixed field. (Formation and destruction of bubbles are discussed later.)

Bubble movement is accomplished using Permalloy (a magnetic material) bars deposited on the insulating cover and supplying a rotating magnetic field surrounding the complete bubble memory chip. The direction of the rotating field is shown in Figure 10-10, and it should be noted that it rotates *perpendicular to* the film field, thereby not affecting the bubble formation. As the magnetic field rotates, it induces fields in the Permalloy "T-bars," causing the bubbles nearby to be either attracted or repulsed, depending on the polarity of the field. Figure 10-10 also shows the movement of a single bubble from left to right under the influence of the rotating magnetic field. Thus, the magnetic bubble memory operates as a *magnetic shift register*.

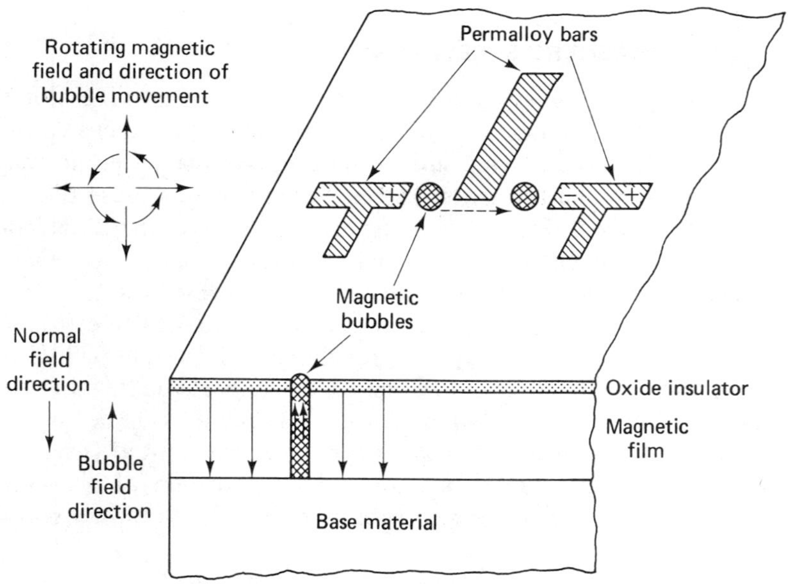

Figure 10-10 Bubble Memory Concepts

Memory Organization and Control. One possible organization of magnetic bubbles is shown in Figure 10-11. A *major loop* is provided in which data input/output operations are performed. The rotating magnetic field discussed above causes magnetic bubbles to constantly circulate around the major loop, passing by the *detector station*, *annihilator station*, and *generator station*. If a data-input operation is to be performed, bubble positions in the major loop are first annihilated so that all positions are clear. Each binary 1 in the input data causes the bubble generator to form a bubble, while a binary 0 leaves the appropriate position without a bubble. The number of bits available in the major loop is a function of design considerations and varies from manufacturer to manufacturer.

Data output is accomplished during the normal recirculation of information around the major loop. As the bubble passes the detector station it is detected and stored in external logic circuitry, such as a shift

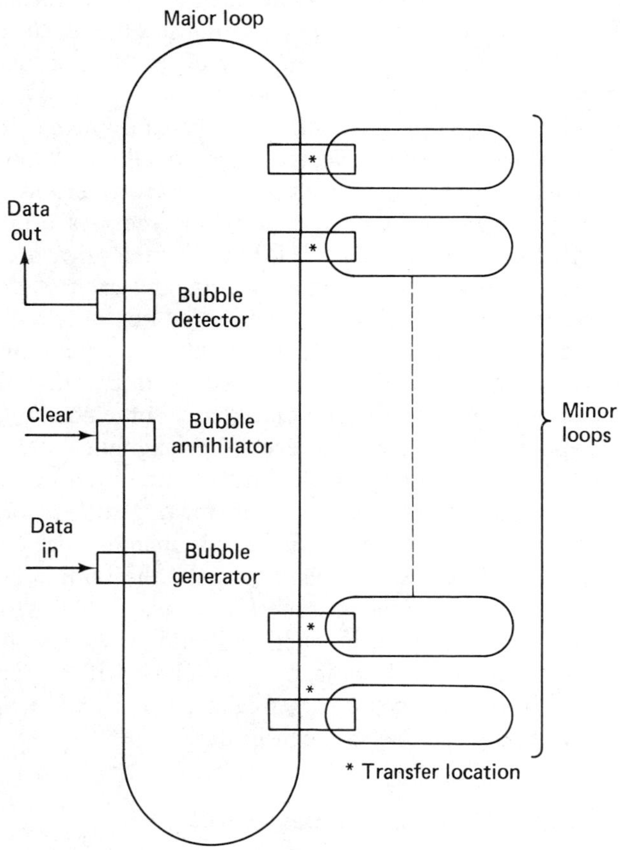

Figure 10-11 Bubble Memory Organization

register. As in most serially oriented storage devices, the serial information is stored in a shift register until the required number of bits are available, and then transferred in parallel to the remainder of the microcomputer. It should be noted that detection of the bubble often results in weakening of the field, and the detector may include a *replicator* to reestablish the bubble. Normally, however, the magnetic bubble memory may be considered to be *nonvolatile* because data remain in the bubble even with loss of power.

With only a major loop in operation, it can be seen that the storage capability of the bubble memory would be quite small. The extremely large storage capacity of the bubble memory is achieved by storing the information not presently being inputted or outputted in a number of *minor loops*. Each minor loop contains a large number of storage locations, and there are as many minor loops as there are data locations in the major loop. A typical example is a small-scale bubble memory with 100,637 storage locations. The major loop contains 157 locations, while each minor loop contains 641 bubble sites. Thus, the capacity of the memory is 641×157, or 100,637 bits.

As with the major loop, information is constantly being circulated in the minor loops. During a read cycle, for example, all minor loops transfer a bit to the major loop simultaneously as the *transfer location* is passed. Likewise, during a write cycle, each minor loop receives a bit from the major loop. Although the concepts of bubble memories are relatively simple, considerable logic circuitry must accompany each of the chips. A clock is needed to provide the critical timing for bubble circulation, and the circuitry required for generation of the rotating magnetic field is quite complex. Read/write control is relatively straightforward, although the actual writing and reading of information requires some nondigital circuits. From the viewpoint of the user, however, the magnetic bubble memory may be treated as any other large-scale memory, and interface circuits are similar. Programming techniques used with moving-surface memory devices are easily applied to the magnetic bubble memory devices, and the absence of mechanical devices promises to make the bubble memory the logical replacement for the higher-cost disk devices.

The theoretical size limit of magnetic bubbles using existing materials is 0.01 micrometer (micron). Bubble memory chips with capacity of 10^7 bits are easily within reach at the present time, and it is anticipated that 10^8 bits in a single package may be seen in the future.

10-5 CHARGE-COUPLED DEVICES (CCD)

Concepts. High-capacity storage may also be accomplished using *charge-coupled-device* (CCD) memories. CCDs are functional memory devices that manipulate information in discrete packets of electrical charge by transfer of the electrical charge from one semiconductor storage element (sometimes called a potential well) to a similar adjacent storage element. Perhaps the easiest way to visualize the CCD is to consider it as a number of water buckets arranged side by side. If the first bucket is filled with water and the remainder are empty, the water (information) is transferred to the next bucket by pouring from one bucket to the other. In the CCD it is only possible to transfer from one bucket to the one immediately adjacent, so the CCD acts very much like a serial shift register. Input to the CCD is accomplished by "filling the bucket," and output is implemented by "emptying the bucket" at the other end of the register.

CCDs operate in much the same manner as the dynamic RAM cells discussed in Chapter 9, but without the use of complete semiconductor cells. A semiconductor base material is overlayed with an insulating cover and "gates" deposited to form the CCD. Potential wells are formed when a HIGH logic level is applied to a gate because the + charge carriers in the base material are forced away from the gate. Thus, an area of charge depletion exists under each gate that has a HIGH logic level applied. If the opposite type of charge carriers (−) are now inserted in the potential well

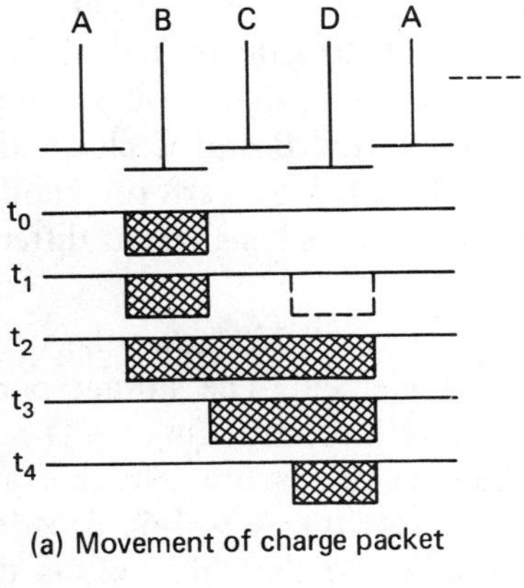

(a) Movement of charge packet

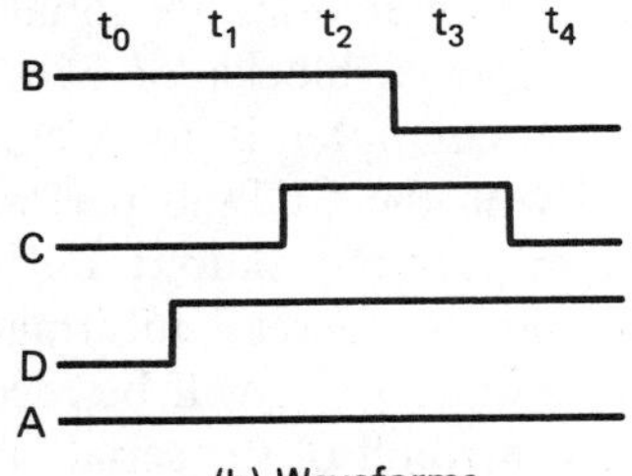

(b) Waveforms

Figure 10-12 CCD Memory Concepts: (a) Movement of Charge Packet; (b) Waveforms

(a binary 1 is represented by a – charge packet) by the input circuitry, the potential well in use can be used to store the representation of the binary digit. The movement of the binary digit represented by the negative charge packet is shown in Figure 10-12a.

Four separate, but related, timing signals are used to transfer information from a CCD cell to an adjacent CCD cell. By sequencing the timing signals in the manner shown in Figure 10-12b, the potential wells generated furnish an easy path for the charge packets to follow. Only the B gate is HIGH at time t_0, forming a storage well under the B gate. (It is assumed that the storage well under the B gate contains the equivalent of a binary 1 as furnished by the input circuitry.) At t_1, both B and D gates are HIGH, forming an additional potential well under gate D. Since no information has been furnished to this newly created potential well, it is shown with no charge packet stored. B, C, and D gates are all HIGH during t_2, and a new potential well is formed overlapping gates B and D. A continuous path now exists from B gate to D gate, and the charge packet representing the input information is dispersed throughout the continuous storage well. At t_3 the B gate input goes to a LOW logic level, and the potential well associated with the B gate disappears. The charges now must regroup to fit into the smaller area. When the C gate goes LOW at t_4, all charges are forced into the potential well at gate D. The information now has been shifted from gate B to gate D (i.e., shifted one bit position). It should be noted that the A timing signal was not used during the explanation above, and that actually the B and C clocks did the shifting. The A and D clocks may also be used to perform another shifting operation, allowing use of all four of the clock signals at different times.

Organization. Figure 10-13 shows the organization of a typical 16K × 1-bit CCD memory element. The storage portion of the CCD chip consists of 64 separate 256-bit recirculating CCD registers with associated refresh amplifiers (CCD memories are dynamic and require occasional refresh to maintain data integrity). A 1-of-64 decoder, in conjunction with the external address supplied to the chip, selects the appropriate 256-bit register to be used. Since CCD memories are serial in nature, the memory element in Figure 10-13 supplies blocks of 256 bits at a time to the microcomputer. If the microcomputer is to operate in parallel, support circuitry must be supplied with the CCD to perform the serial-to-parallel conversion. Conventional input and output buffers are used to write information into the memory or to read information from the selected register. Once again, blocks of 256 bits will be recovered from or written into the memory element as a result of the serial organization of the CCD chip. The four-phase timing signals discussed above are used to circulate information in each of the registers, and to develop timing required for

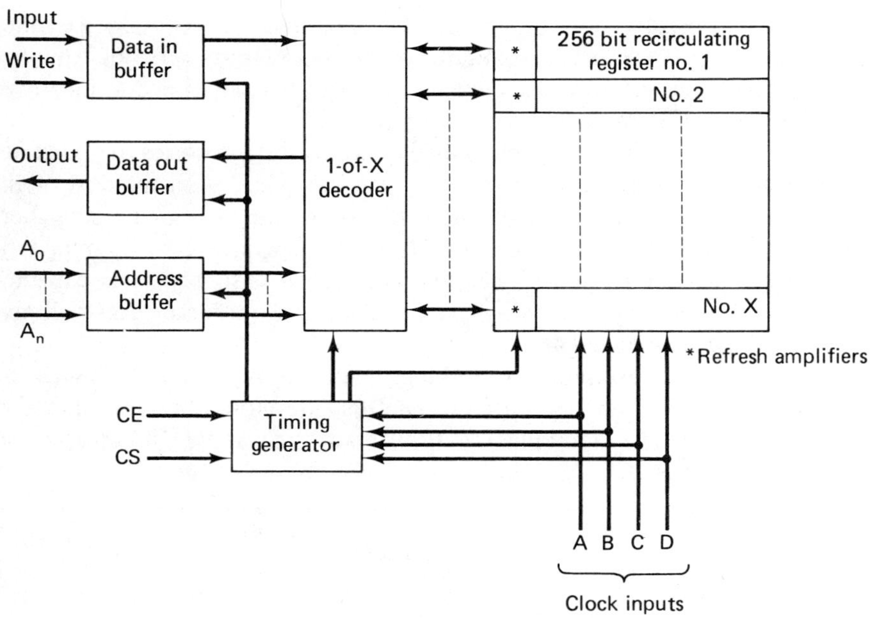

Figure 10-13 CCD Memory Organization

input and output operations. Input/output operations are implemented using the write and chip select/chip enable inputs.

CCD memory chips may be combined with appropriate support chips to form large-scale memories in the megabit category. For example, it takes only 64 of the 16K×1 CCD chips (plus support circuitry) to form a 128K×8 bit memory. Such density is easily achievable on a single printed circuit card. This large capacity places the CCD-based memory in direct contention with moving-surface memory devices. Access time is easily within range of the disk memories typically used with microcomputers, and the reduction of mechanical operations improves reliability greatly. The future may see many of the mechanically operated storage devices replaced with CCD and/or magnetic bubble memories.

10-6 SUMMARY

Even as this text is being prepared, new and innovative ideas are appearing in the field of large-scale memory for microcomputers. The development of static and dynamic memory chips with large capacities is making it more and more possible to approach the capacity of large-scale memories directly within many microcomputers. Recording densities are becoming greater and greater with magnetic-surface types of media, and soon the miniature versions of the floppy disks will contain as much information as the now standard sizes. Numerous versions of tape cartridge and cassette storage devices are appearing, all aimed at increased volume of data at much shorter access times.

Magnetic bubble and charge-coupled-device memories are with us, and although somewhat high in cost when compared with the cost of the microcomputer, the lack of mechanical assemblies may well make them more economical in the long run. Increased experience with magnetic bubbles and charge-coupled devices will reduce cost, and it is expected that many of the larger moving-magnetic-surface storage devices will give way to these new concepts.

The rapidly expanding technology makes it impossible to discuss all storage techniques and be current. This text develops the basic concepts behind the various techniques, and the reader may apply these basics as new devices appear.

QUESTIONS

1. Why is software, rather than hardware, the most common link between the microcomputer and large-scale memories?
2. Explain how information may be stored on a magnetic surface.
3. _______ access is common to all large-scale magnetic storage devices.
4. Discuss the advantages and disadvantages of NRZ, PM, and FM recording techniques.
5. Discuss the advantages and disadvantages of parity, checksum, and CRC error detection.
6. Explain the organization of data stored on a floppy disk using the IBM format.
7. What is the difference between soft sectoring and hard sectoring?
8. How long does it take for a complete rotation of the standard floppy disk?
9. Explain the functions and operation of a typical floppy disk controller.
10. Why would digital data recording be used in preference to audio-frequency data recording with cassette devices?
11. Why would audio-frequency data recording be used in preference to digital data recording with cassette devices?
12. Explain the recovery of digitally recorded cassette tape data.
13. Explain the recovery of audio-frequency recorded cassette tape data.
14. What is the difference between storage of data on moving-surface magnetic devices and magnetic bubble devices?
15. How is a magnetic bubble moved?
16. Explain the organization of a magnetic bubble memory chip.
17. How is information transferred in a charge-coupled device (CCD)?
18. Explain the organization of a charge-coupled-device memory chip.
19. What large-scale memory technology do you think will dominate the microcomputer field in the next five years? Why?

CHAPTER 11

A Representative Microcomputer (Hardware)

As the characteristics of microprocessors and microcomputers have been developed in this text, the reader has probably noted an increasing level of sophistication as new subjects have appeared. In the beginning such simple things as gates and flip-flops seemed quite complex, yet by the time the microprocessor hardware arrived, they seemed quite simple. The microprocessor now seemed complex, and considerable time was spent discussing it from the viewpoint of actual 1s and 0s, and to a good purpose. Without a thorough grounding in what the hardware is doing, working with a microprocessor-based device can be quite difficult.

But as discussions progressed, it may also have been noted that 1s and 0s disappeared and were replaced by more general means of description. Hexadecimal and octal notations were used, and as more and more capability was realized, it should have become apparent that even more general notations, such as programming mnemonics and higher-level programming languages, could describe the workings of the microcomputer.

Actually, when microprocessors first appeared, almost all input was via switches at the binary level, and outputs were via lights arranged in binary form. As mentioned in earlier chapters, such methods of input and output were very slow. More sophisticated applications resulted in keyboards interfaced to the microcomputer for input and printers or cathode ray tubes (CRTs) for output. The microcomputer discussed in this chapter is a natural evolution from the switches and lights of yesterday on the way to yet-unknown heights of tomorrow.

11-1 SYSTEM DESCRIPTION

A typical microcomputer designed for small business and home-management applications is shown in Figure 11-1. The microcomputer measures 18 by 19 by 16 inches and weighs approximately 40 pounds. From the user's viewpoint it consists of a keyboard to input information, a cathode-ray tube (CRT) display for output, and an audio cassette for external storage. Microcomputer functions are implemented as shown in Figure 11-2.

As with most microcomputers, the microprocessor is the center of all activity. The microprocessor operates at a 1-MHz clock rate, which results in 1-microsecond cycle time. Thirteen addressing modes and 56 instructions combine with 64K of addressable 8-bit memory bytes to provide an extremely versatile base for microcomputer operations. Chapter 12 integrates hardware operation with software capabilities.

Input to the microcomputer is via the keyboard, which contains all standard characters, numerals, punctuation marks, and numerous graphic

Figure 11-1 Small Business and Home-Management Microcomputer (Courtesy Commodore)

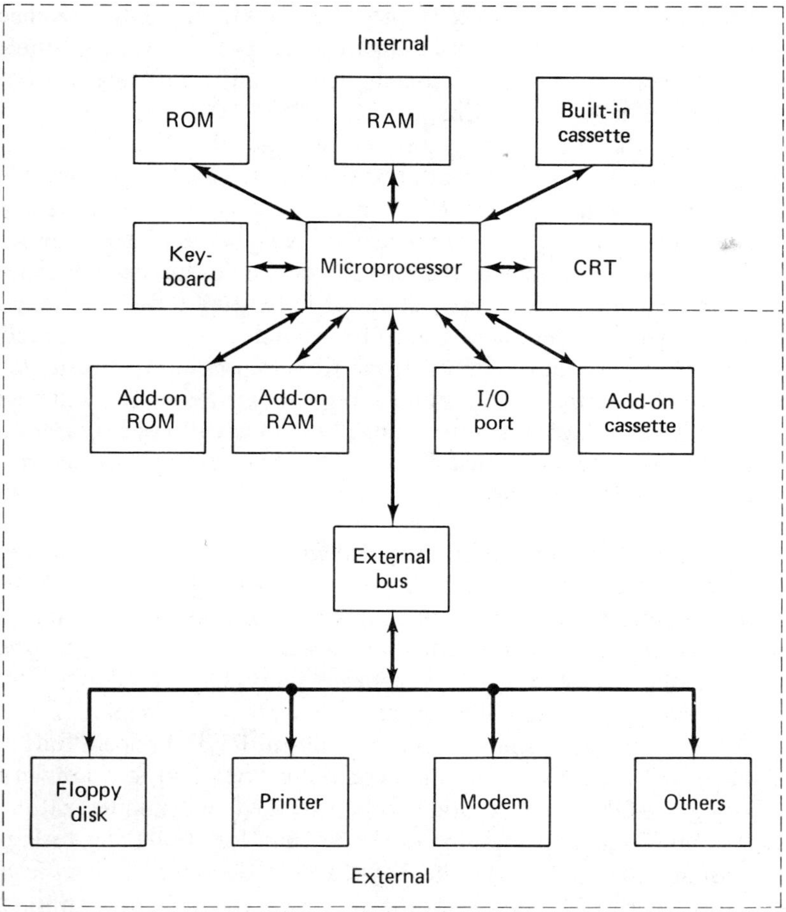

Figure 11-2 System Description

characters. The CRT supplies output in 25 lines of 40 characters each, which may be any combination of the characters input from the keyboard. Thus, it is possible to display not only conventional material, but graphic presentations may also be programmed. Although the audio cassette used to implement program storage and handling provides only relatively slow access time, large amounts of data may be stored at very low cost. The keyboard, CRT, and audio cassette are also discussed in Section 11-5.

Internal memory consists of 8K of 8-bit RAM and 14K of 8-bit ROM. RAM is used conventionally, while ROM contains the data necessary to provide the microcomputer with immediate power-on capability.

The operating system occupies approximately 4K of ROM, 1K each is used for the systems monitor and diagnostics software, and $8\frac{1}{2}$K is used for the primary language of the microcomputer, BASIC. Software characteristics are further explained in Chapter 12.

Considerable flexibility is offered by use of "add-on" capability. Additional memory up to the addressable limit may be connected via a *memory expansion bus*. Either RAM or ROM may be used so that additional operating systems and languages are easy to implement by merely plugging in preprogrammed ROMs. A second audio cassette input is provided, and software is already included in existing ROM. Two methods of interfacing other peripherals are also included. An 8-bit bidirectional bus accepts inputs and outputs from devices oriented toward parallel methods of data transfer. A second input/output bus is implemented by the IEEE-488 standard data interchange bus, which allows easy interfacing of many measurement-oriented devices. Details of external device connection are included in Section 11-6.

Data Flow. The *hardware-oriented* block diagram of the microcomputer being discussed in this chapter is shown in Figure 11-3. Many of the functions are identical with diagrams seen in previous chapters, but a short description of system information flow will be advantageous prior to discussion of each of the major blocks. Figure 11-3 should be used as reference throughout the remainder of this chapter and Chapter 12.

Application of power causes a built-in RESET operation to be performed. All sections of the microcomputer are taken to a known state by causing a ROM-stored routine to be executed. After initialization, the microcomputer enters a monitor mode, waiting for something to happen.

Assume that a key on the keyboard is depressed. The *peripheral interface adapter* (*PIA*) recognizes the keyboard and generates an interrupt signal. A ROM-stored routine is entered upon microprocessor recognition of the interrupt, and the keyboard information is extracted from the PIA by placing the PIA address on the address bus and executing the proper control signals. Keyboard information (now on the data bus) is made available to all sections of the microcomputer. Normally, however, the information will ultimately be placed in internal RAM and in the CRT RAM. Internal RAM will store the information for future use, while CRT RAM will store the information for display on the CRT in a screen location determined by CRT control. Keyboard information may also be routed to external RAM or to any of the external devices via the PIA or the *versatile interface adapter* (*VIA*). Although Figure 11-3 shows specific devices assigned to each PIA/VIA, it will be seen that optimum use of hardware and software will dictate sharing PIA/VIA operations among the input/output devices.

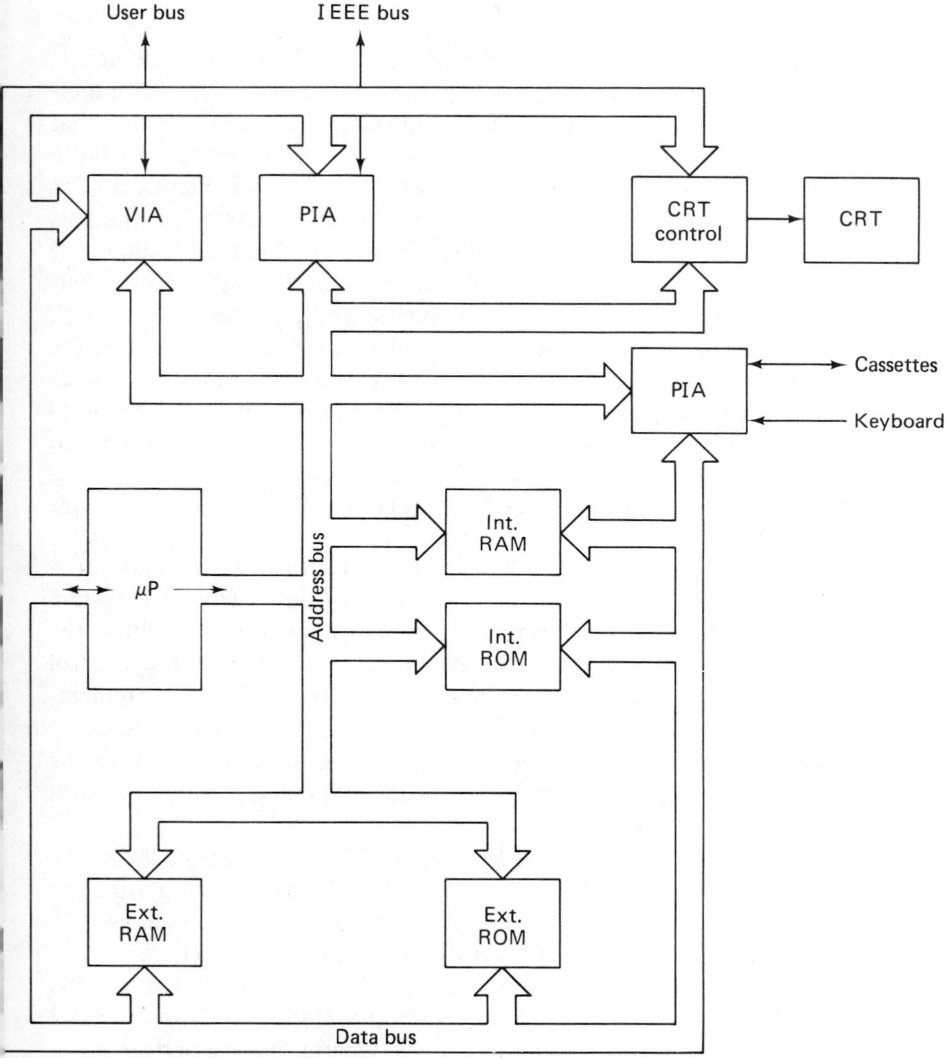

Figure 11-3 Detailed Microcomputer Block Diagram

It should be pointed out, however, that all functions performed by the microcomputer are *directly* under the control of the ROM-stored program. *Every* operation results in the microprocessor accessing ROM for further instructions. Therefore, regardless of the function to be performed, external or internal, the power of the microcomputer is in the efficiency of the ROM-stored program. Chapter 12 integrates the hardware explanations of this chapter with the software capabilities of the microcomputer.

11-2 THE MICROPROCESSOR

In order to point out one of the many variations of the generalized microprocessor discussed in Chapters 6 and 7, a common commercially available microprocessor is used in this chapter. It will be noted that many of the blocks in Figure 11-4 are labeled in the same manner as those of earlier diagrams. When so labeled, the functions are identical. For example, the *accumulator* of Figure 11-4 performs the same functions as the *accumulator* of Figure 6-1; *ALU* functions are the same; and so on.

The microprocessor is organized around an *internal* address bus (separated into 8-bit *high* and 8-bit *low* buses) and an internal 8-bit data bus interconnected by the processor's registers. All information enters the microprocessor via the microcomputer's 8-bit data bus and is routed to either the instruction register for storage and subsequent decoding or to the data bus buffer for transfer to appropriate internal registers. Output from the microprocessor is via the 16-bit address bus for address information and the 8-bit data bus for data. Various control inputs and outputs are also provided.

For purposes of discussion, the microprocessor may be divided into a *register section* and a *control section*. Instructions routed from the system memory are executed by performing a series of data transfers within the registers of the *register section*. The *control section* supplies the control information necessary to sequence the data transfers. It should be remembered that the instructions from system memory must first be decoded, and this operation is performed by temporarily storing the information in the *instruction register*, then performing the decoding and combining with timing signals in the *instruction decoder*.

The timing signals provided by *timing control* are derived from the internal clock generator, which supplies a 1-MHz basic timing function with a resultant 1-microsecond cycle time. In the microprocessor used in the typical microcomputer of this chapter a two-phase clock (Figure 11-5) is required. Both the phase 1 and phase 2 clock signals are about 480 nanoseconds wide. The phase 1 clock occurs first, followed by the phase 2 clock after a delay of about 10 nanoseconds. Operations are performed in specific parts of the microprocessor on *both* the HIGH and LOW portions of each phase. Execution of an instruction always begins with an instruction fetch cycle at phase 1, and continues with as many cycles as are required to complete execution. Actual data transfers take place during phase 2. A typical instruction is analyzed in detail and traced through its operations in Chapter 12.

Other parameters also determine the actual generation of control signals from the *instruction decoder* to the register section. The *interrupt logic* interfaces the instruction decoder to external devices that require the attention of the microprocessor. Interrupts were discussed in Chapter 8,

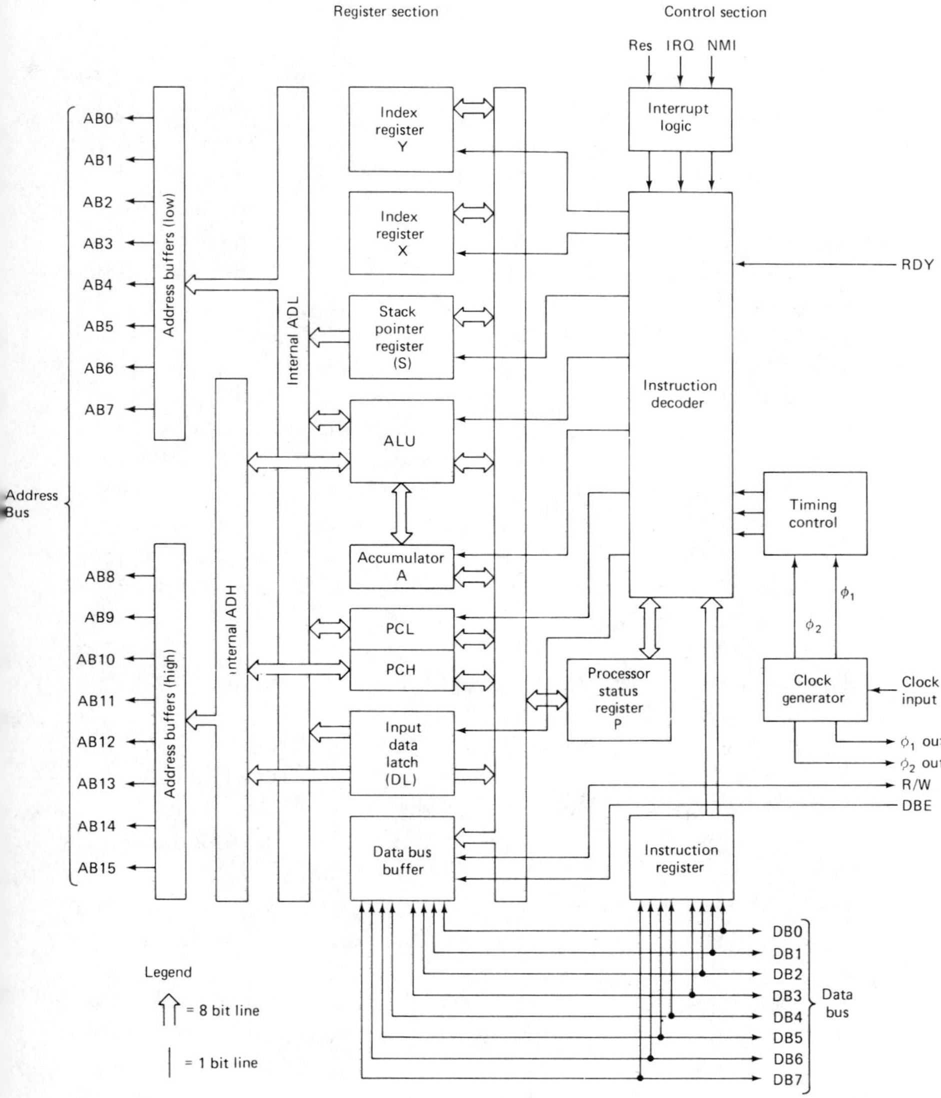

Figure 11-4 Microprocessor Internal Architecture

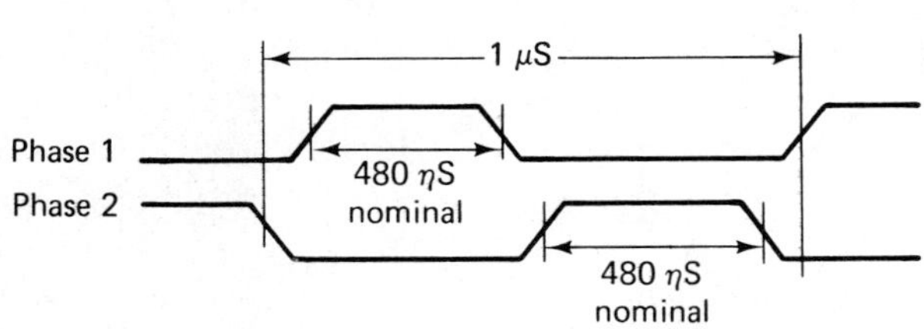

Figure 11-5 Two-Phase Clock Waveforms

and our typical microprocessor functions in the same manner. Furthermore, the *processor status register* stores the results of arithmetic and logic operations and status of internal data operations to supply additional control information to the instruction decoder. Thus, instructions from system memory, timing, interrupt information, and internal processor status are combined in the instruction decoder to develop controls for the data transfers and manipulations that take place in the register section of the microprocessor.

All the remaining operations of the microprocessor take place in the register section. Each of the registers and other components are defined and discussed in the following paragraphs. Since information coming into the processor from system memory or peripheral devices is present on the system data bus for only a very short period of time (as little as 100 nanoseconds), some means of capturing the information for more leisurely processing must be provided. The *input data latch* holds the incoming information that comes from the system data bus via the *data bus buffers* until the processor is ready for it. For example, data transferred from memory into the *accumulator* (*A*) will be placed on the *internal data bus* and thence into the accumulator. Control signals derived from the instruction decoder using the instruction register information and system timing perform the sequencing of these operations.

If an arithmetic or logic operation is to be performed, information in the input data latch will be transferred onto the internal data bus as before. Next, it will be transferred into the *ALU*. At the same time, the contents of the accumulator will be transferred onto an internal register section bus and thence into the second input to the ALU. The results will be transferred back to the accumulator on the next cycle by transferring first onto the bus and then into the accumulator. All these data transfers take place during phase 1 clock pulses.

All logic and arithmetic operations take place in the *ALU*. However, the ALU cannot store data for more than one cycle, and data placed on the inputs to the ALU at the beginning of one cycle must be processed and the results gated into one of the storage registers or to external memory on the next cycle.

The addresses that step the microprocessor through the sequential instructions stored in system memory are provided by the *program counter* (*PCL, PCH*). Each time the processor fetches an instruction from program memory, PCL contents are placed on the low-order 8 bits of the address bus and PCH contents are placed on the high-order 8 bits. The program counter is incremented each time an instruction or data are fetched.

The two index registers (*X* and *Y*) and the stack pointer (SP) are simple 8-bit latches. These registers store data that is to be used in calculating addresses in data memory, and their operations are further discussed in Chapter 12. Finally, the *address bus buffers* store the addresses that are used in accessing peripheral devices such as ROM, RAM, and INPUT/OUTPUT. They are simple latches and drivers for storage and power amplification.

11-3 THE PERIPHERAL INTERFACE

Input/output for modern microprocessors exist in two general configurations. Input/output (I/O) ports may be included as an integral part of the microprocessor chip itself or may be implemented as a separate assembly. In either case, the I/O function is performed in relatively the same manner, using a combination of software and hardware to achieve maximum flexibility.

Figure 11-6 shows implementation of the I/O function as a separate IC assembly. The functions performed are typical of those ICs called *programmable peripheral interface adapters* (*PIAs*). Such devices are termed "programmable" because the I/O functions are software-controlled. That is, the interface may be customized using control words contained in the main program.

The I/O interface of Figure 11-6 contains two separate but identical sections. Each section contains an 8-bit *control register*, an 8-bit *function register*, an 8-bit *output register*, a *gated buffer assembly*, and an *auxiliary I/O control assembly*. A *data bus buffer* supplies microprocessor-generated information to both sections and accepts peripheral device and internally generated information from both sections. The microprocessor data bus is connected to the data bus buffer, while the address and control bus are used by the *chip control* to select the appropriate I/O interface adapter, clear all registers, and initiate the read or write cycles. All registers can receive information on the *interface input bus* and all register outputs appear on the *interface output bus*. Since sections 1 and 2 are identical, only section 1 is discussed.

Control register 1 (CR-1), in conjunction with function register 1 (FR-1), determines the overall configuration and operation of section A of the I/O interface chip. Both CR-1 and FR-1, as well as Output Register 1 (OR-1), are capable of receiving and storing information from the microprocessor data bus through the data bus buffer. Each of the registers is treated as an addressable storage location, and in conjunction with the read/write signal on the microprocessor control bus may be either "written into" or "read from." The usual mode of operation employs only OR-1, except during initialization of the I/O system. At initialization time, CR-1 and FR-1 are addressed (one at a time) and the control word and function designation are supplied from the computer program. The contents of FR-1 determine the flow of information through the selected section of the

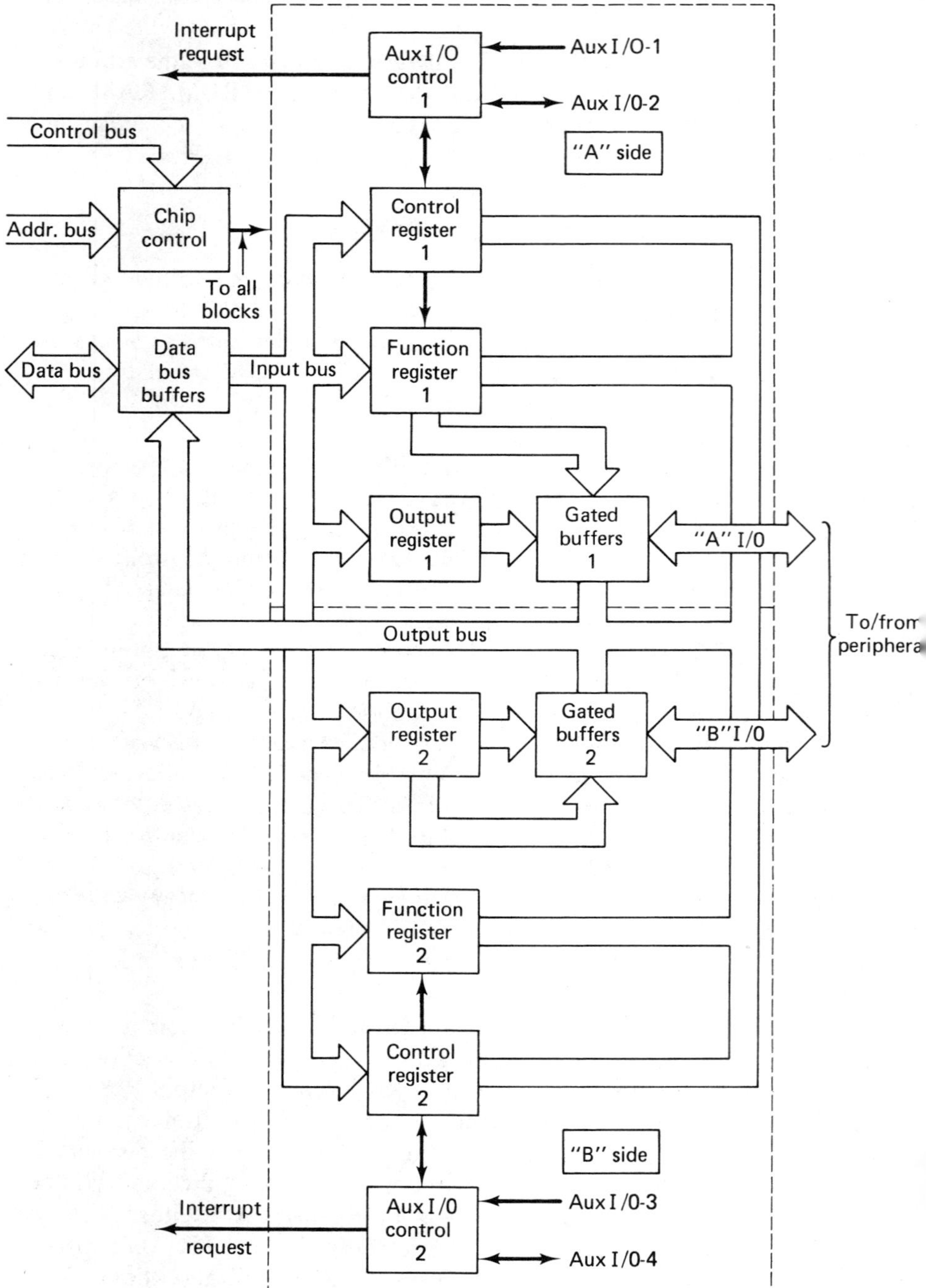

Figure 11-6 Input/Output Interface Device: The Peripheral Interface Adapter (PIA)

I/O interface. For example, if all 8 bits of FR-1 are 1s, the interface outputs information to the peripheral device. If 0s, the input function to the microprocessor is performed. Mixed input and output may be performed under program control; that is, some bits may be treated as output while others are used for input.

CR-1 uses its 8 bits of stored information to enable loading of the function register, control the operation of the two auxiliary I/O lines (Aux I/O-1 and Aux I/O-2), and enable and store interrupt information. Aux I/O-1 is used as an interrupt input for section 1, while Aux I/O-2 may be used as either an interrupt input *or* an output control signal. Functional bit assignments for CR-1 are shown below.

Bit	*Function*
0	Enable interrupt signal for Aux I/O-1 to microprocessor
1	Active transition direction (*H* or *L*) for Aux I/O-1
2	Enable FR-1 access
3	Enable interrupt signal for Aux I/O-2 to microprocessor
4	Active transition direction (*H* or *L*) for Aux I/O-2
5	Aux I/O-2 function (Input or Output)
6	Aux I/O-2 flag
7	Aux I/O-1 flag

As mentioned earlier, system initialization must be performed prior to actual use of the I/O interface. Steps will be required in the main program to

1. Access CR-1 and store the control word required to configure section 1.
2 Access FR-1 and store input or output control word.
3. Access CR-2 and store the control word required to configure section 2.
4. Access FR-2 and store input or output control word.

Assuming that section 1 is to serve as input interface and section 2 as output interface, the register diagrams of Figure 11-7 can be used to explain system operation.

A RESET input on the control bus will reset *all* register bits in the I/O interface to 0 as shown in Figure 11-7a. Examination of Figure 11-7b (after initialization) will show that

1. Section 1 is configured to act as input (FR-1 all 0s), Aux I/O-2 is an output, FR-1 access is enabled, and Aux I/O-1 is an input with interrupt enabled.

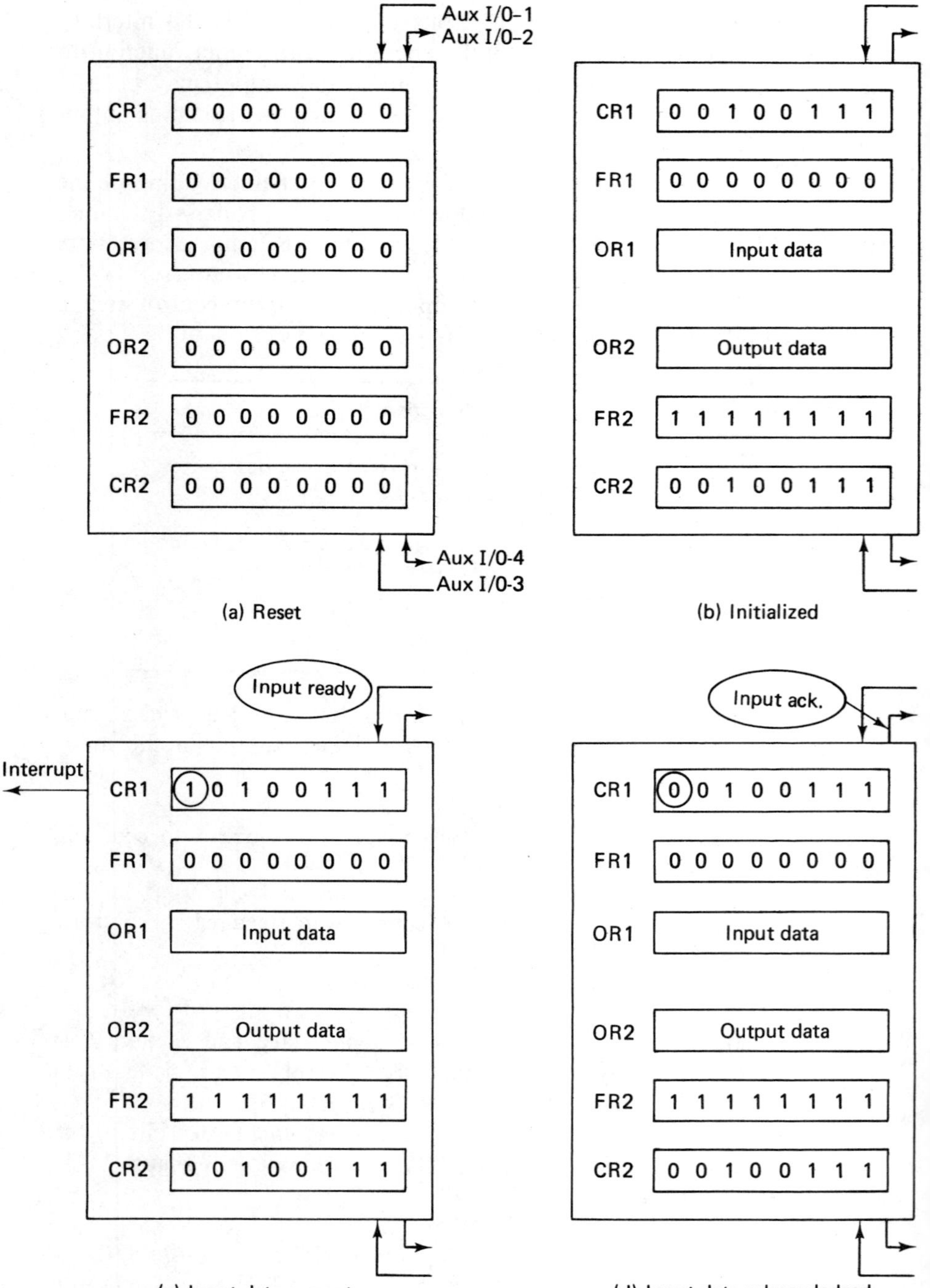

Figure 11-7 PIA Register Diagrams

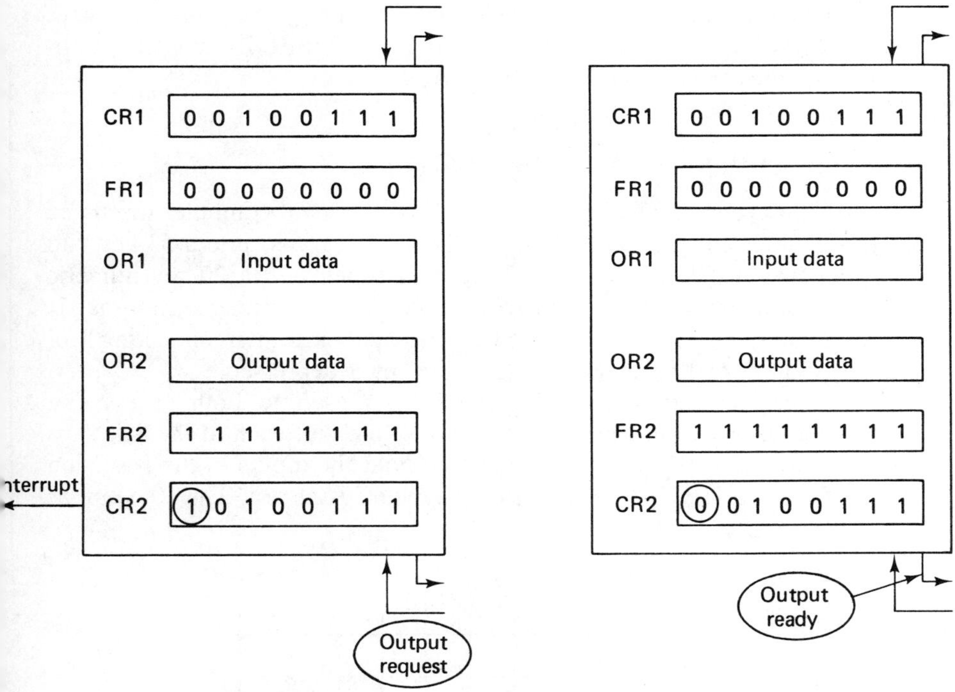

(e) Output device ready for data

(f) Output device received data

Figure 11-7 (Continued)

2. Section 2 is configured to act as output (FR-2 all 1s), Aux I/O-4 is an output, FR-2 access is enabled, and Aux I/O-3 is an input with interrupt enabled.

The configuration for data present at the peripheral I/O of section 1 is seen in Figure 11-7c. When data become available at the peripheral I/O of section A, the inputting device also generates an *input ready* to Aux I/O-1. A status bit (bit 7, CR-1) is set and an interrupt is generated on the microprocessor control bus. Detection of the interrupt will cause the microprocessor to enter an *interrupt service routine* (ISR) which will check the status bit in CR-1 to identify the interrupting device. When the ISR reads the information in OR-1 (Figure 11-7d) the interrupt is cleared and the input device receives an *input acknowledge* signal (Aux I/O-2).

Figure 11-7e assumes that an output device connected to section 2 requests data by supplying an input on Aux I/O-3. The *output request* signal sets a status bit and generates an interrupt on the microprocessor control bus. The ISR identifies the interrupting device by checking the status bit and a read operation clears the interrupt. When information is placed in OR-2, it is also presented to the external device and automatically generates the *output ready* signal (Aux I/O-4).

11-4 INTERNAL COMMUNICATION

Keyboard. The keyboard input for the microcomputer discussed in this chapter consists of a 73-key keyboard, separated into a 53-key pad for standard ASCII character and graphics input, and a 20-key numeric pad for numbers and cursor control. All the keys are arranged in a matrix form as shown in Figure 11-8. Each key is capable of interconnecting 1 of the 10 columns to 1 of the 8 rows when the key is depressed.

A peripheral interface adapter (PIA) is used to both access and collect the information from the keyboard. Note that each of the 8 rows is returned to +5 volts through a resistor to hold the inputs to the B section of the PIA at a HIGH level when the keyboard is at rest. The 10 columns

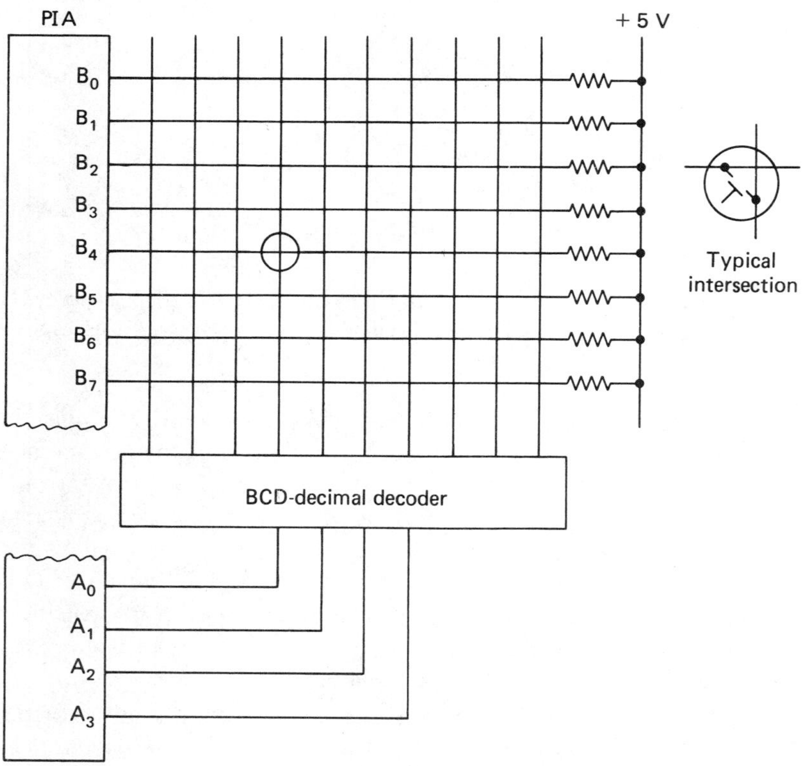

Figure 11-8 Keyboard Operations

are driven by a BCD-to-decimal decoder, acting as a 1-of-10 decoder. Input to the decoder is from the 4 least-significant bit positions of the *A* side of the PIA.

Software stored in ROM causes the *A* side of the PIA to sequentially strobe each of the columns of the keyboard matrix, driving the column line LOW when activated. Key closure is detected by LOW-going signal on one of the eight row lines. The software is also keeping track of which of the columns is being activated, and when a row line goes LOW, the exact key that is closed is identified. Row and column information correlated with time is combined as character identification.

As soon as a character is recognized, the software branches to a "table lookup" routine. The information obtained from the key closure is applied to a ROM stored table, which contains the ASCII equivalent of each of the keyboard characters. The keyboard information essentially serves as an address to the ROM, and the information at the address is the ASCII representation of the character.

Cathode-Ray-Tube Output. The basic indicator is a 9-inch black-and-white CRT (cathode-ray tube), which displays up to 1000 characters arranged in a 25-line by 40-character configuration. The CRT screen is "swept" in a pattern very much like the sweep pattern of a regular TV set (i.e., left to right and top to bottom). As the electron beam sweeps across the screen, its intensity may be varied to increase or decrease its brightness and hence the brightness of the image traced on the screen. The *exact* position of the electron beam is precisely controlled and its position is defined by the state of integrated-circuit counters. Therefore, the screen may be partitioned into a known number of precisely positioned writing locations. The CRT display being discussed has 1000 writing locations, which are further defined as "cells" made up of 64 positions each (8×8 matrix).

Each character that can be displayed on the CRT can be defined by a unique arrangement of dots and spaces in the 64-dot cell. Figure 11-9 shows how some typical letters may be defined. In fact, a ROM can be programmed with the unique pattern for each character and can be addressed to output a specific character on command.

It can be inferred from the preceding discussion that the position of each dot in each cell may be defined. Figure 11-10 shows how the requirements may be implemented. The smallest increment to be defined is the position of a dot, which is a very small percentage of the total length of one horizontal line. When eight dots have been counted, a single line portion of a character has been defined (*cell counter*). Furthermore, when 40 cells have been counted, a complete line has been traversed (*line counter*), and the electron beam must be quickly returned to the left side of

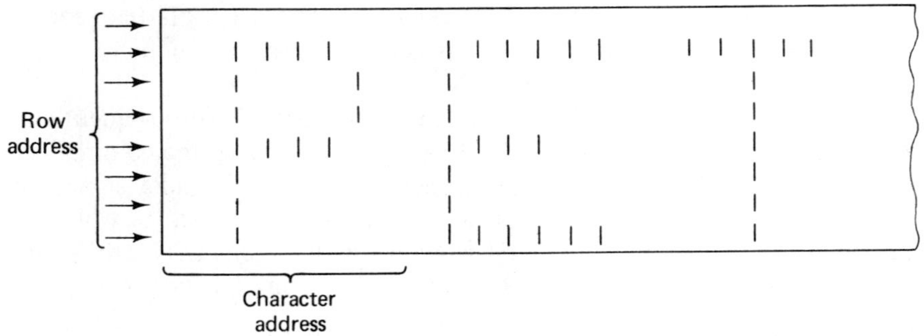

Figure 11-9 CRT Character Display

the screen (*horizontal timing*). When eight lines have been counted, 40 *complete* characters have been defined (*character counter*). *Vertical movement* of the electron beam is derived from the character counter.

With the position of each of the dots in each of the cells defined, it is only necessary to intensify the electron beam when a dot is to be displayed and to "blank" the beam when no dot is to be displayed. A 16K ROM arranged as 256 8×8 cells stores the patterns for the characters to be displayed. A 1 intensifies the beam while a 0 (blank) results in no image for the specific location. Each of the cells must be addressable. Actually, each of the 64 dot locations in each cell is addressed, but in a rather unique manner. The cell itself is addressed conventionally. Assume that only one cell must be written in a greatly simplified CRT display. The cell address defines a *beginning* location for the character, both row and column, and the counter chain causes the beam to reproduce the information for all the dot positions of the first row of the character. The next row is then scanned, and so on, until the complete character is written. Figure 11-10 shows how the rows are developed, but it must be assumed that each time a character is addressed eight shifts must be provided. Note the parallel-to-serial shift register. Each time a character is addressed, the output of the specific *row* of the ROM currently being viewed is placed in the shift register and latched in place. The dot counter then shifts the stored character-generator information out in *serial* form to the video input of the CRT.

The row address of the character generator is generated sequentially by the line counter. How, then, does a character address come about? In most systems, the character(s) to be displayed is stored in RAM. When timing is positioning the electron beam to write at a specific location on the CRT, it is also addressing a like location in RAM. Therefore, RAM is storing whatever is to be written on the CRT; that is, the ASCII representation of the letter *A* stored in a given location in RAM is displayed at a like location on the CRT.

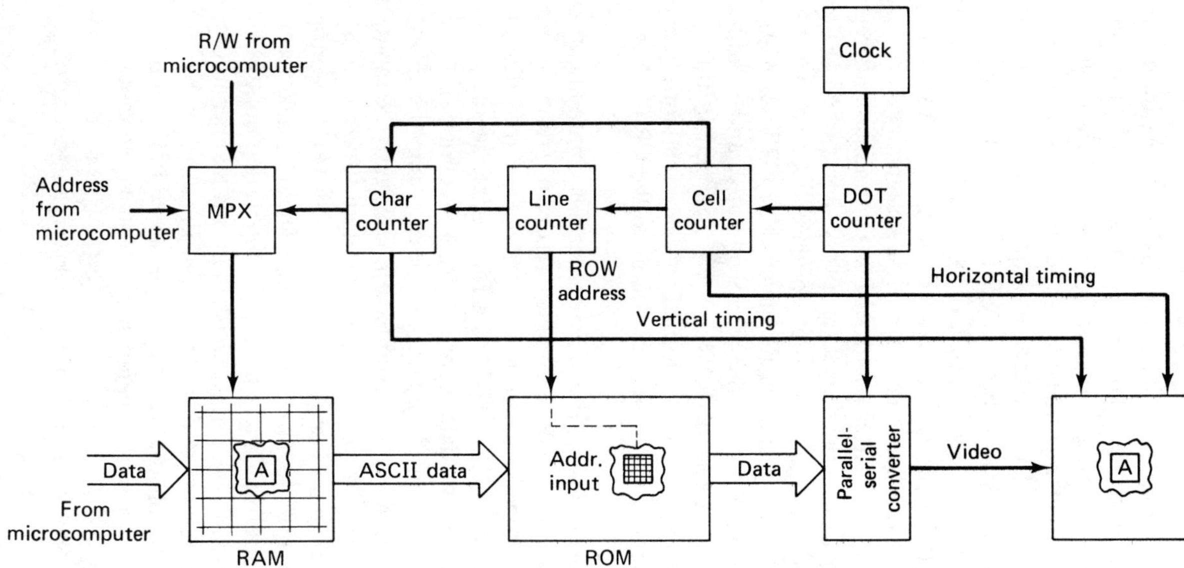

Figure 11-10 CRT Display Block Diagram

The microcomputer is responsible for initially loading the character(s) to be displayed in RAM, and will supply *both* RAM address and ASCII coding of each character. When the microcomputer has information to be displayed, it effectively steals the RAM address input from the CRT timing chain generator to load the information and then returns address control to the CRT timing chain. Data output of RAM, then, is the ASCII code for the character stored at the addressed location in RAM. The address input of ROM uses the data output of RAM to locate the stored pattern of dots and spaces that the CRT will use to display information. Thus, each character stored in ASCII form in RAM causes the ROM to provide the CRT (through the parallel-to-serial shift register) with a pattern of dots and spaces that displays the originally stored character.

Scanning of the CRT and RAM is continuous and repetitive. The CRT image remains on the screen for a very short period of time, so RAM must be stepped through all locations often enough so that it appears that the information is permanently displayed. The CRT RAM is included as part of the overall RAM of the microcomputer, and is addressed in the same manner as the rest of RAM. When arranged in this manner, CRT display is treated as part of the normal software functions, greatly simplifying the programming of output operations to the CRT.

Cassette. Cassette to microprocessor communication also takes place via a PIA. The audio-frequency recording/reproducing techniques discussed in Chapter 10 are used to interface the recording medium with the microprocessor. Figure 11-11 is a functional diagram of the hardware associated with cassette data recording/reproduction.

As in other input/output operations using the PIA, the PIA is selected and a control word is supplied to configure the input/output lines and the control lines for either input or output. Consider that information is to be recorded on a cassette. The A side of the PIA is configured so that bit A_4 and Aux I/O-4 will act as outputs. The VIA also receives a control word to configure bit B_3 as an output. Separating control functions (i.e., read and write controls) is accomplished in this manner to obtain optimum use of both hardware and software. Information to be written on the cassette is then supplied to the PIA to be stored in the output register bit-4 position until the cassette is commanded to start the motor via Aux I/O-4 and then to actually write the data (bit B_3 from the VIA). The R/W electronics perform the function of converting the data from bit A_4 into an audio frequency for recording on the cassette tape.

For reading information from the tape, bit A_4 is reconfigured to act as an input and Aux I/O-1 as an output. When commanded to read information (assuming that the motor has already been turned on by Aux I/O-4), the read/write electronics convert the audio frequency from the tape to digital form for storage in the A_4 position of the PIA. It should be

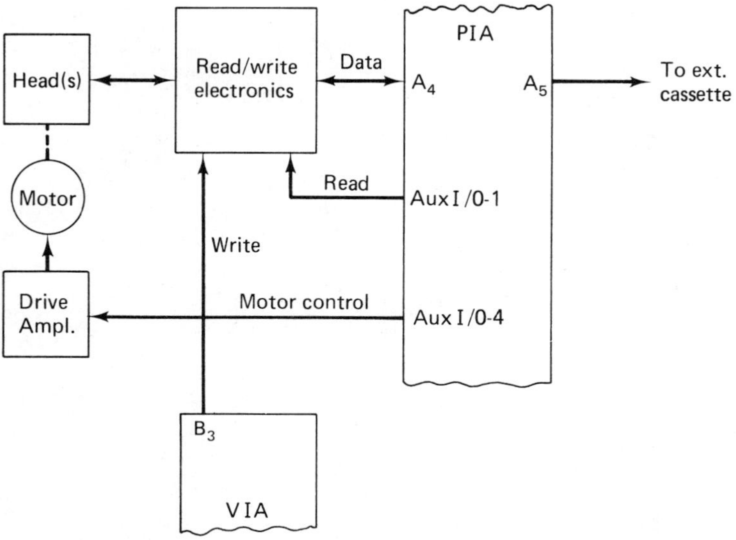

Figure 11-11 Cassette Hardware Functional Block Diagram

noted that only one bit position of the 8-bit PIA *A* register is used. Thus information is effectively written or read in serial form, one bit at a time. The software must assume the responsibility of making up 8-bit bytes from the serial bits received from or to be written on the cassette tape. In fact, the actual interface between the cassette recorder and the microprocessor can be seen to be quite simple. As is often the case, simplicity in hardware results in complexity in software, and the routines for reading and recording data are quite complex. Portions of these routines are discussed in Chapter 12.

11-5 EXTERNAL COMMUNICATION

IEEE-488 Bus. The microcomputer communicates with the outside world using peripheral interface adapters (PIAs) and the versatile interface adapter (VIA). IEEE-488 bus interface is via a PIA identical to that used for communication with the cassettes, keyboard, and CRT control, while the VIA is also used for some of the control functions. The IEEE-488 bus is used to transfer data and commands between the microcomputer and external devices, and consists of 15 signal lines. Eight of the lines (DIO_1 through DIO_8) transfer data bidirectionally, and may be either information or addresses. Data transfer is asynchronous and is coordinated by three "handshake" lines (DAV, data available; NRFD, not ready

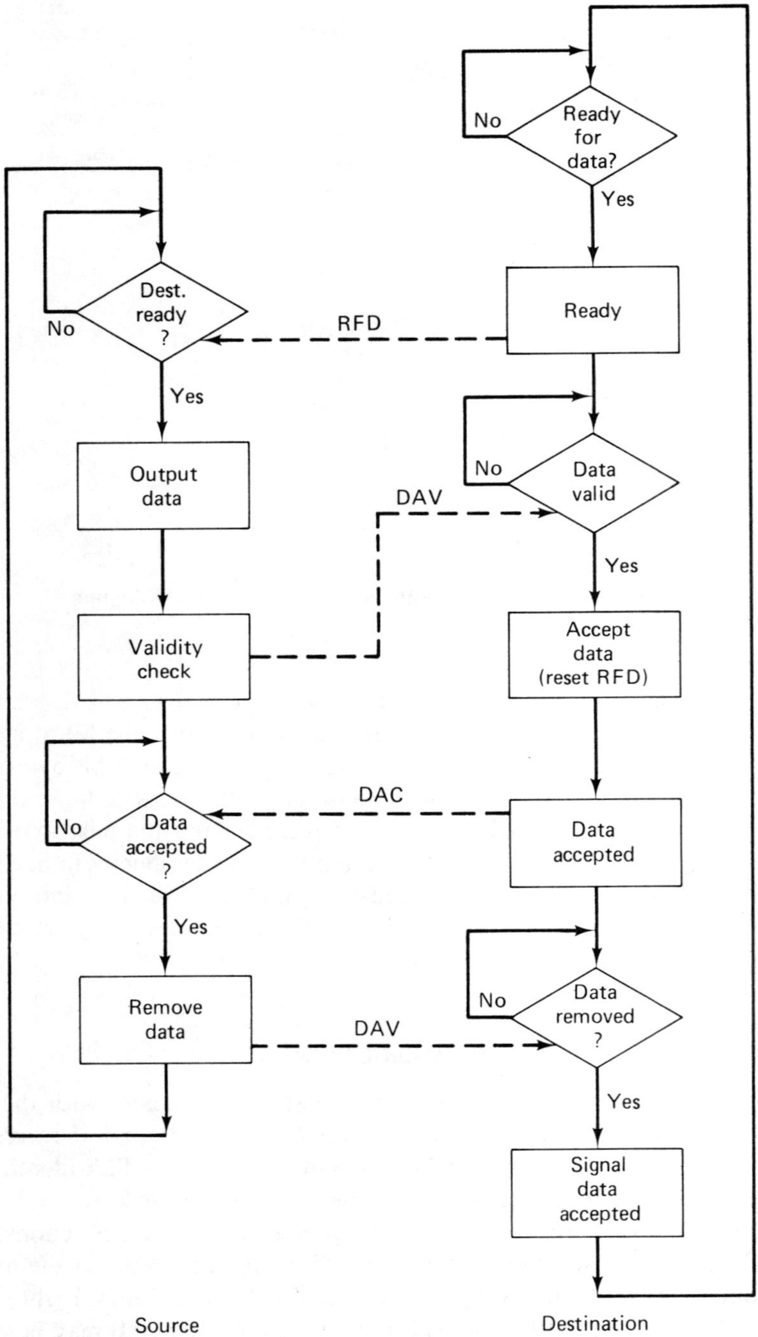

(a) Flowchart

Figure 11-12 IEEE-488 Bus Implementation: (a) Flowchart; (b) Hardware

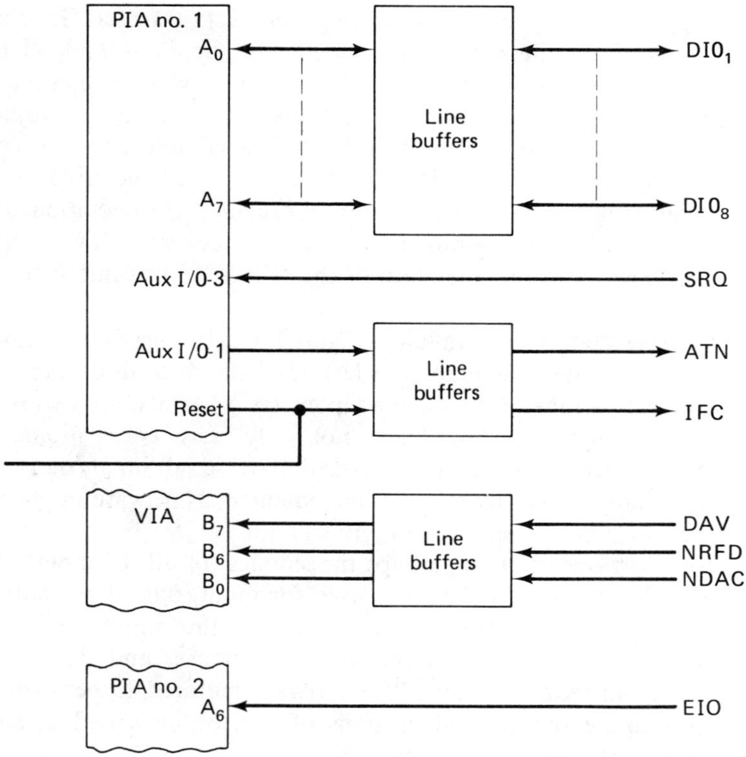

(b) Hardware

Figure 11-12 (Continued)

for data; and NDAC, not data accepted). The remaining four lines control bus activity:

1. IFC—interface clear (places system in a known state).
2. ATN—attention (indicates that address and universal commands are being transmitted).
3. SRQ—service request (indicates that a device requires attention).
4. EOI—end or identify (indicates end of a transfer sequence).

A flow diagram outline of the sequence of events during the transfer of a data byte is shown in Figure 11-12a; the hardware implementation appears in Figure 11-12b. PIA 1 is used for the data input/output function, together with three of the control functions. The *A* side of the PIA stores either received or transmittable data, which are buffered by line transceivers between the PIA and the external IEEE-488 bus. The line transceivers isolate the microcomputer from the outside world and supply adequate power and matching to assure proper data transfer. Once again, the functions of all the inputs and outputs of the PIAs and VIAs are controlled by reading into the PIA or VIA a control word to configure the interface lines as either input or output. Only the data input/output lines need to be reconfigured as data transfers are accomplished. The control lines can be configured at the beginning of the use of the IEEE-488 bus use, and will remain in that configuration throughout the operation. Just as with the previous input/output devices and functions, software plays a very important part in the operation of the IEEE-488 bus interface.

User Port. The parallel-configured user's port is implemented with the versatile interface adapter (VIA). In addition to those capabilities of the standard peripheral interface adapter, the VIA includes two interval timers and a serial register. Thus, not only may conventional 8-bit input/output operations be accomplished, but serial input/output and specialized timing functions may be implemented. Hardware implementation of the user's port is shown in Figure 11-13.

User's port operation is perhaps the simplest of all the input/output operations. The *A* side of the VIA is used for input/output as configured by the ever-present control word. Likewise, the input and output handshake lines are configured for proper response, and data transfer proceeds one full byte at a time. The software required to perform these operations is quite simple, and portions of the routines used in parallel data transfer are shown in Chapter 12.

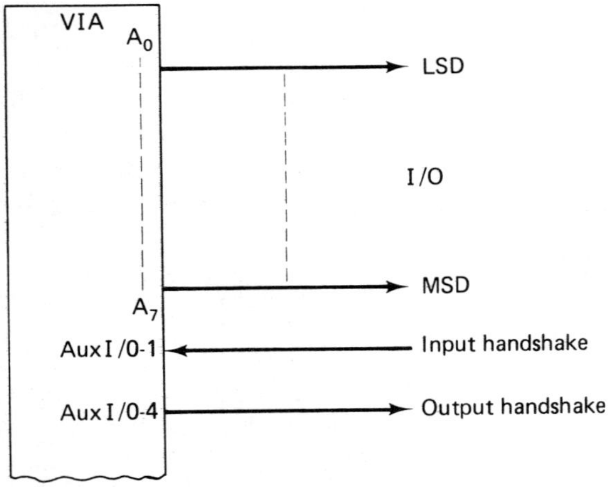

Figure 11-13 User Input/Output Port Block Diagram

Table 11-1 Microcomputer Memory Map

Block	Location Address	Contents	Type of memory
	65536		
16		I/O, diagnostics, monitor	ROM
	61440		
15		I /O ports and expansion I/O	RAM
	59391		
		Screen editor	
	57344		
14		"Basic" interpreter	ROM
	53248		
13			
	49152		
12		Expansion	
	45056		
11		Expansion	
	40960		
10		Expansion	
	36864		
9	33792		
		Display storage	RAM
	32768		
8			
	28672		
7			
	24576		
6			
	20480		
5			
	16384		
4			
	12288		
3			
	8192		
2			
	4096		
1	1024		
		Working storage, stack, etc	
	0		

11-6 MEMORY

The available 64K of memory in the typical microcomputer discussed in this chapter is arranged as shown in Table 11-1. Although the first 33,792 bytes of storage are RAM, only certain portions are available for general use. For example, the first 1024 bytes are generally reserved for such functions as stack pointer storage, cassette tape input working storage, operating system working storage, and so on. The last 1024 bytes (32,768 to 33,792) are reserved for screen storage operations. RAM addresses between 1024 and 32,768, however, may be used for any type of temporary storage desired. Addresses 33,792 to 59,391 are ROM, and specific functions are assigned to certain locations. The screen editor program resides from location 59,391 downward to 57,344, while the BASIC interpreter program occupies 57,344 downward to 49,152. The remainder of the ROM downward to 33,792 is reserved for expansion.

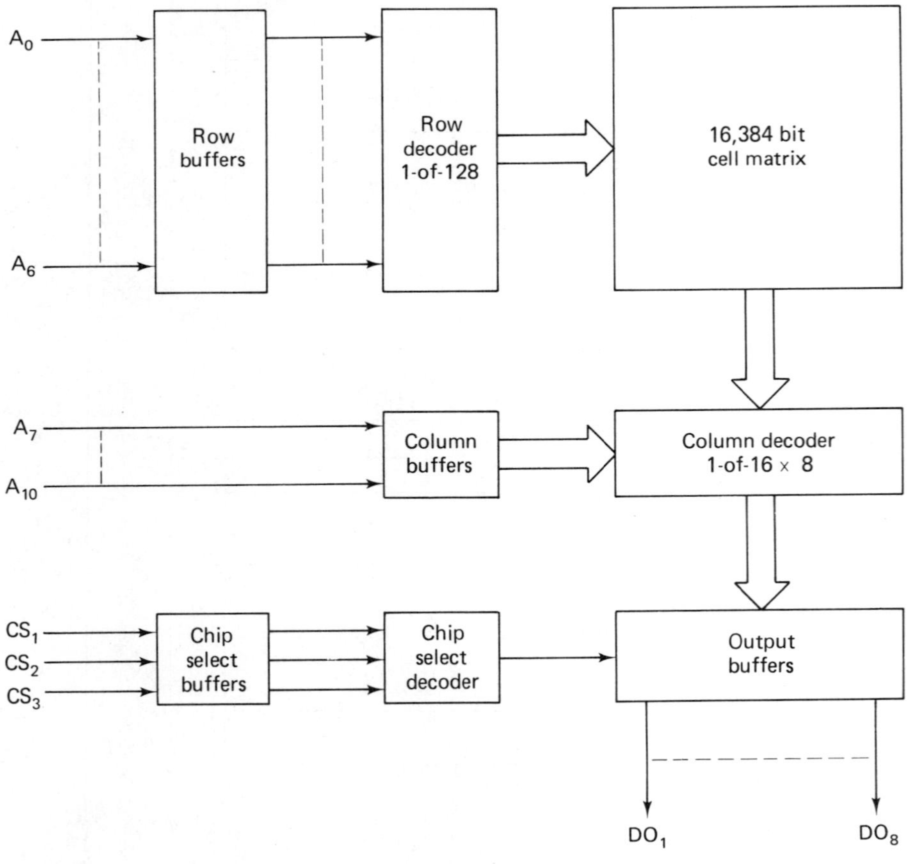

Figure 11-14 ROM Chip Organization

A small area of RAM is located from location 59,391 to 61,440 and is used for temporary storage of input and output data, and for expansion with additional input/output capabilities. The final 4K of memory is ROM and is used to store input/output routines, diagnostics, and the monitor routine.

ROM. Internal ROM is implemented in seven 2K×8-bit chips. Organization is shown in Figure 11-14. ROM operation has been discussed in detail in Chapter 9, and ROM chips used in the typical microcomputer of this chapter operate in the same manner.

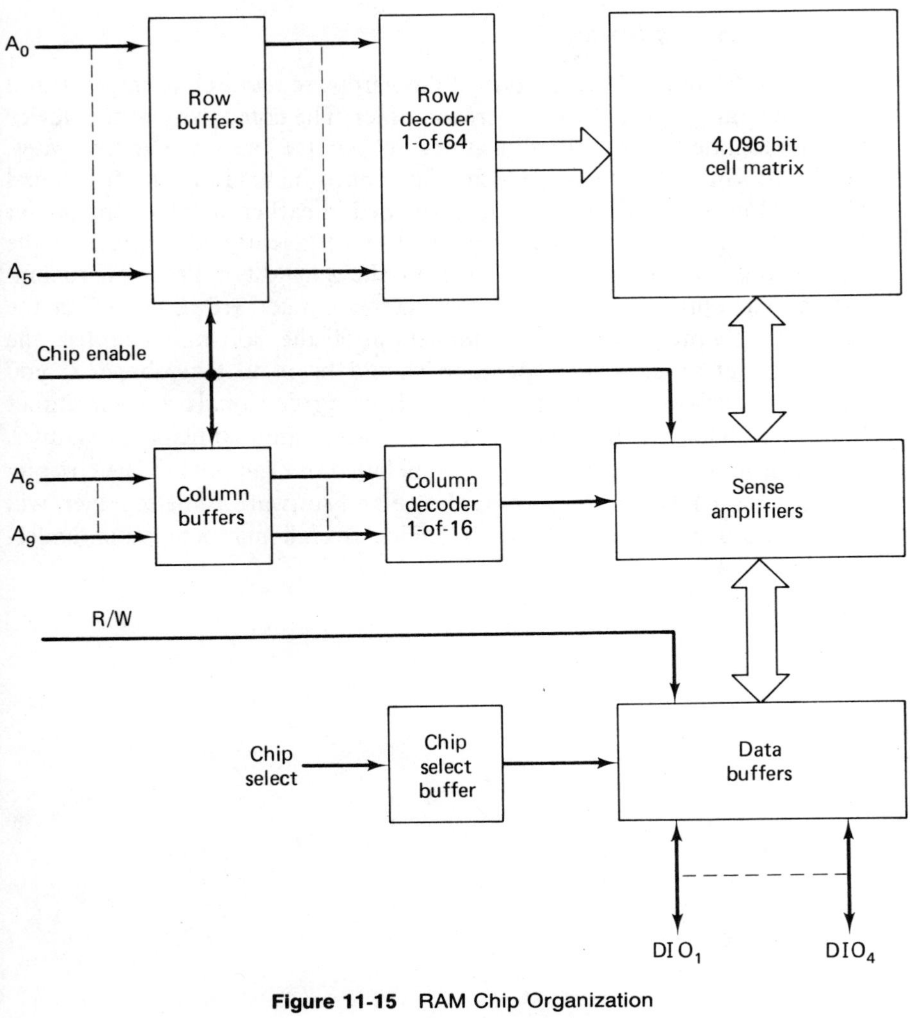

Figure 11-15 RAM Chip Organization

The 14K of ROM is used to provide permanent storage of the operating system, system monitor, diagnostic software, and the BASIC interpreter. Thus, the microcomputer is configured so that initial start-up results in immediate availability of usage without the need to load any bootstrap programs or start-up routines.

RAM. RAM consists of two blocks of 4K each, accomplished with conventional 1K×4 static RAMs. Expansion to 32K of RAM is accomplished by addition of blocks of 4K. Each RAM chip is organized as shown in Figure 11-15, and it will be noted that its organization is very similar to those RAM chips discussed in Chapter 9. The reader is referred to Section 9-2 if a review of RAM operation is desired.

11-7 SUMMARY

Chapter 11 has described the hardware required to implement a small general-purpose digital microcomputer. The complexity of the device has necessitated some simplifications, of course, and the many gates, amplifiers, and discrete components have been included in the functional blocks. However, with the knowledge gained in earlier chapters, the reader should at this time be able to generalize sufficiently to appreciate the tremendous capability existing in even the smallest modern microcomputers. The software aspects of the microcomputer are discussed in the following chapter, and upon completion of the software chapter, the microcomputer's complete operation should begin to take shape. If you still have questions, don't feel alone. Even professionals are sometimes bewildered when the microcomputer responds in an unanticipated manner.

Questions and problems are not included in Chapter 11. The completion of Chapter 12, where both hardware and software come together, will be a more appropriate time to determine overall microcomputer understanding.

CHAPTER 12

A Representative Microcomputer (Software)

Chapter 11 discussed a representative microcomputer from the hardware viewpoint. This chapter discusses the same microcomputer from the software viewpoint. The general organization of this chapter is similar to Chapter 7, where a microprocessor was discussed from the software side. Preliminary background such as instruction format, addressing modes, and a programmer's view of the microcomputer are discussed first. The instruction set is then examined in detail, followed by some examples of use of the instruction set.

Once the reader is familiar with the instructions available for the microcomputer, it becomes possible to examine more complex programming schemes. Assembly language programming is discussed as an extension of use of the primitive instruction set, followed by an introduction to a conversational language, BASIC. Correlation between assembly language and BASIC is shown by a simple example.

The basic concepts of computer programming were discussed in Chapter 7, where *analysis*, *organization*, *coding*, *testing*, and *documentation* were explained. Each of these steps must be performed, regardless of the microprocessor or microcomputer being used. Chapter 7 was concerned with the *machine language* aspects of programming for a "universal" microprocessor. In Chapter 12, other aspects of programming are considered, from machine language programming of our representative microcomputer through the use of *high-level languages*. The *hierarchy* of programming languages is shown in Figure 12-1. At the lowest and most detailed level is the machine language, the only language that the computer can really understand. A program at this level is called an *object* program and, as explained in Chapter 7, requires separate steps for even such primitive operations as moving a byte from one register to another.

The next higher level is a symbolic system of programming which uses a language that is mnemonically related to the operations to be performed. The symbolic language *source* program is converted to a machine language object program by a special program called an

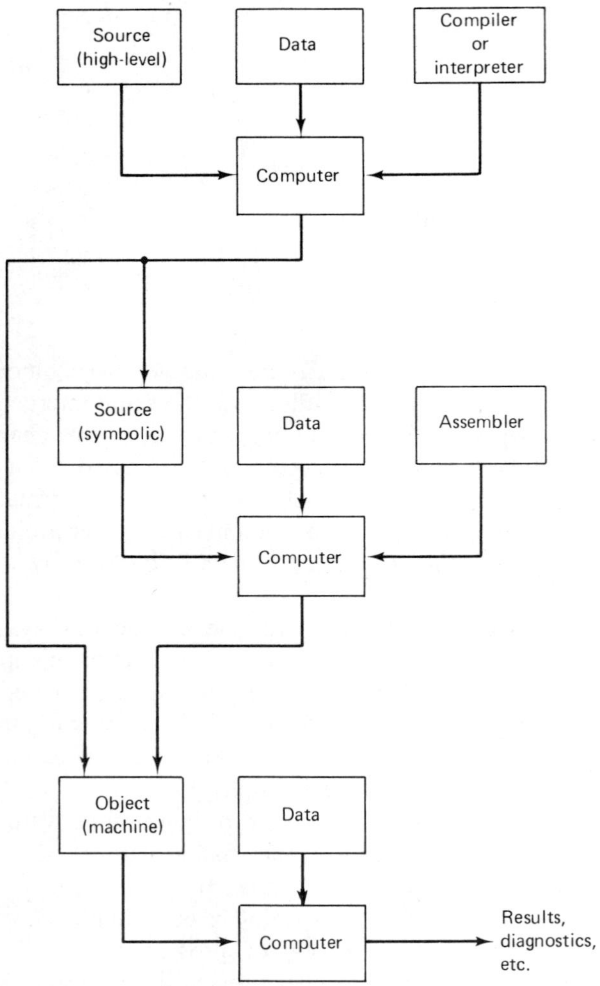

Figure 12-1 Hierarchy of Programming Languages

assembler. Assembly language programming also requires separate steps for each operation, but is easier to use because the instruction abbreviations resemble English-language operations (i.e., STA for *st*ore *a*ccumulator).

At the highest level of programming is the *compiler-based* system. The source program is written in a language that very closely resembles the type of problem being solved, and is translated by a special program called a compiler. The computer outputs either a symbolic assembly language program, which must be further converted to a machine language program, or a machine language program directly. At the same level as the compiler, although slower in operation, is the *interpretive* system. This language is problem-related and in a conversational mode. An interpreter program converts each statement into either a symbolic or a machine language program. Both compilers and interpreters convert single statements into *many* symbolic or machine language statements.

In Chapter 12 examples of machine language, assembly language, and interpreter language programs are shown following descriptions of the microcomputer's instruction set. The interpretive language used is BASIC, which is a powerful set of statements that allows use of the microcomputer without the details of machine and/or assembly language operations.

12-1 MICROCOMPUTER INSTRUCTIONS

Instruction Format. As noted in Chapter 7, most microprocessors and microcomputers store and manipulate information in groups of 8 binary digits (bits), called bytes. Our representative microcomputer is no exception, and as with other 8-bit microcomputers, uses as many bytes as are necessary to adequately define the instruction and other information. Single-byte, two-byte, and three-byte instructions are encountered.

All the instructions in our microcomputer's repertoire can easily be described with 8 bits. The requirement to refer to other locations for information, however, means that one or two more bytes must be attached to adequately describe what is to be done and where to go for additional data. The instruction formats discussed in Chapter 7 are identical to the instruction formats to be used in this chapter (although the list of instructions is different), and the reader is referred to Figure 7-1 as a review of 8-bit instruction formats.

Addressing Modes. Thirteen addressing modes are available to the microprocessor. When each of the 56 instructions is correlated with applicable addressing modes, 151 very powerful address-oriented instructions result. The instruction set is shown in Table 12-1, while each of the addressing modes is explained below. It should be noted in all cases that

Table 12-1 Instruction Set

ADC	Add Memory to Accumulator with Carry	**JSR**	Jump to New Location Saving Return Address
AND	AND Memory with Accumulator	**LDA**	Load Accumulator with Memory
ASL	Shift Left One Bit (Memory or Accumulator)	**LDX**	Load Index X with Memory
		LDY	Load Index Y with Memory
BCC	Branch on Carry Clear	**LSR**	Shift One Bit Right (Memory or Accumulator)
BCS	Branch on Carry Set		
BEQ	Branch on Result Zero	**NOP**	No Operation
BIT	Test Bits in Memory with Accumulator	**ORA**	OR Memory with Accumulator
BMI	Branch on Result Minus	**PHA**	Push Accumulator on Stack
BNE	Branch on Result Not Zero	**PHP**	Push Processor Status on Stack
BPL	Branch on Result Plus	**PLA**	Pull Accumulator from Stack
BRK	Force Break	**PLP**	Pull Processor Status from Stack
BVC	Branch on Overflow Clear	**ROL**	Rotate One Bit Left (Memory or Accumulator)
BVS	Branch on Overflow Set	**ROR**	Rotate One Bit Right (Memory or Accumulator)
CLC	Clear Carry Flag	**RTI**	Return From Interrupt
CLD	Clear Decimal Mode	**RTS**	Return From Subroutine
CLI	Clear Interrupt Disable Bit	**SBC**	Subtract Memory from Accumulator with Borrow
CLV	Clear Overflow Flag	**SEC**	Set Carry Flag
CMP	Compare Memory and Accumulator	**SED**	Set Decimal Mode
CPX	Compare Memory and Index X	**SEI**	Set Interrupt Disable Status
CPY	Compare Memory and Index Y	**STA**	Store Accumulator in Memory
DEC	Decrement Memory by One	**STX**	Store Index X in Memory
DEX	Decrement Index X by One	**STY**	Store Index Y in Memory
DEY	Decrement Index Y by One	**TAX**	Transfer Accumulator to Index X
EOR	EXCLUSIVE-OR Memory with Accumulator	**TAY**	Transfer Accumulator to Index Y
INC	Increment Memory by One	**TSX**	Transfer Stack Pointer to Index X
INX	Increment X by One	**TXA**	Transfer Index X to Accumulator
INY	Increment Y by One	**TXS**	Transfer Index X to Stack Pointer
JMP	Jump to New Location	**TYA**	Transfer Index Y to Accumulator

the first byte of each instruction is the operation code and specifies both the instruction and the addressing mode.

Accumulator addressing is a one-byte instruction. An operation on the accumulator is implied. In *immediate addressing*, the operand is contained in the second byte of the instruction. However, in *absolute addressing*, the second byte of the instruction specifies the 8 low-order bits of the effective address while the third byte specifies the 8 high-order bits. Thus, the absolute addressing mode allows access to the entire 64K bytes of addressable memory. These three modes resemble the register, immediate, and direct modes discussed and shown in Section 7-1.

Zero-page addressing allows for shorter code and execution times by fetching only the second byte of the instruction and assuming a zero high address byte. Note the similarity to the absolute addressing mode. Careful use of the zero page can result in a significant increase in programming efficiency. *Indexed zero-page addressing* (X, Y indexing) is used in conjunction with the index register and is referred to as "zero page, X" or "zero page, Y." The effective address is calculated by adding the second byte to the contents of the index register. Since this is a form of zero-page addressing, the content of the second byte references a location in page zero. Indexed zero-page addressing is the same as the indexed mode shown in Figure 7-2, without use of byte 3.

Indexed absolute addressing (X, Y indexing) is used in conjunction with X and Y index registers and is referred to as "absolute, X" and "absolute, Y." The effective address is formed by adding the contents of X or Y to the address contained in the second and third bytes of the instruction. This mode, which is identical to the indexed mode discussed in Section 7-1, allows the index register to contain the index or count values and the instruction to contain the base address. This type of indexing allows referencing of any location *and* index modification of multiple fields, resulting in reduced coding and execution time. In the *implied addressing* mode the address containing the operand is implicitly stated in the operation code of the instruction. *Relative addressing* is used only with branch instructions and establishes a destination for the conditional branch. The second byte of the instruction becomes the operand, which is an "offset" added to the contents of the lower 8 bits of the program counter when the counter is set at the next instruction. The range of the offset is -128 to $+127$ bytes from the next instruction. See *relative addressing* in Figure 7-2.

In *indexed indirect addressing* (sometimes called "indirect, X"), the second byte of the instruction is added to the contents of the X index register, discarding the carry. The result of this addition points to a memory location on page zero whose contents is the low-order 8 bits of the effective address. The next memory location in page zero contains the

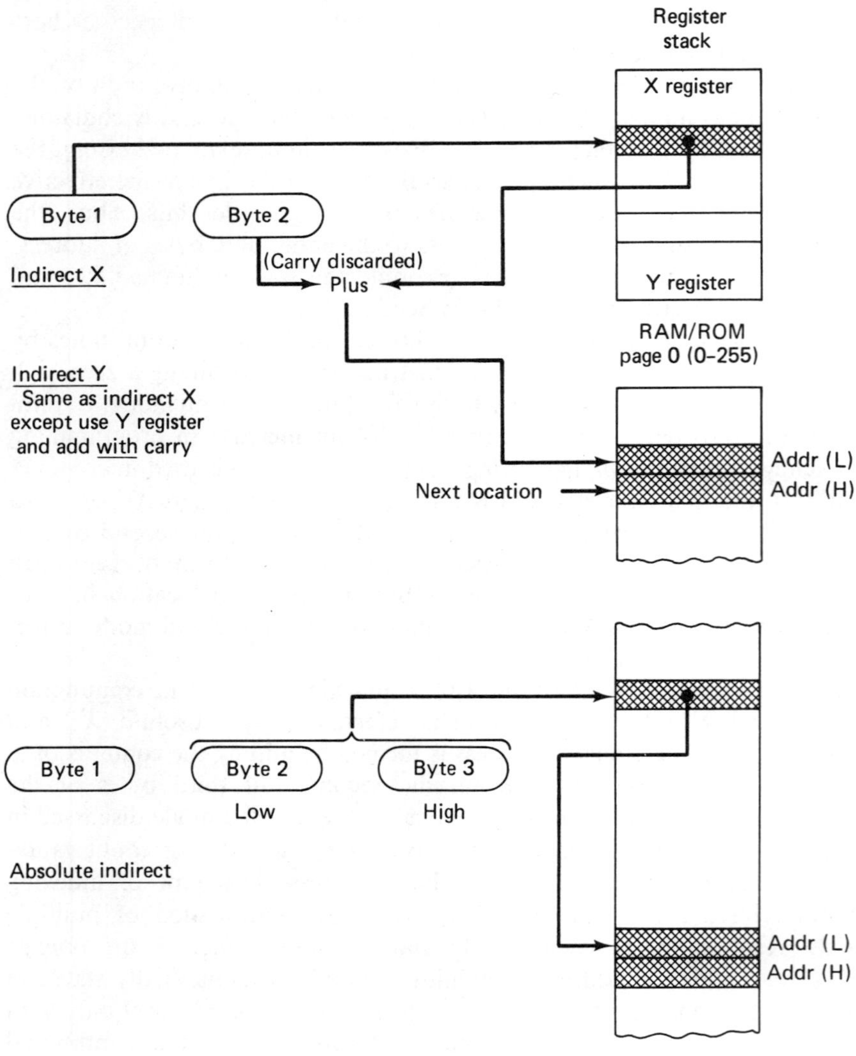

Figure 12-2 Special Addressing Modes

high-order 8 bits of the effective address. Both memory locations specifying the high- and low-order bytes of the effective address must be in page zero. The second byte of the instruction points to a memory location in page zero in *indirect addressing* (referred to as "indirect, *Y*"). The contents of this memory location are added to the contents of the *Y* index register, the result being the low-order 8 bits of the effective address. The carry from this addition is added to the contents of the next page-zero memory location, the result being the high-order 8 bits of the effective address (see Figure 12-2).

When *absolute indirect addressing* is used, the second byte of the instruction contains the low-order 8 bits of a memory location. The high-order 8 bits of that memory location are contained in the third byte of the instruction. The contents of the fully specified memory location are the low-order byte of the effective address. The next memory location contains the high-order byte of the effective address, which is loaded into the 16 bits of the program counter. Figure 12-2 shows absolute indirect addressing.

The Instruction Set. As mentioned in Chapter 7, the instruction set is the means of connecting one part of the microprocessor and/or the microcomputer to another. The concept of instruction set design was developed in Chapter 7 and is not repeated here. It should be remembered, however, that a certain amount of correlation can be found between groups of instructions if the instruction set is examined closely. The instruction set for our representative microcomputer is developed in the remainder of this section.

Instruction Classification

Group 1. These instructions are all *single-byte instructions*, requiring two cycles for execution. It should be noted that all these instructions are applicable *only* in the implied and accumulator addressing modes.

Group 2. *Internal execution on memory data* is the common thread among Group 2 instructions. They require two to six cycles for execution, and are two bytes in length for all addressing modes except for absolute modes which are three bytes long.

Group 3. Group 3 instructions perform *storing operations*. They are two bytes in length for all addressing modes except the absolute modes, which are three bytes long. Three to six cycles are required for execution.

Group 4. These instructions are the READ, MODIFY, and WRITE instructions. They require the same amount of bytes as Group 2 and Group 3 instructions, and need five to seven cycles for execution.

Table 12-2 Instruction Classifications

* SINGLE BYTE	** INTERNAL EXECUTION ON MEMORY DATA	** STORE	** READ/MODIFY/WRITE	** MISCELLANEOUS
ASL(accum)	ADC	STA	ASL	BCC
CLC	AND	STX	DEC	BCS
CLD	BIT	STY	INC	BEQ
CLI	CMP		LSR	BMI
CLV	CPX	3 to 6	ROL	BNE
DEX	CPY	cycles	ROR	BPL
DEY	EOR			*** BRK (7 cycles)
INX	LDA		5 to 7	BVC
INY	LDX		cycles	BVS
LSR(accum)	LDY			JMP
NOP	ORA			JSR
ROL(accum)	SBC			*** PHA (3 cycles)
ROR(accum)				*** PHP (3 cycles)
SEC	2 to 6			*** PLA (4 cycles)
SED	cycles			*** PLP (4 cycles)
SEI				*** RTI (6 cycles)
TAX				*** RTS (6 cycles)
TAY				3 to 7
TSX				cycles
TXA				
TSX				
TYA				
2 cycles				

Notes: First cycle always OP CODE fetch
- * Accumulator or implied addressing mode only
- ** 2-byte for all addressing modes except 3-byte for absolute modes
- *** Single-byte but more than 2 cycles

Group 5. The MISCELLANEOUS instructions may be one, two, or three bytes in length, requiring up to seven cycles to execute. The STACK, JUMP, BRANCH, and RETURN instructions are included in this group.

Table 12-1 lists the 56 instructions available for use by the microprocessor, while Table 12-2 groups the instructions into the five previous classifications.

There are numerous methods used to display the complete operation code table for a microprocessor. They can be as simple as Table 12-1, which merely defines the meaning of each mnemonic (abbreviation). However, such a table does not supply the actual code used when programming at machine language. A very detailed table may be constructed, supplying such information as addressing mode, the actual op code, the number of bytes required for each operation, and the number of computer cycles. Such a table is most commonly found in manufacturers' specification sheets. For our purposes, a matrix (Figure 12-3) will be formed, showing the op code, the mnemonic code for each operation, and the addressing mode. When used with Tables 12-1 and 12-2, adequate information is supplied to actually design and construct small programs, as will be noted shortly.

Programmer's Model of the Microcomputer. The programmer's model of the microcomputer is seen in Figure 12-4, while the instructions available for use with the microcomputer were shown in Table 12-1. Those portions of the microcomputer that perform data-manipulation functions *and* are available to the programmer are included in the model. In addition, the arithmetic/logic unit (ALU) is also shown because of its almost constant use during data manipulations.

Most of the programmer's model of the microcomputer centers around the microprocessor and its registers. The hardware description of the microprocessor in Chapter 11 defined the purpose of the registers in Figure 12-4, and the reader should refer to Section 11-3 if a review is required. In general, the programmer's model is used to depict the flow of information throughout the microcomputer during execution of instructions. A detailed description of each instruction is required to fully employ the programmer's model.

Since the use of programmers' models was detailed in Chapter 7, additional coverage is not provided in this chapter. The details of specific instruction flow are included in Section 12-2, where machine language programming is discussed.

MSD \ LSD	0	1	2	3	4	5	6	7	8	9	A	B	C	D	E	F
0	**BRK** *imp* 5	**ORA** *inx* 2				**ORA** *z* 2	**ASL** *z* 4		**PHP** *imp* 5	**ORA** *im* 2	**ASL** *acc* 1			**ORA** *ab* 2	**ASL** *ab* 4	
1	**BPL** *rel* 5	**ORA** *iny* 2				**ORA** *zpx* 2	**ASL** *zpx* 4		**CLC** *imp* 1	**ORA** *aby* 2				**ORA** *abx* 2	**ASL** *abx* 4	
2	**JSR** *ab* 5	**AND** *inx* 2			**BIT** *z* 2	**AND** *z* 2	**ROL** *z* 4		**PLP** *imp* 5	**AND** *im* 2	**ROL** *acc* 1		**BIT** *ab* 2	**AND** *ab* 2	**ROL** *ab* 4	
3	**BMI** *rel* 5	**AND** *iny* 2				**AND** *zpx* 2	**ROL** *zpx* 4		**SEC** *imp* 1	**AND** *aby* 2				**AND** *abx* 2	**ROL** *abx* 4	
4	**RTI** *imp* 5	**EOR** *inx* 2				**EOR** *z* 2	**LSR** *z* 4		**PHA** *imp* 5	**EOR** *im* 2	**LSR** *acc* 1		**JMP** *ab* 5	**EOR** *ab* 2	**LSR** *ab* 4	
5	**BVC** *rel* 5	**EOR** *iny* 2				**EOR** *zpx* 2	**LSR** *zpx* 4		**CLI** *imp* 1	**EOR** *aby* 2				**EOR** *abx* 2	**LSR** *abx* 4	
6	**RTS** *imp* 5	**ADC** *inx* 2				**ADC** *z* 2	**ROR** *z* 4		**PLA** *imp* 5	**ADC** *im* 2	**ROR** *acc* 1		**JMP** *abi* 5	**ADC** *ab* 2	**ROR** *ab* 4	
7	**BVS** *rel* 5	**ADC** *iny* 2				**ADC** *zpx* 2	**ROR** *zpx* 4		**SEI** *imp* 1	**ADC** *aby* 2				**ADC** *abx* 2	**ROR** *abx* 4	
8		**STA** *inx* 3			**STY** *z* 3	**STA** *z* 3	**STX** *z* 3		**DEY** *imp* 1		**TXA** *imp* 1		**STY** *ab* 3	**STA** *ab* 3	**STX** *ab* 3	

Figure 12-3.

9	**BCC** *rel* 5	**STA** *iny* 3			**STY** *zpx* 3	**STA** *zpx* 3	**STX** *zpy* 3		**TYA** *imp* 1	**STA** *aby* 3	**TXS** *imp* 1			**STA** *abx* 3		
A	**LDY** *im* 2	**LDA** *inx* 2	**LDX** *im* 2		**LDY** *z* 2	**LDA** *z* 2	**LDX** *z* 2		**TAY** *imp* 1	**LDA** *im* 2	**TAX** *imp* 1		**LDY** *ab* 2	**LDA** *ab* 2	**LDX** *ab* 2	
B	**BCS** *rel* 5	**LDA** *iny* 2			**LDY** *zpx* 2	**LDA** *zpx* 2	**LDX** *zpy* 2		**CLV** *imp* 1	**LDY** *aby* 2	**TSX** *imp* 1		**LDY** *abx* 2	**LDA** *abx* 2	**LDX** *aby* 2	
C	**CPY** *im* 2	**CMP** *inx* 2			**CPY** *z* 2	**CMP** *z* 2	**DEC** *z* 4		**INY** *imp* 1	**CMP** *im* 2	**DEX** *imp* 1		**CPY** *ab* 2	**CMP** *ab* 2	**DEC** *ab* 4	
D	**BNE** *rel* 5	**CMP** *iny* 2				**CMP** *zpx* 2	**DEC** *zpx* 4		**CLD** *imp* 1	**CMP** *aby* 2				**CMP** *abx* 2	**DEC** *abx* 4	
E	**CPX** *im* 2	**SBC** *inx* 2			**CPX** *z* 2	**SBC** *z* 2	**INC** *z* 4		**INX** *imp* 1	**SBC** *im* 2	**NOP** *imp* 1		**CPX** *ab* 2	**SBC** *ab* 2	**INC** *ab* 4	
F	**BEQ** *rel* 5	**SBC** *iny* 2				**SBC** *zpx* 2	**INC** *zpx* 4		**SED** *imp* 1	**SBC** *aby* 2				**SBC** *abx* 2	**INC** *abx* 4	

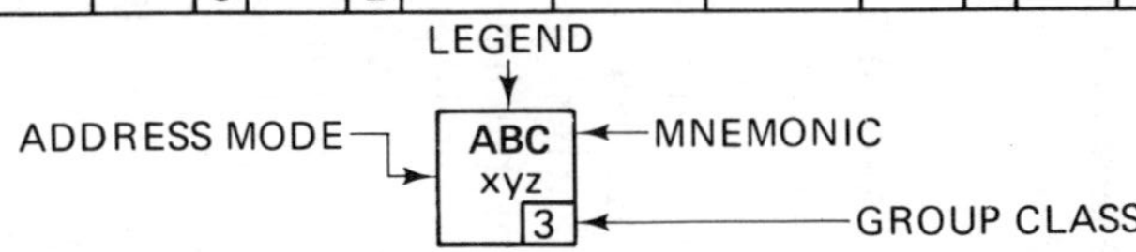

ADDRESS MODE

ab—absolute
abx—x indexed absolute
aby—y indexed absolute
acc—accumulator
im—immediate
imp—implied
inx—x indexed indirect
iny—y indexed indirect
abi—absolute indirect
rel—relative
z—zero page
zpx—x indexed zero page
zpy—y indexed zero page

GROUP CLASS

1—Single byte
2—Internal execution on memory data
3—Store operations
4—Read, modify and write
5—Miscellaneous

Note: Blank indicate no instruction implemented.

Figure 12-3 Op-Code Matrix

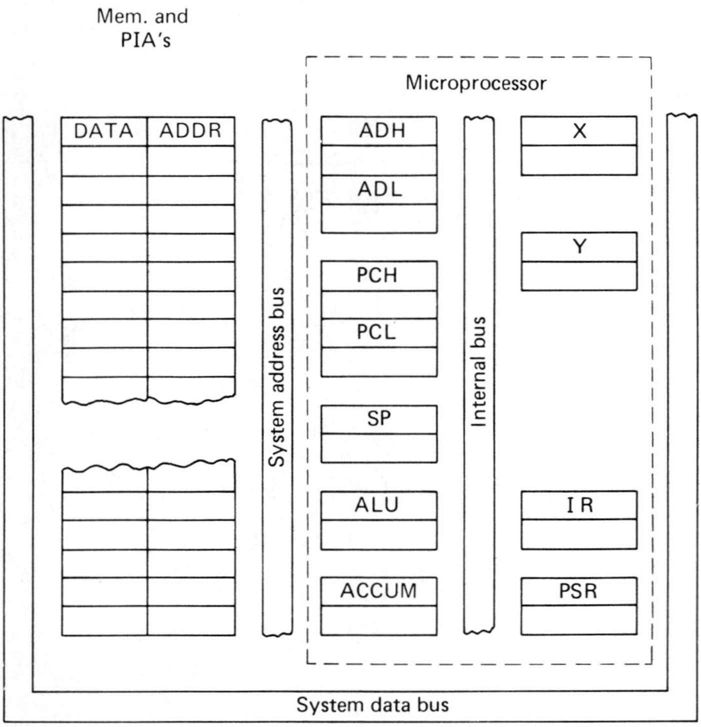

Figure 12-4 Programmer's Model of Microcomputer

12-2 MACHINE LANGUAGE PROGRAMMING

Characteristics. In machine language programming, a knowledge of the programmer's model of the microcomputer and the details of the instruction set is required. Furthermore, it is the programmer's responsibility to keep accurate records of all storage locations and to assure that all program "branches" are to the proper location.

The use of machine language and assembly language are closely related. Actually, most programmers who work with machine languages employ the mnemonics of the assembly language during preparation of their programs. It is much easier to use the mnemonics because of their resemblance to the actual operations to be performed. However, the program must still be entered into the microcomputer in machine code. An example of a machine-language-coded routine that uses the instruction set and mnemonics of Table 12-1 is shown in Figure 12-5. Decimal notation is initially included to aid understanding of the concepts of machine language programming. The flowchart describes an operation that provides a controlled delay between microcomputer operations. It is often necessary to delay operations in a microcomputer. For example, when a key on the input keyboard is depressed, it is necessary to delay storing the information from the key due to the "bounce" that occurs upon original key depression. A delay routine is often used to hold off reading the key switch until the mechanical operations have settled down and no further bounce is occurring. The flowchart steps are discussed below.

Sample Routine. The concept of the delay routine is that of loading a constant into one or more of the microcomputer's registers, and using the time that it takes to decrement the register(s) to zero as the delay. In the delay routine that serves as an example of machine language programming, index register *X* stores the master delay constant, while index register *Y* stores the multiplier constant. Figure 12-5 includes the mnemonic development of the delay routine.

The master delay constant is determined in the following manner. The DEX and BNE instructions form a loop that is executed the number of times determined by the master delay constant loaded in register *X*. DEX, a single-byte instruction, requires two processor cycles for execution, hence it takes 2 microseconds to complete (a 1-MHz cycle time is assumed). The BNE instruction requires three cycles, or 3 microseconds. Each time through the loop, then, requires 5 microseconds. A 1-millisecond delay may be obtained by setting the *X* register to 200_{10}, since each decrement and test operation requires 5 microseconds ($5 \times 200 = 1000$ microseconds, or 1 millisecond). The multiplier loop consists of the DEY, BNE, and LDX instructions, which provide a 7-microsecond delay. Since the multiplier loop contains the delay loop, each time through the multiplier loop takes the selected delay time *plus* the multiplier time (e.g., for a 1000-microsecond delay loop, the overall time is 1007 microseconds). Total routine delay is then the selected delay time plus 7 microseconds times the multiplier stored in the *Y* register. If 10 is stored in the *Y* register, the total delay is then 1007×10 microseconds, or 10,070 microseconds (10.07 milliseconds). It should be noted that in long delays such as shown here, the delay in executing the multiplier loop is of little consequence. For general estimates, total delay is the delay loop time multiplied by the number of times the routine executes the delay loop (i.e., the multiplier constant).

Discussion of each of the instructions will provide an insight into application of the instruction set and its many variations. It should be noted that each of the bytes in the routine occupies one location in memory. To assure accurate record keeping, each of the locations is assigned an address, and as the program progresses, the locations are incremented for each new byte. Since most programs require returns to certain portions of their routines for repeat operations, it is very important

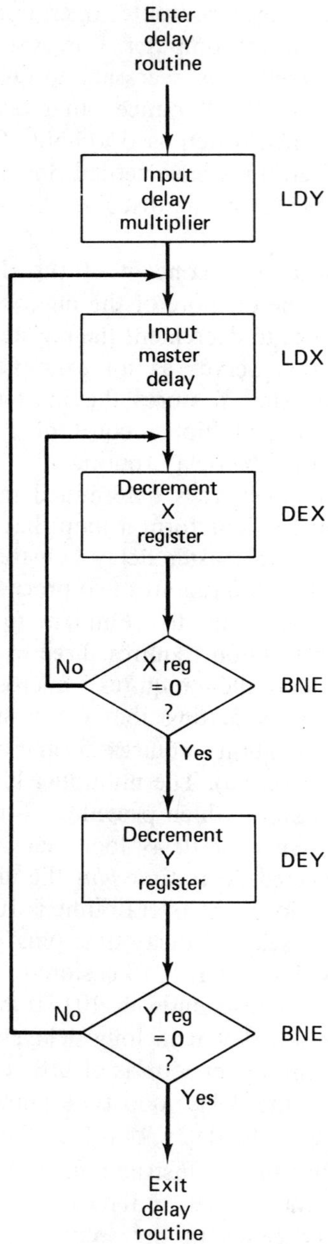

(a) Flow chart

Figure 12-5 Machine Language Routine: (a) Flowchart; (b) Routine

Address			Data			
EX	Decimal		HEX	Decimal	Mnemonic	Remarks
800	2048		AØ	16Ø	LDY	Load multiplier constant in Y
801	2049		ØA	1Ø		
802	2050		A2	162	LDX	Load delay constant in Y
803	2051		C8	2ØØ		
804	2052		CA	2Ø2	DEX	Decrement X
805	2053	Multiplier loop / Delay loop	DØ	2Ø8	BNE	X = 0?
806	2054		FD	253		
807	2055		88	136	DEY	Decrement Y
808	2056		DØ	2Ø8	BNE	y = 0?
809	2057		F8	248		
80A	2058		60	96	RTS	Return

ote: The machine language program consists of only the hex address and hex data columns. (see below) The remainder is shown as an aid to understanding.

tarting address 0800 H	AØ	ØA	A2	C8	CA	DØ	FC	Actual machine-language listing
	88	DØ	F8	96				

nd address 080A H

(b) Routine

Figure 12-5 (Continued)

that the programmer maintain good records, and that the routines and programs are well documented even in the early development phases. The delay routine of Figure 12-5 starts at location 2048_{10}.

Both the two-byte LDY and LDX instructions are straightforward. They employ the *immediate* addressing mode, so the byte following each instruction is comprised of the data to be loaded into registers *Y* and *X*, respectively. Thus, register *Y* is set to $10_{10}(0A_{16})$ and register *X* is set to $200_{10}(C8_{16})$ as a result of the first two instructions. The DEX instruction requires only one byte, since the register to be decremented is implicit within the instruction. Each time DEX is executed, the *X* register is decremented by 1 (i.e., 200 to 199 to 198, etc.).

The BNE instruction is not quite so simple. Its function is to test the Processor Status Register "*Z*" bit to see if it indicates that the result of the previously executed instruction was zero. If *not* zero, the byte following the instruction is the *relative address* of the next instruction to be executed. As noted in Section 12-2, the second byte is an "offset" that is to be added to the contents of the lower 8 bits of the program counter when the counter is set at the next instruction. It should be noted that the offset range is -128 to $+127$, and if the branch must be greater than that amount, other addressing techniques must be employed. In the delay routine being discussed, it is necessary to *go back* to the DEX instruction and repeat if the previous DEX instruction did not reduce the *X* register value to zero. Branching forward is no problem, because it is merely necessary to add the offset to the program counter contents to reach the next required step. Since there is no convenient way to tell the program to branch backward, it is merely told to branch forward by an amount that is greater than the +127 limit. When the BNE instruction detects that the branch address has exceeded the allowable limit, it goes back to the beginning of the 256 byte "page" that has its upper address at the upper allowable limit and continues counting. In other words, addition without carry is performed, and the branch takes place as though subtraction had actually been performed. Thus, to branch backward from 2054 to 2052, the routine is told to branch forward 253 locations.

When the BNE instruction recognizes that the *X* register has been decremented to zero, the loop is exited and the next instruction (DEY) is performed. The next BNE instruction then begins to look for the *Y* register's contents to become zero. If not zero, control is transferred back to the LDX instruction, and the *X* register is reloaded and redecremented to zero before the *Y* register is redecremented. The multiplier loop is repeated until the *Y* register's contents reach zero, and the loop is exited to the final instruction (RTS). The RETURN FROM SUBROUTINE instruction then transfers control back to the portion of the program that initially called for the delay routine.

12-3 ASSEMBLY LANGUAGE PROGRAMMING

Characteristics. The transition from machine language to assembly language is a small step from the viewpoint of the programmer. Assembly mnemonics have already been used to ease the burden of machine language programming, and the programmer should already be familiar with them. The major advantage of using an assembler is that it is no longer necessary to work with binary or hexadecimal number systems unless desired, and the translation from mnemonics to machine language is a function of the assembler, not the programmer.

An *assembler is a collection of tables and routines that performs much of the bookkeeping normally required of the programmer.* Each of the instructions is correlated in tables with its machine language equivalent and addressing modes so that it is only necessary to input the mnemonic describing the instruction. The matching entries in the tables are recovered and entered as part of the routine being programmed. All characters, letters, numbers, symbols, and so on are defined so that a keyboard entry automatically results in the proper machine language code being generated from applicable tables. Routines for conversion between the various number systems are provided and identifier symbols assigned so that the assembler knows with which number system it is working.

Sample Routine. The assembler requires a strict format of information presented to it in order to perform its functions. Table 12-3 shows a typical assembly input/output format that displays the same program used in the preceding section. Each line, which may or may not have a line number attached, represents *one* machine language instruction. As such, it should be noticed that assembly language programming does not present any savings in terms of numbers of instructions required to perform a specific function. In fact, if one considers the assembler program required to obtain the benefits of mnemonic programming, considerably more memory space may be required for small programs in an assembly language.

The address column contains the hexadecimal address of the first byte of the instruction on the line. This address is assigned by the assembler, relieving the programmer of the bookkeeping task of keeping track of address assignments. Up to three bytes may appear in the next column, depending on the instruction. These codes are also assigned by the assembler and are based on the operator-entered information in the *operation* and the *operand* columns. A label column is also provided to identify the beginning of a specific section or subroutine of a larger routine. These points are identified so that the assembler can assign an address to the label, providing a location for return when another operation requires access to the same subroutine.

The operation column contains the mnemonic that describes the specific operation to be performed. It is the operation that causes assignment of values to the address and byte one of the coding column. As noted earlier, this information is contained in a table, and when a valid operation

Table 12-3 Assembly Language Routine

Line No.	*Address*	*Coding*	*Label*	*Operation*	*Operand*	*Remarks*
01	0800	A0 0A	DELAY	LDY	# $ 0A	Load Multiplier Constant
02	0802	A2 C8	LOOP 2	LDX	# 200	Load Delay Constant
03	0804	CA	LOOP 1	DEX		Decrement *X*
04	0805	D0 FD		BNE	LOOP 1	*X*=0?
05	0807	88		DEY		Decrement *Y*
06	0808	D0 F8		BNE	LOOP 2	*Y*=0?
07	080A	96		RTS		Return

code is input to the assembler, a search of the table is made to identify and assign the proper hexadecimal code. Bytes 2 and 3 of the coding column are identified by yet another table. Consider line 01 in the sample assembler output. The operand #$0A provides considerable information. Immediate addressing mode is identified by the # sign, while the $ sign denotes that the numbers that follow are hexadecimal in nature. Thus, an operand such as #$0A indicates that 0A hexadecimal is to be acted upon by the operation indicated in the adjacent column.

Operator input of LDY, #$0A would result in a complete line of information such as line 01. An address would be assigned, LDY would be decoded to 0A as a machine language operation, and 0A would be placed in the second byte of the instruction. The label would be recognized, and an address for storage of DELAY's address would be assigned. Thus, it can be seen that much less operator intervention is required when using an assembler.

Using an Assembler. An assembler is used in the following manner. The programmer constructs a program in the conventional manner, implementing each of the operations required by the flowchart. Mnemonics are used to identify each instruction and arbitrary names are assigned as labels for specific routines. Variables are identified by letters or combinations of letters and listed as operands for applicable operations. Thus, the label, operation, and operand columns in the assembled listing of Table 12-3 are the responsibility of the programmer. Remarks may be added to remind the reader of the instructions' functions, but they are not processed by the microcomputer.

When the programmer is satisfied that the written program will function as designed, the label, operation, and operand information is entered, in the order written, into the microcomputer. The assembler initially scans the input information for labels, assigning addresses to each. Instructions are decoded and addresses assigned to variables, although no codes are generated at this time. Any errors, such as invalid codes, are detected and brought to the operator's attention. "Pass 1" is considered complete at this time, and if no errors are found, "pass 2" may be initiated (either manually or automatically). Instruction and operand codes, together with address information, are generated during pass 2. The program is executed, and the results are displayed in the form called out in the program. A complete listing, such as shown in Table 12-3, may also be displayed if desired.

The reader is urged to correlate Table 12-3 with the machine language version of the same program. It should be apparent that assembled programs are much easier to work with. Hexadecimal or binary input of a machine language program is tedious and error-prone, supplying another "plus" for assembled programs. About the only "plus" for machine language programs is that considerably more memory space is required to obtain assembler capability. Applications exist for both, and the reader is urged to develop capability with both machine and assembly language programming if even moderate use of microprocessors and microcomputers is anticipated.

12-4 HIGH-LEVEL PROGRAMMING

Characteristics. Both machine language and assembly language programming tend to be unique to a specific model of microprocessor or microcomputer. As such, their use with machines developed by other manufacturers is limited, if not impossible. Numerous problem-oriented rather than machine-oriented languages are in use, and programs written in these languages are transferrable to other machines. In addition, problem-oriented programming languages tend to use instructions that more closely resemble the English language.

High-level programming languages may be categorized as *compilers* or *interpreters*. Compilers and interpreters are collections of machine or assembly language routines that convert statements in the language into machine language instructions. In very broad terms, a *compiler collects all the programming statements provided, translates them all at one time, and runs the program*. An *interpreter translates and runs statement by statement without having to wait for the complete program to be input*. Thus compilers are usually faster than interpreters, but interpreters tend to appear more "interactive" to the user, since results of executing a program statement are almost immediately available.

The representative microcomputer discussed in Chapters 11 and 12 uses an interpretive language called *BASIC*. Developed in the early sixties at Dartmouth College in cooperation with General Electric, BASIC was designed to be a very simple math and science language usable by students without special training. The version used in this chapter's microcomputer requires 8K of ROM to store the routines necessary to convert the relatively few statements of the BASIC programming language into machine language instructions. An additional 6K of ROM is employed to handle input/output routines, diagnostic error routines, and so on. It is perhaps apparent by now that as programming languages grow in complexity, more and more memory space is required just to do the translating from "easier-to-use" language into the machine language required by the microprocessor. However, as languages become easier and easier to use, more and more people can use the facilities of the microcomputer. Furthermore, as the demand rises, mass-production techniques can be applied to manufacture, and prices fall. It can be anticipated that as the microprocessor becomes applicable to more and more everyday tasks, programming languages will be further simplified to attract the potential consumer. Additional memory will be required, of course, but memory costs are plummeting, and the future may well see the cost of supplying the translating software far exceeding the cost of the hardware.

BASIC. A list of the BASIC instructions and operations common to many small-scale microcomputers is shown in Table 12-4. An example and the general meaning of each of the instructions are also included. It is not anticipated that the reader will become a proficient programmer as a result of this chapter, but several examples of simple BASIC programs are included to acquaint the reader with the characteristics of the language.

Each line in a program is identified by a *line number*, used to determine the actual sequence of execution. Line numbers are assigned by the programmer. A BASIC *statement* is written on each line to describe the

Table 12-4 Some Typical Basic Statements, Functions, and Operations

Statements/ Functions/ Operations	*Example*	*Purpose*
DATA	10 DATA 1,2,3,4	Specifies data to be read, left to right
END	999 END	Terminates program
INPUT	20 INPUT *A*	Accepts value of *A* from input device
PRINT	30 PRINT *A*	Prints value of *A* on output device
READ	40 READ *B*	Obtains value of *B* from DATA statement
FOR...NEXT	25 FOR *A*=1 TO 10 100 NEXT *A*	Performs all instructions between FOR and NEXT as many times as specified by index value *A*, i.e., loop control statement
IF...THEN	50 IF *B*=2 THEN 10	If condition is TRUE, instruction following THEN is executed
GOTO	70 GOTO 100	Jumps to statement number 100
RETURN	200 RETURN	Return from subroutine; control is transferred to main program.
COS	20 *X*=COS(*B*)	Returns cosine of argument *B*
INT	30 *Y*=INT(*X*)	Returns largest integer less than or equal to argument *X*
SQR	40 *B*=SQR(*A*)	Returns square root of argument *A*
=	10 *X*=Y	Assigns a value to a variable
*	20 *A*=B*7	Multiplication
/	30 *X*=Y/Z	Division

Note: Only a *partial* list. Consult manual for microcomputer in use for complete list of BASIC language vocabulary.

operation to be performed. BASIC statements, like English-language sentences, are made up of *words* from the *vocabulary* of the BASIC language. A number of statements are combined to make up a *program*. An example of a BASIC statement is

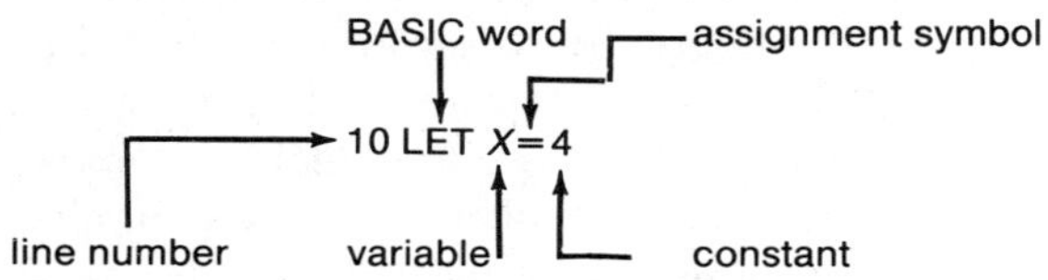

The statement on line 10 contains the line number (10), the BASIC word LET, a variable X, the = sign that is used in this case as an assignment symbol, and the number 4 that is assigned by statement number 10 to the variable X. When statement 10 is executed, the interpreter sets aside room in memory for the variable X, and places 4 in the assigned location. The LET statement is one way to input information to the microcomputer.

By combining a few simple BASIC statements, a program may be developed that performs a simple arithmetic operation on two variables. The microcomputer is relatively useless unless information can be supplied to it and recovered after processing. Program 1 below demonstrates some of the simple input and output statements allowed in BASIC, and at the same time provides the product of two variables.

Program 1, Product of Two Variables

		sample of output after program executed
10	LET X=4	
20	LET Y=6	RUN
30	LET Z=X*Y	4 6 24
40	PRINT X, Y, Z	READY
50	END	

Both statements 10 and 20 perform the same function (i.e., assigning numerical values to variables). Other methods of supplying input to the microcomputer exist and are shown in later sample programs. Also note that, although the line numbers are in sequence, they are separated by a numerical gap. The gap is left so that additional statements may be inserted if desired. When the total program is completed, the interpreter will place all lines in numerical order before execution, thus allowing additions and deletions to the program without completely reordering each step. This is just another of the many bookkeeping functions performed by the interpreter to relieve the user of the numerous menial tasks associated with programming.

Statement 30 performs two functions. First, it sets aside a location to store the value of the variable Z. It then causes the value of the stored variable X to be multiplied by the value of the stored variable Y. (Note the use of * to denote the multiplication operation.) The result is placed in the location set aside for the variable Z. The stored values of X, Y, and Z are output from the microcomputer by statement 40, one of the many forms of output in BASIC. PRINT will thus cause the numbers 4, 6, and 24 (see Program 1) to be displayed on the output device, separated by a fixed number of spaces between each number. The actual number of spaces is a function of the microcomputer *and* the commas separating each of the variables in the PRINT statement. Other spacing symbols are allowed, resulting in either less space between numbers, or printing of numbers on separate lines. Statement 50, END, is a signal to the microcomputer that no more statements are forthcoming in this program.

Program 2 below demonstrates some additional BASIC words by providing the statements required to prepare a list of numbers starting at A and continuing to B.

Program 2

```
10      INPUT A, B
20      FOR N=A TO B
30      PRINT N
40      NEXT N
50      END
```

Statement 10 is another example of inputting information to the microcomputer. Upon execution, statement 10 causes some type of prompting character (usually a ?) to be printed. The microcomputer then waits until a value for each of the variables in the INPUT statement is entered and then proceeds to the next line. Statements 20 and 40 establish a loop, which is repeated as many times as required by the parameters A and B. The PRINT statement displays each number in sequence as the loop executes. FOR $N=A$ TO B establishes the smallest number (at the beginning of the list) by obtaining the number stored at location A by the INPUT statement. The last number in the list (the largest number) is likewise recovered from location B.

Under normal circumstances statements are executed sequentially. Thus when statement 20 establishes A as, for example, 10 and B as 20, N will initially become equal to A and statement 30 is executed. When N is printed, the program proceeds to statement 40. N is incremented by 1 (to 11) and compared with B (20) by returning to statement 20. A is still less than B, so the program executes statements 30 and 40 again. The loop continues until B is reached, at which time the program jumps to the step immediately following statement 40. Statement 50 causes cessation of program execution. The final output of the program will be a list of numbers from 10 to 20, sequentially arranged.

As a final example, Program 3 performs an operation similar to the delay routine discussed in the machine language and assembly language sections. The microcomputer under discussion includes a number of timers,

Program 3, Delay Routine

```
10    X=TI
20    IF (TI-X)<60 THEN 20
30    END
```

one of which increments every $\frac{1}{60}$ second (approximately 16.7 milliseconds). A storage location permanently designated as TI accumulates the number of $\frac{1}{60}$-second intervals. If, for example, a 1-second delay was required, it would only be necessary to start the delay, accumulate 60 timing intervals, and then terminate the delay. Statement 10 stores the *current* value of TI in a temporary location called X. In statement 20 the IF··· THEN operation in BASIC is introduced. If the condition following the IF word is true [i.e., $(\mathrm{TI}-X)$ *is* less than 60], the statement following the THEN word is executed. If not true, the next statement is executed. Thus, following the storage of the current value of TI in X, the program continues to repeat statement 20 until 60 timing intervals have been accumulated. When $(\mathrm{TI}-X)$ is no longer less than 60, the control transfers to statement 30, which ends program execution.

Although three simple programs using the BASIC programming language have been shown, only the "tip of the iceberg" has appeared. It should be apparent, however, that many advantages exist with the use of an interpretive language. The reader is urged to explore the use of BASIC and other high-level languages by reference to the many excellent texts now available.

12-5 SUMMARY

Chapter 12 has followed the software of a typical microcomputer from its primitive instruction set to the implementation of a high-level language. At the machine language level it should have been noted that operations were closely allied to the hardware discussed in Chapter 11. Therefore, the machine language programmer must be quite aware of the microprocessor and microcomputer components and their functions. In fact, even at the assembly language level, considerable interaction with the hardware exists. As indicated, both machine language and assembly language programming are usually machine-dependent.

When the high-level languages such as BASIC are used, however, machine dependency fades, giving way to dependency on knowledge of the problem being programmed and the relatively simple grammar and vocabulary of the language itself. Programs may be transferred from microcomputer to microcomputer with a minimum of modification. Programmers can communicate easily, and program ideas flow freely, even to relatively inexperienced persons.

A price has been paid for ease of programming, however. Even as programming changed from machine language to assembly language, memory-space requirements zoomed. Where only two or three bytes were required to perform a machine language instruction, up to 100 bytes were needed to handle the same assembly language step. Higher-level languages appear even more demanding. As noted, 14K of ROM space was required to implement BASIC language and its associated monitoring, diagnostic, and input/output operations in the typical microcomputer of this chapter.

The combination of Chapters 11 and 12 has provided an in-depth look at a small-scale microcomputer. As the exercises that follow are performed, refer often to the preceding chapters. They contain a wealth of information and will materially aid solution to the exercises and acquisition of information concerning modern microcomputers.

QUESTIONS

1. Discuss the characteristics of machine languages. Relate your discussion to the microcomputer of Chapter 11, if desired.
2. Discuss the characteristics of assembly languages. Compare to those characteristics described in Question 1.
3. Discuss the characteristics of high-level languages, including both compilers and interpreters.
4. Explain the purpose of each of the 13 addressing modes discussed in Section 12-1 and show how each could be used. Figure 7–2, the functional diagrams of Chapter 11 or the programmer's model of Chapter 12 may be used if desired.
5. Construct a flowchart and write a machine language routine that will provide a 0.25-second delay. Use the instruction set shown in this chapter.
6. Repeat Question 5 using an assembly language routine.
7. Using the instruction set of this chapter, write a machine language and an assembly language routine that multiplies two unsigned 8-bit data bytes.
8. Using the instruction set of this chapter, write a machine language and an assembly language routine that converts an 8-bit serial byte into an 8-bit parallel byte.
9. Modify Program 2 of this chapter to list selected numbers in descending order. Use BASIC.
10. Modify Program 3 of this chapter to provide a print of time every 10 seconds. Use BASIC.

CHAPTER 13

Applications

The traditional uses of the microcomputer are not discussed in this chapter. Such applications are immediately evident in daily life. In this final chapter, two unusual microcomputer/microprocessor applications are shown. A household appliance, the microwave oven, is controlled by a small microprocessor to efficiently manage time and energy. Second, a microprocessor-based "intelligent typewriter" is explained, showing the vast improvement possible in office procedures and tasks. Finally, although not an actual application, a 16-bit microprocessor is discussed as an introduction to the expanding field of larger microcomputers.

13-1 MICROWAVE OVEN CONTROL

Functional Operation. Adaptation of microcomputer principles to high-sales-volume household devices is seen in Figures 1-8 and 13-1. The microwave oven has become a common appliance, and the microprocessor and associated supporting hardware and software make it a versatile addition to the kitchen.

Control of the microwave oven is accomplished using a 4-bit microcomputer contained on a single chip. The only additional hardware required is that necessary to tailor the input/output circuits to the microcomputer chip circuits. Review of Figure 13-1 will show that all traditional computer functions are present. *Input* and *output* devices interface the

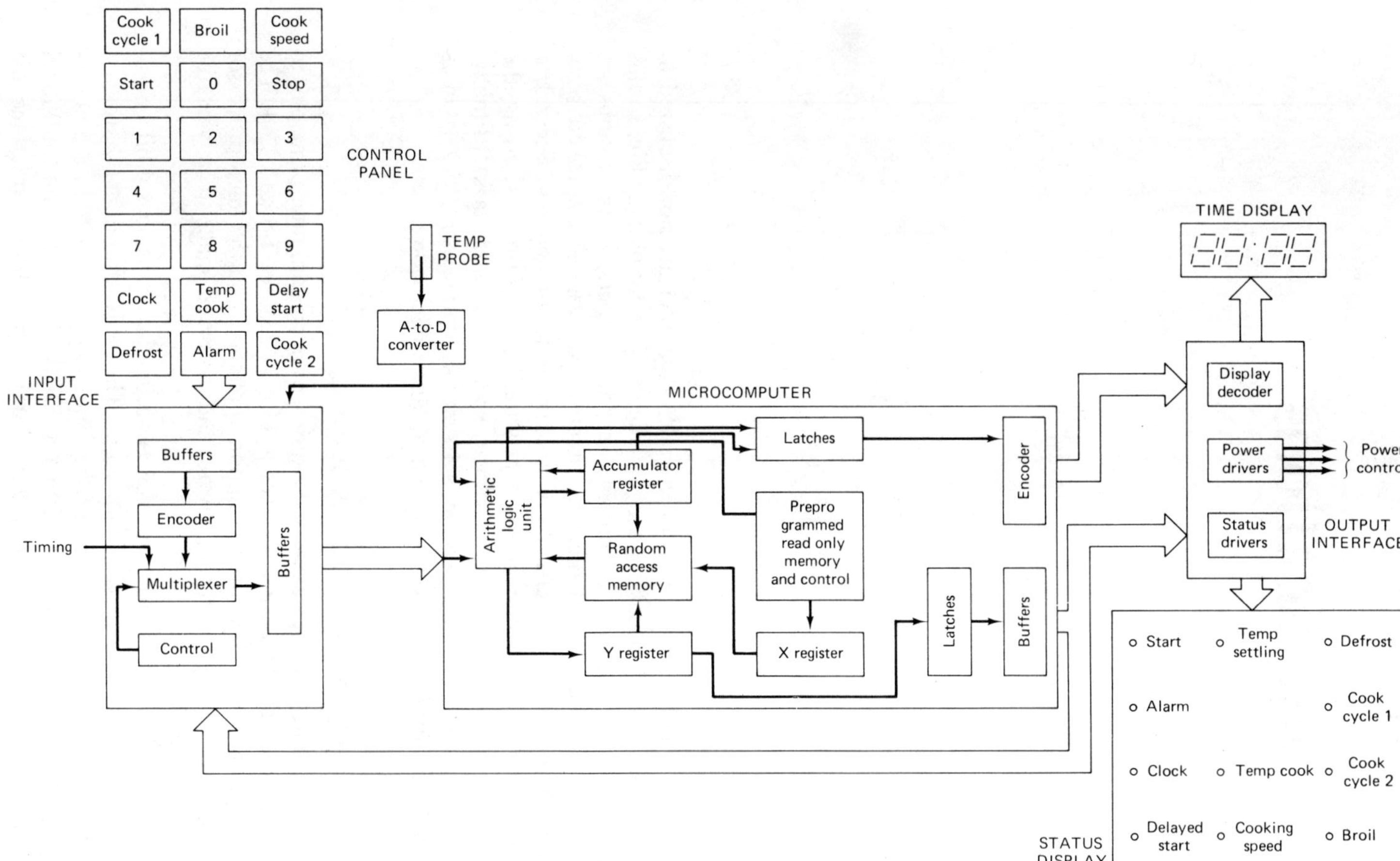

Figure 13-1 Microwave Oven Control (Courtesy Texas Instruments)

microcomputer with the outside world. *Memory* and *arithmetic / logic* functions are easily identified as part of the microcomputer chip, while the remainder of the chip accomplishes the *control* function. It should be noted that both RAM and ROM exist as part of the microcomputer chip, thereby making applications limited only to the software that is stored in ROM. Actually, this microcomputer chip is a standard item programmed by the manufacturer to function as a microwave oven controller. The same chip, with different software in ROM, is used in many other applications.

The complete controller acts as a special-purpose digital computer, performing its control functions as directed by the ROM-stored program. Software implements cooking by supplying seven separate timers (four cooking timers, a delayed-start timer, an alarm timer, and a time-of-day clock). All timers are "adjustable" from the control panel. The timing related to the actual operation in process is available on the time display, while the mode of operation is shown on the status display. Temperature cooking is also available by use of a temperature-sensing probe. Self-test programming is included for the convenience of service personnel. Previous chapters have detailed operations of all of the components shown in Figure 13-1, and the reader should readily be able to follow the flow of information within the controller.

The *control panel* is used to enter all numerical information (time-of-day, cycle duration, cooking speeds, etc.), select specific cycles, and reset alarms and errors. When a key on the control panel is pressed, the resultant switch closure is encoded by the *encoder* in the *input interface*. A 4-bit word is generated by the interface and routed to the microcomputer for processing. Timing and temperature probe signals are also converted to 4-bit words and switched by the *multiplexer* under program control for use by the microcomputer.

Subfunctions within the microcomputer operate in much the same manner as those functions in the microprocessors and microcomputers discussed in earlier chapters. All inputs to the microcomputer result from actions in the *input interface*, and all outputs are fed to the *output interface* as either control signals from internal *buffers* to actuate *power* and *status drivers* or time-related information from an internal *encoder* to be decoded and displayed on the *time display*. Input information is processed in the *ALU* and *accumulator register* under control of the preprogrammed *ROM*. Results of processing are routed to *latches* via either the accumulator or the *Y register* and thence to the *output buffers* and *encoder*. The *X register* and *RAM* are used for routing and temporary storage of program information and data.

Detailed Operation. An example of a "software" timer is seen in Figure 13-2. Each of the timers could have been implemented in hardware, but it will be seen that the software timer is also quite practical. The

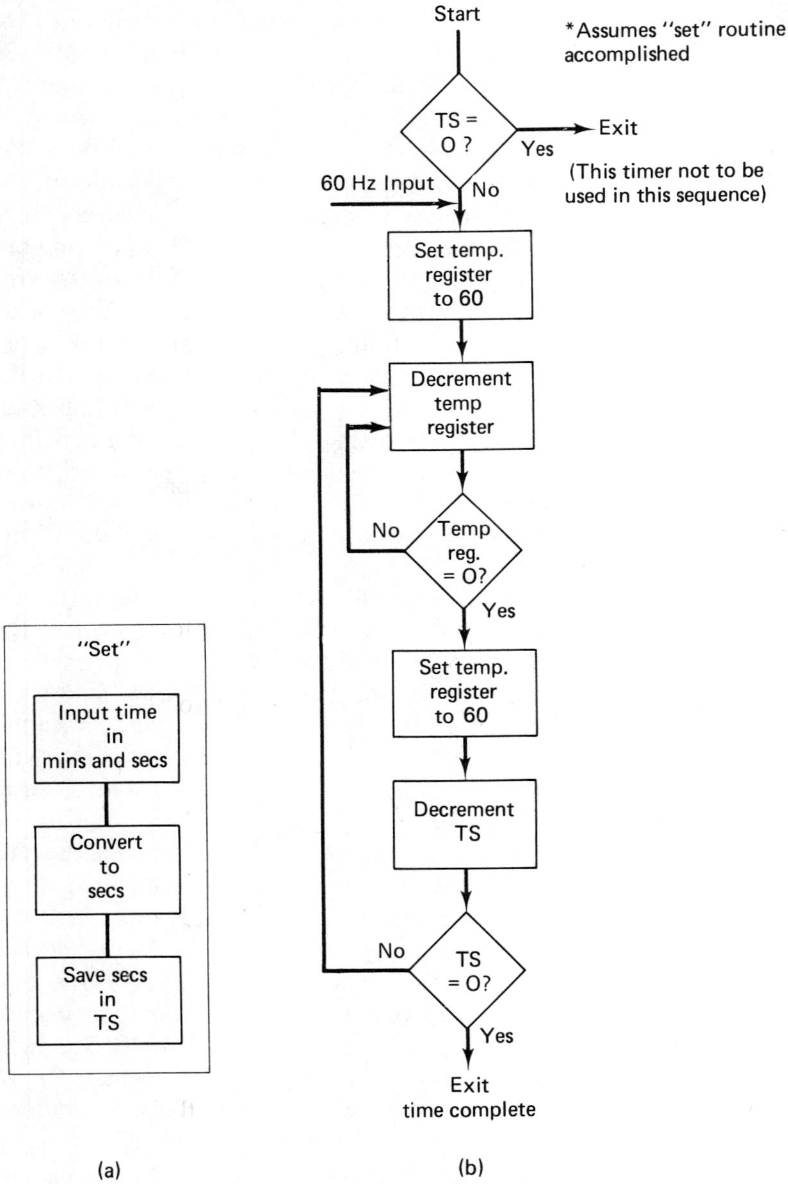

Figure 13-2 Timer Flowchart

scheme employed is much like the delay routines used in earlier discussions of serial input/output.

A common subroutine that can be used to set any of the timers is assumed to be available. Such a routine is quite simple and may be seen in outline form in Figure 13-2a. Although it is possible to perform the timing functions directly in minutes and seconds, the set routine used in this discussion converts minutes and seconds into a total number of seconds and stores it in a temporary storage location that is mnemonically named "TS." The timer routine will constantly refer to the number stored in TS.

Actual use of the timer routine is accomplished by depressing the START button on the control panel of the microwave oven. The first step in the routine (Figure 13-2b) determines whether TS is 0. If TS is 0, no timing is necessary during this cooking sequence, and the routine exits to the next timing sequence. If TS is not 0, however, a temporary register is loaded with the decimal number 60 and the 60-Hz power-line frequency is applied to the input of the timer routine. Each time the 60-Hz input generates a timing pulse (60 times per second), the temporary register is decremented by 1 and its contents compared to 0. If the temporary register has not yet reached 0, the routine returns to the decrement step to await the next 60-Hz input. If 0 has been reached (occurs only once each second), the temporary register is reset to 60 and TS is decremented by 1.

TS is examined to determine if it has reached 0, and if not, control is returned to the temporary register-decrementing step to continue counting. If 0 has been reached, the timer has completed its function, and the routine is exited with an indication that cooking time has expired.

The microwave oven executes its five cooking programs in a fixed sequence, as shown in Figure 13-3. Each of the programs has an associated timer, and their times may be set in any order. However, when the oven door is closed and the START control is depressed, execution will proceed in the order shown in Figure 13-3.

Up to 60 minutes is available from the *defrost timer*, followed by an equal amount of *automatic temperature settling time*. During the defrosting period, the oven's magnetron (source of cooking heat) is cycled on and off at 15-second intervals to provide maximum defrosting with minimum cooking of the food. The 50% duty cycle is an integral part of the defrost timer's ROM-stored program.

Cook cycle 1 can exist for up to 60 minutes maximum, with selectable duty cycle from 100% to 10% in 10% intervals. Upon completion of cook cycle 1, *Cook cycle 2* is entered. Selectable duty cycle is also provided in cook cycle 2. However, maximum cooking time is limited to 30 minutes. A *broil timer* (up to 30 minutes) is also provided by the microwave oven controller. A separate heating element is used by the broil timer to brown the food. Upon completion of *all* cycles, an alarm is sounded to inform the operator.

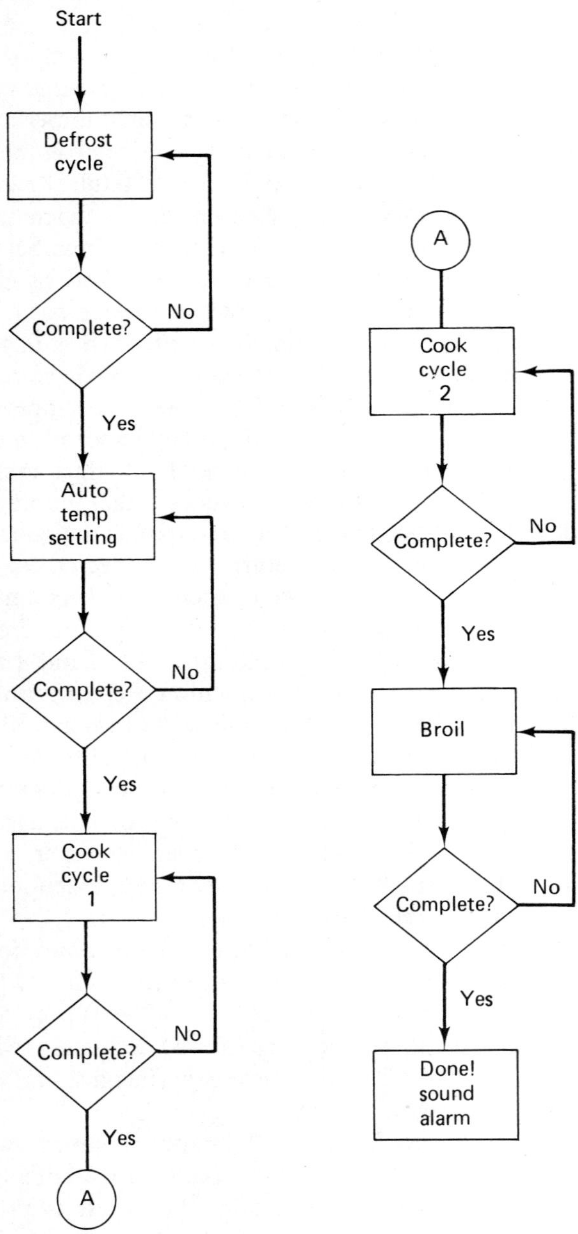

Figure 13-3 Sequence Flowchart

A delayed start capability is also provided by the microwave oven controller. The microcomputer computes the time to start the cooking sequence by subtracting the total of all cooking times from the desired end time. Thus, a complete cooking program may be entered at any time during the day, the food placed in the oven, and the operator may leave the unit unattended. At the proper time, the oven goes into action and prepares the food so that it will be ready at the calculated time.

Temperature-controlled cooking is also a feature of the microwave oven controller. A temperature probe is placed in the food to be cooked, and the temperature of the probe converted by the microcomputer into a digital value that may be used to determine completion of the cooking cycle. The programmed temperature is only approximate, however, and the operator must depend on experience to obtain maximum results.

Any of the timer's contents are accessible during the cooking cycle. Depression of the appropriate control button will display the timer's contents for 3 seconds and the cooking cycle will continue without interruption. Furthermore, if the oven door is opened during cooking, the cooking cycle halts and the time remaining is displayed. When the door is closed, cooking resumes where it was interrupted upon depression of the START button. The program is lost only due to unanticipated power failure or use of the STOP button.

The microprocessor controlled microwave oven, then, has been perhaps one of the first consumer-oriented appliances to appear in the home. Microprocessor versatility has provided the homemaker with precise control of the cooking process. This is but one of the labor- and time-saving applications now available to the consumer. As the old saying goes—"The best is yet to come!"

13-2 AN "INTELLIGENT TYPEWRITER"

Functional Operation. Another interesting application of the microprocessor/microcomputer may be seen in a business-oriented requirement. It is often necessary for a business to send copies of the same letter to different addresses, but with "personal" information included within the body of each letter. If the major part of the letter could be stored in some sort of memory, it would be necessary to type in only the personal message while the "intelligent typewriter" took care of the remainder of the letter. Such a system is described in this section.

A block diagram of the intelligent typewriter is shown in Figure 13-4. Microprocessor implementation of the typewriter is relatively simple, requiring only the microprocessor, RAM, ROM, and a few miscellaneous support ICs. The major emphasis is on software, not hardware. A serially

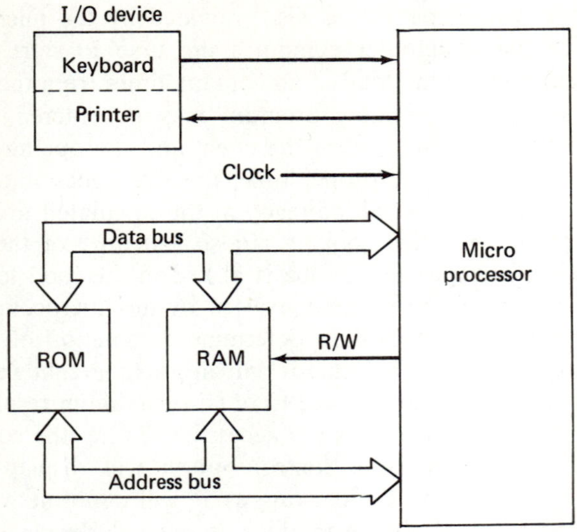

Figure 13-4 Intelligent Typewriter Functional Block Diagram (Courtesy Signetics)

encoded keyboard and printer similar to the conventional Teletype* terminal are used for input/output. RAM stores the information to be typed on the letter, while ROM supplies the software to operate the system. The microprocessor could be one of many available, and since the intelligent typewriter should not be device-dependent in this discussion, only flowcharts of specific routines are shown rather than actual detailed program segments.

The serial output of the keyboard (see Chapter 8 for discussion of serial keyboards) is accepted by a microprocessor input port and stored, via software, as a parallel character. Within the microprocessor all information is processed in parallel. When information is printed, it must be converted to serial form for transfer to the printer. Once again this task is accomplished by software.

ROM stores the routines required to input, process, and output information, along with all control functions. Information and control data required to be stored are placed in RAM during input and transferred to the printer during output. ROM and RAM capacities are determined by control complexity and amount of information to be printed. The flowcharts and sample program segments that follow describe system operation.

*Trademark registered by Teletype Corporation.

Detailed Operation. Figure 13-5 explains the overall functional operation of the intelligent typewriter. When the intelligent typewriter is started, an initialization routine is performed to set the keyboard, hardware, and printer to a known state. Assuming that a character has already been stored in the buffer, the ROM-stored program goes to the buffer and retrieves the information. It is first checked to determine if it is a valid character. If not, control is returned to the INPUT routine. If it was valid, it is next checked to see if it is a control character. The control function is performed if detected, and the program loops back to the INPUT routine. If not a control character, it is assumed to be a printable character. The operational mode is checked and if PRINT-ONLY mode is encountered, the OUT routine is entered and the character printed. If not in PRINT-ONLY mode, the character is stored and the program checks to make sure there is adequate room in memory. An alarm is sounded if storage is full, and the program branches back to the beginning. Characters are printed under control of the OUT routine.

It should be noted that the overall program is merely made up of a number of separate routines tied together by logical decisions. This is a common method of composing programs, and the reader should consider this approach as programs are constructed.

As an example of one of the routines in the general flowchart of Figure 13-5, consult the detailed flowchart of Figure 13-6, which depicts the *input character routine*. The input routine is entered following initialization, and the following sequence of events occurs. The serial input line from the keyboard is continuously checked to determine if a START bit has been received (refer to Chapter 8 for a description of the typical serial Teletype character). As soon as the START bit is detected, a $\frac{1}{2}$-bit delay is initiated to check for noise or an invalid START bit. If the START bit is still present when the delay runs out, it is assumed that the START bit is valid. Registers to be used in the routine are loaded to their required values, and a full-bit time delay is initiated. Upon completion of the full-bit delay, the information present on the input is stored in a temporary register. The storage operation occurs during the middle of the bit to assure that any logic-level transitions have settled. The middle of the first bit was reached in the following manner. The START bit beginning was initially detected and the $\frac{1}{2}$-bit delay was initiated, locating the middle of the START bit. The full-bit delay thus delayed sampling of the input beyond the beginning of the first bit by $\frac{1}{2}$ bit, locating the middle of the first bit.

The information stored in the temporary register is moved to a permanent register immediately after sampling, and a determination is made as to how many information bits have been received. When all 7 bits have been stored (i.e., the loop executed 7 times), one more full-bit delay is

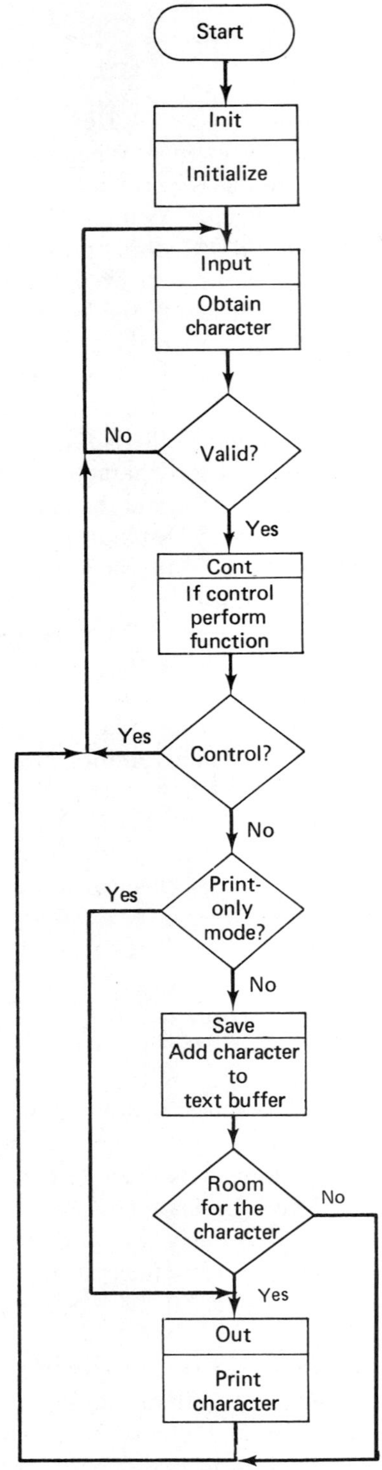

Figure 13-5 Main Flowchart, Intelligent Typewriter (Courtesy Signetics)

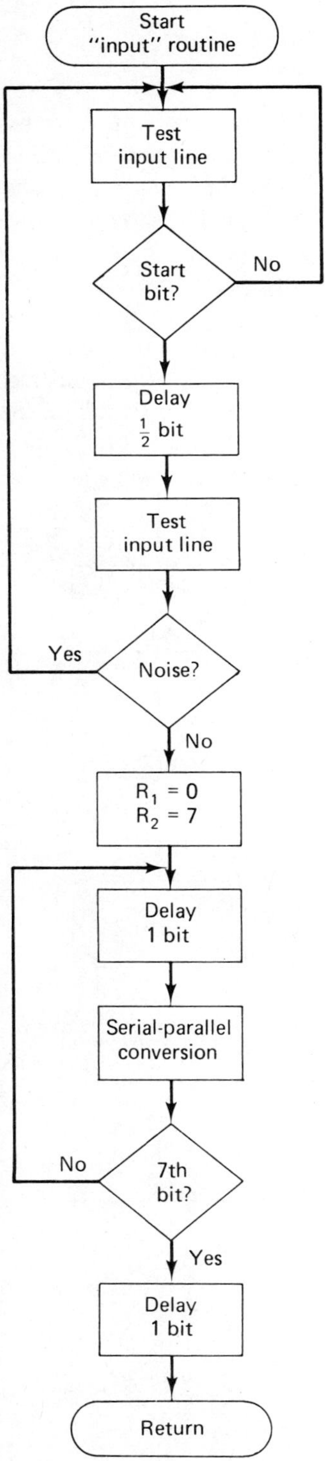

Figure 13-6 Input Character Routine

initiated, and the control of the program returned to the main program. The 7 information bits have now been recovered and are stored in the character buffer mentioned in the general flow chart.

Although this chapter is geared generally toward system discussion, it might be enlightening to look at some of the implementations of the INPUT flowchart. The delay routines have already been discussed in detail in Chapter 12 and may be implemented easily using those techniques. Programming the serial-to-parallel conversion block in the flowchart demonstrates some of the additional techniques available to the programmer. Figure 13-7 provides the details of the analysis of the conver-

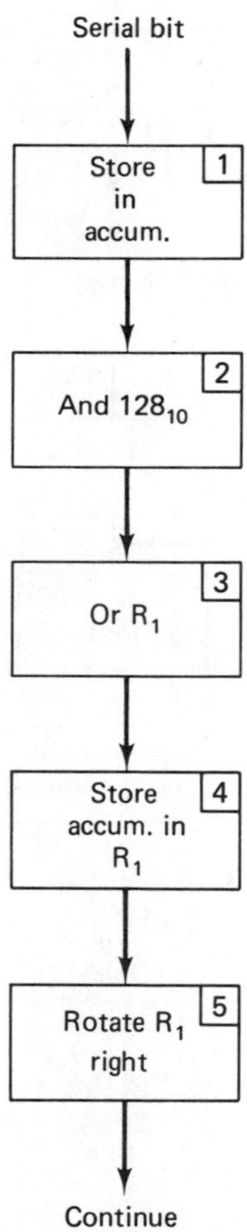

(a) Flow chart

Figure 13-7 Serial-to-Parallel Conversion: (a) Flowchart; (b) Bit Manipulation

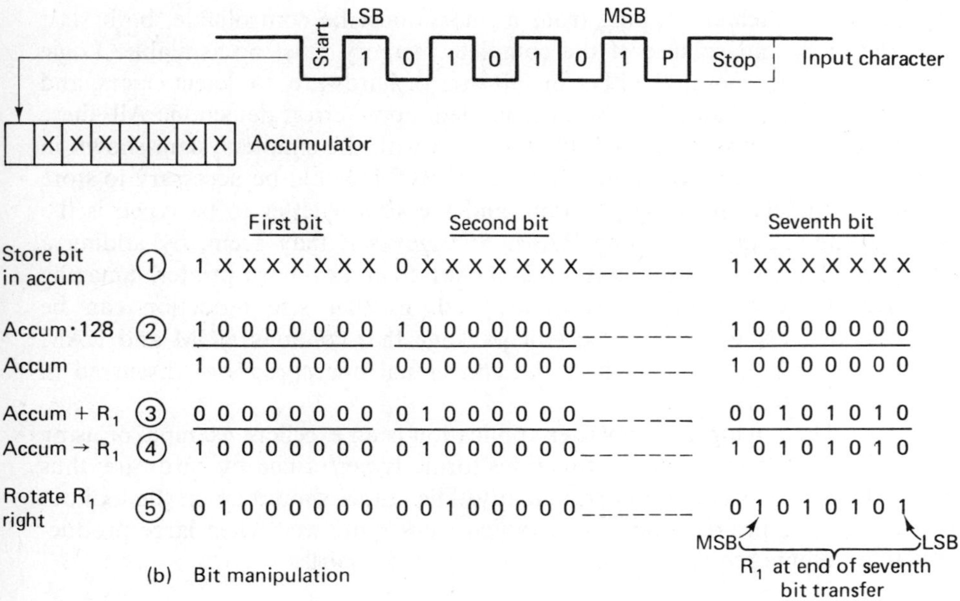

Figure 13-7 (Continued)

sion process. Actual coding is not shown, since its form would depend on the microprocessor selected for the intelligent typewriter implementation. The initial step in the serial-to-parallel conversion for each character is accomplished by transferring the information at the serial input to the most-significant position in the accumulator (block 1). When the accumulator is ANDed with 128_{10} (block 2), all but the most-significant position will be masked out. The result of block 2 is placed back in the accumulator, which is then ORed with R_1 (block 3). Since R_1 has been initialized to 00000000 prior to transferring the first bit of the character, the result will continue to be that in the accumulator, which is transferred to R_1 (block 4). R_1 is now rotated 1 bit right (block 5) to complete the operation. As a result of these five blocks, the LSB of the serial character is stored in R_1, one location to the right of the most-significant position.

According to the main flow chart, if 7 bits have not been received, the loop repeats to read in additional bits. Figure 13-7b shows the manipulation for the first, second, and seventh bit. Note that upon completion of the seventh bit transfer, the serially formed character from the keyboard is now stored in parallel form, MSB first in R_1.

Obviously, many of the details of both hardware and software have been omitted from this discussion. Provision must be made to delete an error from memory if an improper key has been pressed while placing the information into memory. When the location in the letter is reached where the personal information is to be inserted, the operator must be able to stop the machine. Typing from memory must be controllable, both start and stop, and erasure of the complete memory must be available. Logic must be incorporated, either in software or hardware, to detect errors, and some kind of alarm must be sounded upon error detection. All these functions, however, are easily attainable with the simple system shown in Figure 13-4. No more than 1K bytes of ROM should be necessary to store the complete operating program, and the size of letter to be typed is the limiting factor concerning RAM. Strange as it may seem, by adding a printed circuit card about 2×3 inches to a keyboard-printer, amazing improvement in capability is attained. Further size reduction can be obtained using a microprocessor package that contains ROM and RAM directly on the chip, such as the conceptual microprocessor discussed in Chapters 6 and 7.

The intelligent typewriter application is an excellent example of using software to perform many functions formerly performed by hardware, thus reducing parts count and size greatly. The reader should be on the lookout for these types of applications, which work quite well when large production volume requires minimum cost outlay for parts.

13-3 FROM MICRO TO MINI

Characteristics. As microprocessors and microcomputers have evolved, the dividing line between *micro*computers and *mini*computers has become less distinct. Reduced hardware costs have allowed microcomputers to take on memory capacities approaching that of minicomputers, and many tasks formerly assigned to minicomputers are performed by microcomputers. The "gray area" between microcomputers and minicomputers is now being bridged by applications of *16-bit microprocessors.*

One of the major hurdles to be overcome was the 8-bit microprocessor requirement of two or more bytes to identify both the operation and the operand. Each byte requires a memory cycle, and as data-handling requirements became greater, large amounts of microprocessor time were expended just obtaining information from memory. Furthermore, most interrupt-type operations, such as input and output, necessitate memory operations to store microprocessor register data while the interrupts are being serviced. Also, much valuable time is lost servicing interrupts.

The solutions to these shortcomings are many and varied, just as with their 8-bit predecessors. Many of the approaches seem to be tied to a 40-pin package requirement, in an attempt to maintain the minimum number of external connections. When the microprocessor processes 16 bits, the address and data buses must have 16-bit capability also. Continued use of the 40-pin package then requires multiplexing (time sharing) of the external buses, with a resultant increase in overall execution time. A

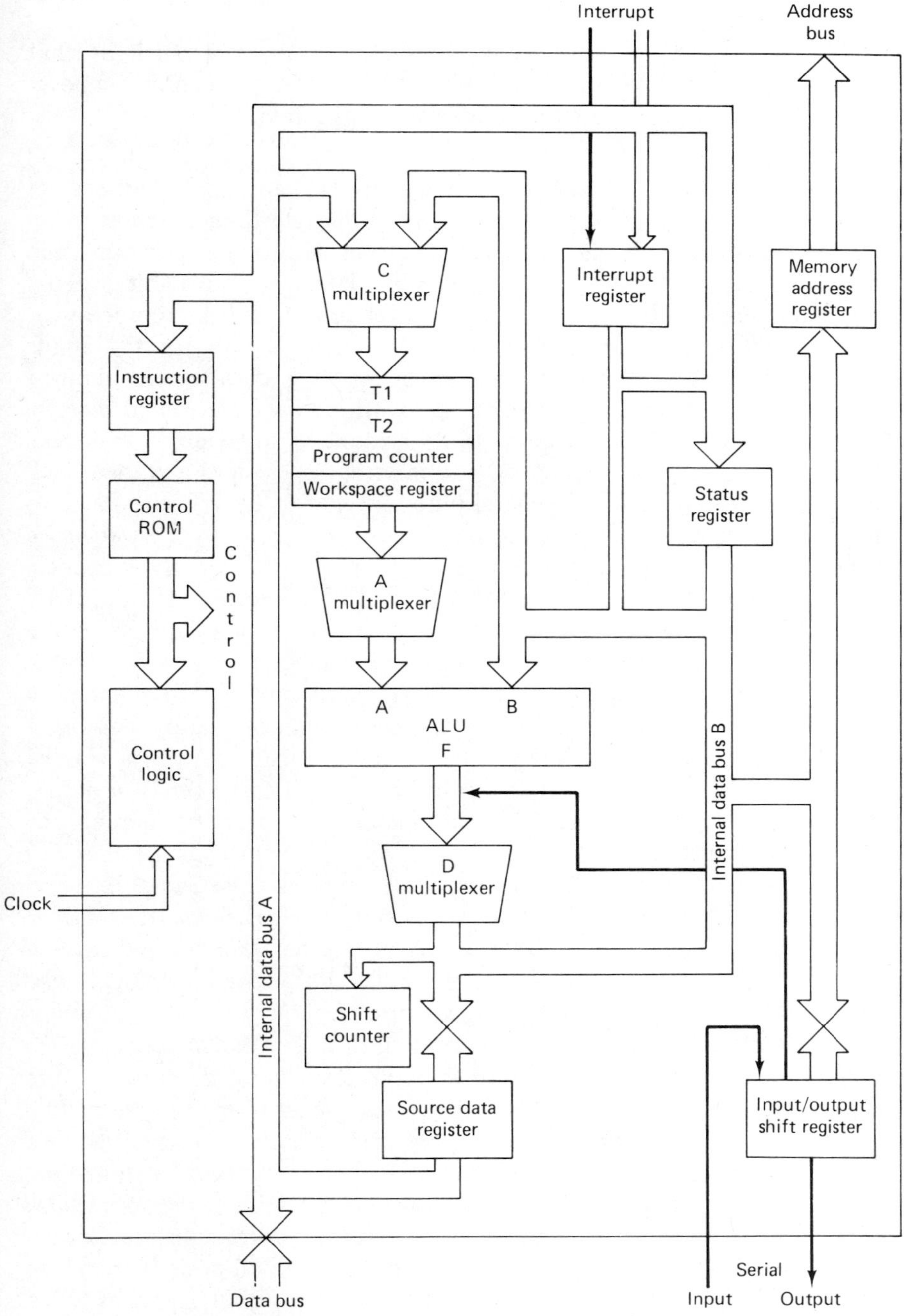

Figure 13-8 16-Bit Microprocessor Block Diagram (Courtesy Texas Instruments)

16-bit microprocessor with *minicomputer-like architecture*, such as that shown in Figure 13-8, solves many of the problems mentioned above. However, an increase to a 64-pin package was required.

Functional Description. Although all the traditional functions are found in the 16-bit microprocessor, some rather startling differences are noted. The solution to the extra memory cycle to obtain operation code and operand is, of course, to use a 16-bit data bus and 16-bit internal registers/memory. But, unlike many conventional 8-bit microprocessors, the internal register complement is quite small. Except for the status register and program counter, no actual working data are maintained directly on the microprocessor chip. All working data are routed directly from memory to the microprocessor's arithmetic/logic unit (or to the special-purpose registers for status and interrupt operation) and then back to memory. Effectively, the general-purpose registers have been moved to memory. The general-purpose register stack is addressed by the *workspace pointer*, which addresses the first of 16 consecutive memory locations. Thus, the processor has access to sixteen 16-bit registers. If additional registers are needed, it is only necessary to load a new address in the workspace pointer and 16 more registers are available.

Microprocessor control is conventional. Instructions from memory are temporarily stored in the *instruction register* and decoded by the *control ROM*. Routing of decoded instructions and external controls is via the *control logic* block. Input/output operations are primarily serial in nature and are accomplished via the *input/output shift register*. Timing for input/output operations is under microprocessor control using the *shift counter*. On-chip memory is for all practical purposes nonexistent. External memory is addressed via the address bus and the *memory address register*, while memory data input/output are via the external data bus and *internal data bus A*. Separate *interrupt logic* with up to 17 separate interrupts is also incorporated to furnish maximum flexibility.

The versatility of this 16-bit microprocessor is in the *arithmetic/logic unit* and its associated registers and multiplexers. One of the inputs to the arithmetic/logic unit is selected by the *A multiplexer* from the register file, which contains two *temporary registers*, the *program counter*, and the *workspace register*. The *temporary registers* contain data used for short-time storage during execution of microprocessor instructions. *Program counter* functions are conventional (i.e., keeping track of instruction sequences), while the *workspace register* points to the current 16-position working register stack in memory. Input to the register file is from either *data bus A* or *data bus B* via the *C multiplexer*.

Information on *data bus B* supplies data to the second arithmetic/logic unit input. Typical sources are the *input/output shift register*, the *status register*, and the *source data register*. The *input/output shift register*, as previously noted, is the primary source of data from the outside world. Flag bits indicating results of most recent arithmetic or logic operations performed plus interrupt information are contained in the *status register*. The *source data register* contains data going to or coming from memory.

Arithmetic/logic unit output is switched via the *D multiplexer* to *data bus B*, which can feed the *shift counter* and the *input/output shift register* for serial data input/output. All other registers connected to *data bus B* may obtain data from the arithmetic/logic unit as determined by the information on the internal control bus.

Working Model. The working model of the 16-bit microprocessor and associated memory is shown in Figure 13-9. Note that as far as the microprocessor is concerned, only three internal registers are available to the programmer: the program counter, the workspace register, and the status register. Each register has already been discussed, and will now be integrated with memory and program operation from the programmer's viewpoint. The program counter contains the address of the instruction following the instruction currently being executed. Program counter contents are referenced by the processor to obtain the next instruction from memory and then automatically incremented. The status register contains the present state of the processor and is used to determine method of response to conditional tests and interrupts. Finally, the workspace register stores the address of the first word in the currently active set of workspace registers. It is used to establish the register stack employed in normal instruction execution.

A workspace register file occupies 16 contiguous memory words in the area of memory not set aside for special functions. It should be noted that memory is actually organized as $64K \times 8$ bits but is addressable in the two-byte mode ($32K \times 16$ bits). Each workspace register may store data or addresses and may function as operand registers, accumulators, address registers, or index registers. The processor may address any register in the workspace during execution of instructions merely by adding the register number to the contents of the workspace pointer and initiating a memory request for the word.

When the processor must change from one environment to another, such as during interrupt servicing or subroutine operations, the workspace concept becomes very important. With conventional 8-bit microprocessors, many internal registers must store their contents in a register stack so that they may return to normal operation upon completion of the interrupted operation. With the 16-bit microprocessor, only three registers as a maximum must be stored, since all other registers are already a part of memory.

As a normal part of its operation, the 16-bit microprocessor stores the program counter, status register, and workspace register during interrupted

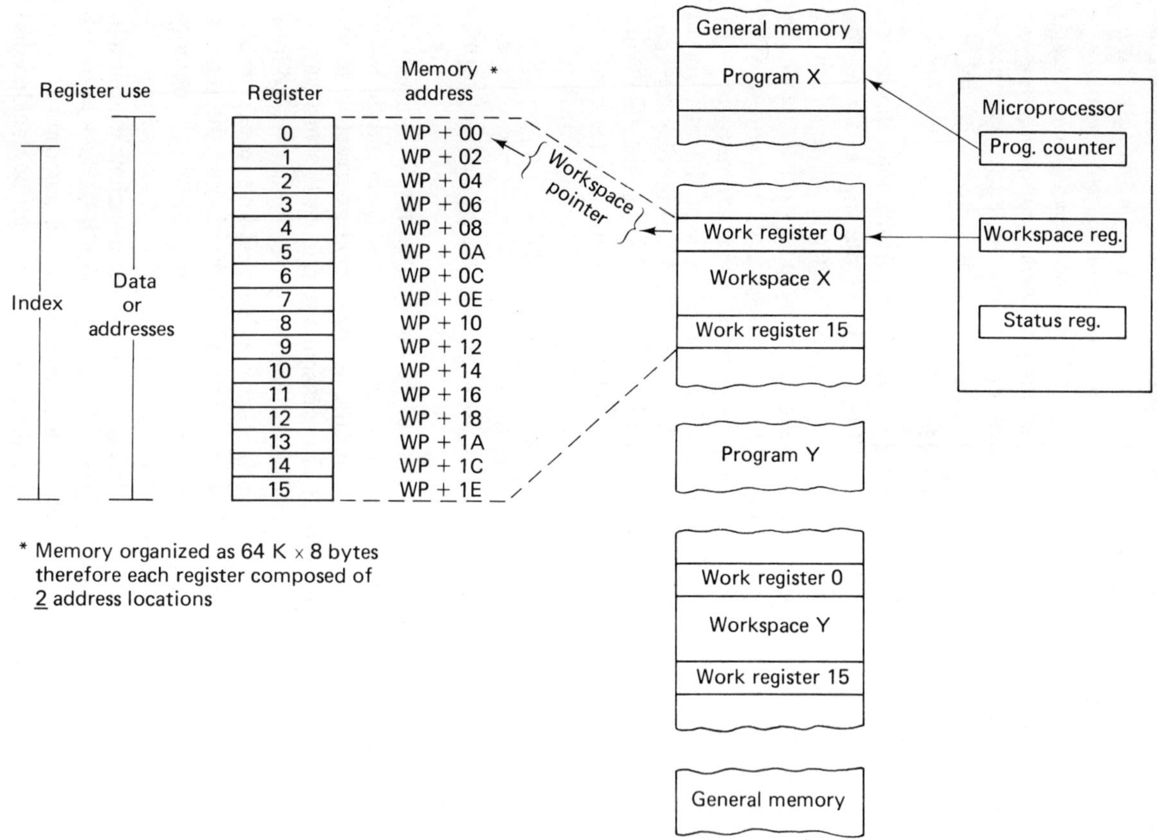

Figure 13-9 Working Model of 16-Bit Microprocessor

operation, and updates the workspace register to a completely new 16-word stack. Thus, the new routine has its own workspace register file even as the original register file is saved for return operations. If program *X* is considered to be the normal program being executed and program *Y* the interrupting program, Figure 13-9 shows how memory could be arranged. The workspace register file concept results in a greatly reduced time requirement for servicing of interrupts and other operations requiring deviation from normal program sequence.

Addressing Modes. Seven addressing modes are available in the 16-bit microprocessor. Each is listed and explained below.

1. Workspace Register	The contents of the indicated workspace register in the word are the operand.
2. Workspace Register (Indirect)	The contents of the indicated workspace register contain the memory address of the operand.
3. Indexed	The contents of the indicated workspace register are added to the address contained in the second command word.
4. Direct	The word following the instruction contains the memory address of the operand.
5. Workspace Register (Indirect with Auto Increment)	The contents of the indicated workspace register contain the memory address of the operand, which is automatically incremented after the access.
6. Immediate	The word following the instruction contains the operand.
7. Relative	The 8-bit displacement of the instruction is added to the updated program counter in jump instructions or to the base address in single-bit input instructions.

When these addressing modes are correlated with the instruction formats and example instructions that follow, it will be seen that the overall capability of the 16-bit microprocessor is impressive.

Instruction Set. The instruction set contains 62 single-word instructions and seven two-word instructions. Each instruction performs one of the following operations:

1. Arithmetic, logical, comparison, or manipulation operations on data.

2. Loading or storage of internal registers.
3. Data transfer between memory and external devices.
4. Control functions.

Samples of each of the categories are shown in Figure 13-10.

The four examples shown in Figure 13-10 dramatically indicate the power and versatility of the 16-bit microprocessor. A simple operation such as moving information from one register to another can take numerous steps in an 8-bit processor. Note, for example, that the MOV instruction in the 16-bit processor required only one step. When such

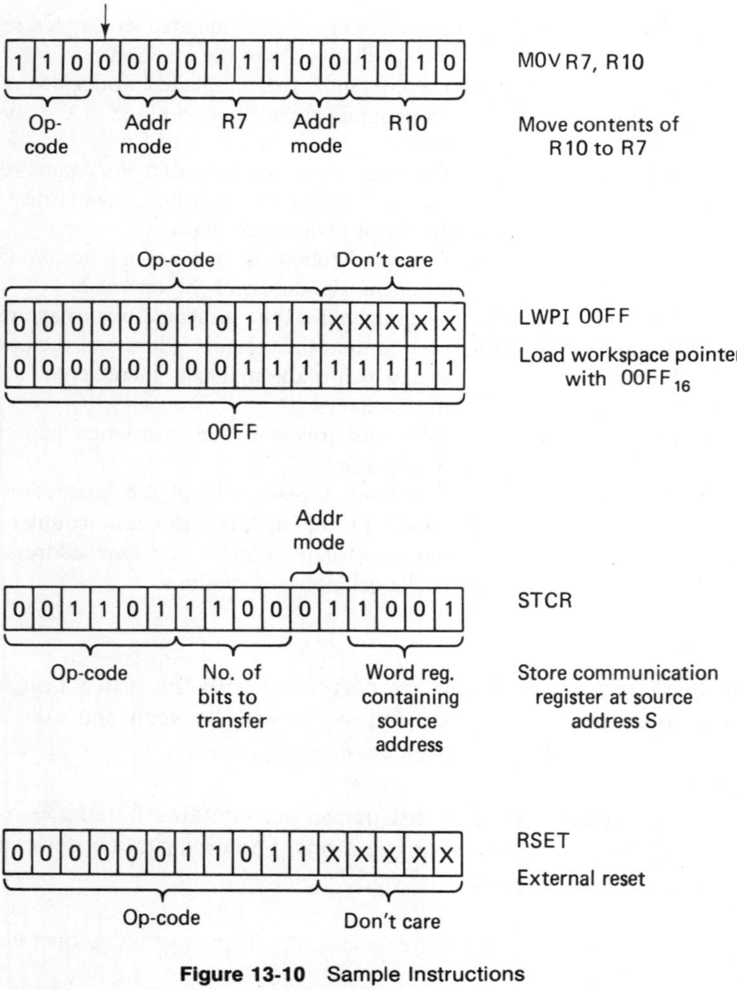

Figure 13-10 Sample Instructions

savings are multiplied by the hundreds of operations required in even the simplest of data-handling programs, the time savings are obvious.

Many applications do not require the speed and versatility of a 16-bit microprocessor. The reader should be alert as various applications are investigated, and should use only the level of complexity required. It is surely not cost-effective to use a 16-bit processor for an application such as the microwave oven control discussed earlier in this chapter. Similarly, it is not practical to employ a 4-bit processor in a large-scale data-handling microcomputer which requires numerous inputs and many interrupts.

13-4 THE END

The full gamut has been run. From the very basic logic assembly to the 16-bit microprocessor, the concepts and applications of microprocessors and microcomputers have been discussed. Where possible, generalization has been attempted. Each manufacturer supplies detailed information on his product, and the reader can refer to the data that meet specific requirements. Except for Chapters 11 and 12, hardware specified is either a composite of existing hardware, or is composed of units that are available from more than one source. Therefore, the reader should, as a result of using this text, possess a well-rounded knowledge of the overall microprocessor/microcomputer field—not enough knowledge to be an expert, but enough to know what the field is doing and how to search for new information.

It is the sincere wish of the author that each reader will find something of interest in this text. Regardless of individual interests or requirements, the reader should find that little bit of motivating force that will cause continuance in this thoroughly fascinating field.

Good luck, and happy microcomputing!

Appendix A

Laws of Boolean Algebra

The Laws of Identity:	$A=A,$	$\overline{A}=\overline{A}$
The Commutative Laws:	$AB=BA,$	$A+B=B+A$
The Associative Laws:	$A(BC)=(AB)C,$	$A+(B+C)=(A+B)+C$
The Idempotent Laws:	$AA=A$	$A+A=A$
The Distributive Laws:	$A(B+C)=AB+AC,$	$A+BC=(A+B)(A+C)$
The Laws of Absorption:	$A+AB=A,$	$A(A+B)=A$
The Laws of Expansion:	$AB+A\overline{B}=A,$	$(A+B)(A+\overline{B})=A$
DeMorgan's Laws:	$\overline{AB}=\overline{A}+\overline{B},$	$\overline{A+B}=\overline{A}\overline{B}$
Logical Multiplication	$(\overline{A}+B)(A+\overline{B})=\overline{A}A+\overline{A}\overline{B}+AB+B\overline{B}$	

Common Identities of Boolean Algebra

$$A(A+B)=AB$$

$$A+\overline{A}B=A+B$$

$$(AB)(A+B)=AB$$

$$(\overline{AB})(A+B)=A\overline{B}+\overline{A}B$$

$$\overline{A\overline{B}+\overline{A}B}=AB+\overline{A}\overline{B}$$

$$(A+B)(B+C)(A+C)=AB+BC+AC$$

$$(A+B)(\overline{A}+C)=AC+\overline{A}B$$

$$AC+AB+B\overline{C}=AC+B\overline{C}$$

$$(A+B)(B+C)(\overline{A}+C)=(A+B)(\overline{A}+C)$$

Appendix B

HEXADECIMAL-TO-DECIMAL CONVERSION

1. Consider the hexadecimal number in the following form:

$$WXYZ_{16} = (W \times 16^3) + (X \times 16^2) + (Y \times 16^1) + (Z \times 16^0)$$

2. Use the table below to determine the value of each term and sum the terms. For example:

$$357B_{16} = 12288 + 1280 + 112 + 11 = 13691_{10}$$

16^3		16^2		16^1		16^0	
Hex	*Decimal*	*Hex*	*Decimal*	*Hex*	*Decimal*	*Hex*	*Decimal*
0	0	0	0	0	0	0	0
1	4096	1	256	1	16	1	1
2	8192	2	512	2	32	2	2
3	12288	3	768	3	48	3	3
4	16384	4	1024	4	64	4	4
5	20480	5	1280	5	80	5	5
6	24576	6	1536	6	96	6	6
7	28672	7	1792	7	112	7	7
8	32768	8	2048	8	128	8	8
9	36864	9	2304	9	144	9	9
A	40960	A	2560	A	160	A	10
B	45056	B	2816	B	176	B	11
C	49152	C	3072	C	192	C	12
D	53248	D	3328	D	208	D	13
E	57344	E	3584	E	224	E	14
F	61440	F	3840	F	240	F	15

DECIMAL-TO-HEXADECIMAL CONVERSION

1. Subtract the largest possible decimal number in the table above from the decimal number to be converted. Write down the hexadecimal equivalent. Continue as shown below.

$$
\begin{array}{rrll}
13691_{10} = & 13691 & & \\
- & \underline{12288} & = 3 & \text{MSD} \\
 & 1403 & & \\
- & \underline{1280} & = 5 & \\
 & 123 & & \\
- & \underline{112} & = 7 & \\
 & 11 & & \\
- & \underline{11} & = \text{B} & \text{LSD} \\
 & 0 & &
\end{array}
$$

Therefore, $13691_{10} = 357B_{16}$.

Bibliography

BOYCE, JEFFERSON C. *Digital Computer Fundamentals.* Englewood Cliffs, N.J.: Prentice-Hall, 1977.

BOYCE, JEFFERSON C. *Digital Logic and Switching Circuits.* Englewood Cliffs, N.J.: Prentice-Hall, 1975.

Data Manual. Signetics Corporation, Sunnyvale, Calif., 1976.

DAVENPORT, WILLIAM P. *Modern Data Communication.* New York: Hayden, 1971.

Designing with Microcomputers. Signetics Corporation, Sunnyvale, Calif., 1976.

KARNAUGH, M. "The Map Method for Synthesis of Combinational Logic Circuits." *AIEE Proceedings*, November 1953, p. 593.

KLINGMAN, EDWIN E. *Microprocessor Systems Design.* Englewood Cliffs, N.J.: Prentice-Hall, 1977.

MCS 48 User's Manual. Intel Corporation, Santa Clara, Calif., 1977.

MCS 85 User's Manual. Intel Corporation, Santa Clara, Calif., 1977.

Memory Design Handbook. Intel Corporation, Santa Clara, Calif., 1975.

SOUCEK, BRANKO. *Microprocessors and Microcomputers.* New York: John Wiley & Sons, 1976.

TMS 1117 Microwave Oven Controller Manual. Texas Instruments, Dallas, Tex., 1976.

TMS 9900 Family System Development Manual. Texas Instruments, Dallas, Tex., 1977.

TMS 9900 Microprocessor Design Manual. Texas Instruments, Dallas, Tex., 1976.

6502 Hardware Manual. MOS Technology, Norristown, Pa., 1975.

6502 Programming Manual. MOS Technology, Norristown, Pa., 1975.

8080 Microcomputer Systems User's Manual. Intel Corporation, Santa Clara, Calif., 1976.

Index